DATE DUE	
NOV 28 1978	
APR 15 1980	
1983	

TOPICS IN MATHEMATICAL GEOLOGY
VOPROSY MATEMATICHESKOI GEOLOGII
ВОПРОСЫ МАТЕМАТИЧЕСКОЙ ГЕОЛОГИИ

ANDREI BORISOVICH VISTELIUS

TOPICS IN MATHEMATICAL GEOLOGY

Edited by
M. A. Romanova and O. V. Sarmanov
Laboratory of Mathematical Geology
Leningrad, USSR

Russsian translations by
J. P. Fitzsimmons
Department of Geology
University of New Mexico
Albuquerque, New Mexico

 CONSULTANTS BUREAU • NEW YORK - LONDON • 1970

ACKNOWLEDGMENTS

Nine of the papers in this collection were originally written in English; one was written in French. The authors of each of these papers were kind enough to provide us with a copy of their manuscript. This allowed us either to print them directly in English or to translate from the French, rather than retranslate from the Russian.

We extend our thanks to the following authors for having made these manuscripts available.

Papers submitted in English:

Gerard V. Middleton, Generation of the Log-Normal Frequency Distribution in Sediments (p. 34)

Felix Chayes, On Locating Field Boundaries in Simple Phase Diagrams by Means of Discriminant Functions (p. 85)

D.M. Shaw, Two-Cluster Discrimination in Analytical Geochemistry using the Distance Coefficient (p. 93)

W. Schwarzacher, The Vertical and Lateral Variation of a Carboniferous Limestone Area Near Sligo (Ireland) (p. 151)

William T. Fox, Use of the Computer for Quantitative Analysis of Fossil Distribution (p. 175)

F. Hori, Heat Conduction Calculations on the Thermal History of Contact Aureoles (p. 217)

E.H. Timothy Whitten, Variance of Some Selected Attributes in Granitic Rocks (p. 231)

D.G. Krige, The Role of Mathematical Statistics in Improved Ore Valuation Techniques in South African Gold Mines (p. 243)

Daniel F. Merriam, Visual Display of Computer Output Aids Geological Interpretation (p. 264)

Papers submitted in French:

B. A. Choubert, Geochemical Behavior of Elements in the Lithosphere (p. 13)

Library of Congress Catalog Card Number 69-12516
SBN 306-10832-1

The original text, published by Nauka Press in Leningrad in 1968 for the V. A. Steklov Mathematical Institute, Leningrad, of the Academy of Sciences of the USSR, has been corrected by the editors for this edition. This translation is published under an agreement with Mezhdunarodnaya Kniga, the Soviet book export agency.

ВОПРОСЫ МАТЕМАТИЧЕСКОЙ ГЕОЛОГИИ

М. А. Романова, О. В. Сарманов

© 1970 Consultants Bureau, New York
A Division of Plenum Publishing Corporation
227 West 17th Street, New York, N. Y. 10011

United Kingdom edition published by Consultants Bureau, London
A Division of Plenum Publishing Company, Ltd.
Donington House, 30 Norfolk Street, London, W.C. 2, England

All rights reserved

No part of this publication may be reproduced in any
form without written permission from the publisher

Printed in the United States of America

FOREWORD

Collections of this sort are a regular publication feature of the Laboratory of Mathematical Geology of the Order of Lenin V. A. Steklov Mathematical Institute of the Academy of Sciences of the USSR. In the future it is intended that further collections and monographs reflecting the activity of the Laboratory be issued.

In this present collection, in addition to workers of the Laboratory of Mathematical Geology, specialists of both Russia and many foreign countries participated. This has permitted us to display the general level of mathematization of geology in 1966. In order to enhance the overall view, the editors have included a section "Chronicle and Bibliography" in which information is given on the most important actions relating to mathematization of geology taking place in 1965 and the first half of 1966, and which includes a bibliography on two-dimensional regressions having great practical value in geology but little known to us in the Soviet Union.

The mathematical level of papers in the collection differs widely. Some papers are investigations of mathematical geology proper, i.e., of the discipline of probability modeling of geological processes, and they are based on the current structure of probability theory. Other papers are devoted to the simplest applications. On the whole the editors have tried to show that successful mathematization of geology is possible only with thorough understanding of the essence of geologic problems and with choice of the mathematical technique corresponding to the specific features of the geologic data. In this respect even the simplest mathematical methods may be of great use if the problem has been well set up geologically.

The material in the collection has been arranged according to type of problem as defined by A. B. Vistelius in a number of publications. Works on the investigation of a single independent random value are placed at the beginning; then come works on several values, and last, investigations using the ideas of random processes and means of approximating random fields.

Within each section the papers are arranged from the more general and more characteristic for this group to the less general and less characteristic. Geologically the papers embrace a wide range of topics, from experimental petrology and regional lithology to tectonics, methods of computing reserves, and selection of computer.

CONTENTS

The Life and Work of Andrei Borisovich Vistelius 1

The Published Works of A. B. Vistelius 6

I. GEOLOGIC HYPOTHESIS AND PROBABILITY DISTRIBUTIONS

Geochemical Behavior of Elements in the Lithosphere 13
   B. A. Choubert

Generation of the Log-Normal Frequency Distribution in Sediments 34
   G. V. Middleton

Correlation of Joint Trends with the Elements of Tectonic Structures 43
   L. D. Knoring

II. USE OF THE SPECIFIC PROPERTIES OF MULTIDIMENSIONAL SPACE IN SOLVING GEOLOGICAL PROBLEMS

The Origin of Clastic Mineral Associations in the Aptian-Cenomanian Rocks of the Southwestern Ural and Mugodzhary Region 63
   M. E. Demina and O. M. Kalinin

On Locating Field Boundaries in Simple Phase Diagrams by Means of Discriminant Functions 85
   F. Chayes

Two-Cluster Discrimination in Analytical Geochemistry using the Distance Coefficient 93
   D. M. Shaw

III. PARAGENETIC ANALYSIS

Distribution of Percentage Values 104
   A. V. Faas and O. V. Sarmanov

Niobium in Metagranites of the Polar Urals 112
   E. P. Kalinin, M. V. Fishman, and B. A. Goldin

Linear Paragenetic Associations in Rocks and Minerals of the Ladoga Formation . 125
   Yu. V. Nagaitsev and Yu. V. Podol'skii

Processes of Magmatic Differentiation in Connection with Paragenetic Features among Rock-Forming Elements in Natural Glass 136
   V. V. Gruza

IV. ANALYSIS OF GEOLOGIC SECTIONS

A Stochastic Model of Stratification (the Case of Unlimited Interstratal Erosion) .. 142
  T. S. Rivlina

The Vertical and Lateral Variation of a Carboniferous Limestone Area Near Sligo (Ireland) 151
  W. Schwarzacher

Use of a Harmonic Model for Analysis of the Dynamic System of Sedimentation in the Jatulian of Central Karelia 160
  K. I. Kheiskanen

Analysis of Sequences of Mineral Grains in Granites of the Kyzyltas Massif (Central Kazakhstan) as a Manifestation of the Markov Process 165
  D. N. Ivanov

Use of the Computer for Quantitative Analysis of Fossil Distribution. 175
  W. T. Fox

V. MAPPING GEOLOGICAL CHARACTERISTICS

Geometrical Properties of the Surface of the Alekseevka Uplift in the Kuibyshev District 186
  M. D. Belonin and I. M. Zhukov

Sorting of Clastic Material in Eolian Deposits of Central Kara Kum 200
  M. A. Romanova

VI. VARIOUS GEOLOGICAL PROBLEMS

Heat Conduction Calculations on the Thermal History of Contact Aureoles 217
  F. Hori

Variance of Some Selected Attributes in Granitic Rocks 231
  E. H. T. Whitten

The Role of Mathematical Statistics in Improved Ore Valuation Techniques in South African Gold Mines 243
  D. C. Krige

Fundamental Problems of Computing Reserves of Mineral Resources 262
  V. Nemeč

Visual Display of Computer Output Aids Geological Interpretation 264
  D. F. Merriam

VII. CHRONICLE AND BIBLIOGRAPHY

Trend Analysis of Geologic Data (Basic Literature) 273
  M. A. Romanova

Mathematical Methods in Geology (Chronicle for the Period from September 1964 to September 1966) 279
  M. E. Demina

THE LIFE AND WORK OF
ANDREI BORISOVICH VISTELIUS

A. B. Vistelius was born in Petrograd in 1915. In 1939 he graduated from the university in mineralogy in the Department of Geology. In 1942 he defended his candidacy, and in 1947 his doctoral dissertation. The first was devoted to the mineralogy and petrography of the intrusive and metamorphic rocks of the Balkhash region, the second to the lithology of upper Paleozoic carbonate rocks of the central Volga and trans-Volga region.

After 1942, Vistelius worked as a senior petrographer and then as director of the Branch of Operating Methods of the All-Union Petroleum Institute (1942-50), senior scientist at the Physicochemical Laboratory and Director of the Mining-Geological Laboratory of the All-Union Salt Institute (1950-52), senior scientist, head, and scientific director of a series of expeditions of the Laboratory of Aerial Methods of the Academy of Sciences of the USSR (1952-61), and, finally, director of the Laboratory of Mathematical Geology at the Order of Lenin V. A. Steklov Mathematical Institute of the Academy of Sciences of the USSR.

Vistelius began to publish when still a student, and since that time has published about 130 works, including seven monographs and manuals on method. Many of the works of Vistelius have been translated into foreign languages (English, French, German, Dutch, and Romanian) and have made the author widely known in both the Soviet Union and foreign countries.

Studies published by Vistelius indicate three principal trends of his scientific endeavors.

1. Regional problems. In this field Vistelius has made fundamental investigations on problems of paleogeography, lithology, and lithostratigraphy of oil basins and also on problems of regional geochemistry.

His lithologic—paleogeographic works represent a great contribution to the understanding of the formational history of upper Paleozoic sedimentary rocks along the middle Volga, the productive beds of the Apsheron Peninsula, the red beds of Turkmenia, and the Middle Miocene deposits of the Caspian region. Here belongs his large-scale, and still unfinished, investigation of the Aptian-Cenomanian strata of southeastern USSR and the adjoining regions of Iran and Afghanistan. His regional works also include studies of phosphorus distribution in the granitoidal rocks of Tien Shan, North America, and Brazil. The significance of phosphorus and a whole series of disseminated elements in the formation of granitic rocks has been shown in a new way, and new data have been thus supplied on a cardinal problem of geology: the origin of granite.

2. Numerous investigations in mineralogy, petrography, geochemistry, and, in part, stratigraphy. All these have been devoted to systematic problems, and each of them has touched necessarily on two sides of the problem: the actual specific problem is solved, on the one hand, and the method of approaching the solution is refined, on the other. The normal method of the author is formulation of an actual problem with emphasis on the fact that this problem is interesting as a whole, since systematic difficulties are encountered in solving it. The range of questions covered in these works is very great. There are papers devoted to tourmaline in the pegmatites of northern Karelia and remarks on the stratigraphy of the Konka horizon of Turkmenia, the migration of molybdenum in the supergene zone, and the properties of structural diagrams. Vistelius has also analyzed already known problems and, in introducing new methods, has found convincing solutions. In this regard, interest is aroused by his notes on enantiomorphism in quartz, and analysis of data from the literature. Vistelius so ably analyzed published data that his notes were included in the latest edition of Dana's manual, known for its exacting demands in the selection of material incorporated in it. Without exaggeration, it may be stated that it would be difficult to find a geologist at the present time with such a wide range of interests, manifested by actual investigations, as we find in the works of A. B. Vistelius.

3. The two fields indicated above, despite their considerable significance, are but ancillary to his main activity, which has been pursued by Vistelius with rare persistence and systematic effort over the course of the last 25 years. This main trend of his interest is the introduction of mathematical methods into geology. This trend appeared in the studies of Vistelius in 1942, and it has been continuously developed by him since that time. This field of endeavor has also brought Vistelius wide renown, although it overshadows significant and purely geological results that were obtained by him in purely geological studies, of which even a few would have been entirely sufficient to bring their author to the fore among the leading representatives of geologic thought.

The works of Vistelius in the mathematization of geology astonish one with their singleness of purpose and systematic approach. At the very beginning of his efforts in this direction, in 1944 and 1945, the author advanced two theses, to the proof of which he has devoted twenty years of energetic work. One can but wonder how he was able to advance such theses at a time when there appeared absolutely no grounds for expecting them to be formulated, to say nothing of the fact that surely no geologist could possibly imagine that after 20 years they would be accepted by a huge number of investigators. The theses are the following.

1. Mathematical methods in geology lead to the development of a specific object of investigation which then becomes the object of a new discipline, mathematical in method and geological in purpose. As early as 1944 Vistelius called this new discipline analytical geology.

2. Mathematical methods are necessary in geology for constructing models of geologic processes, i.e., for testing agreement between geological constructions and observational data.

The idea of mathematical (stochastic) modeling has been ahead of the development of mathematical methods in geology by almost an entire stage. Whereas a few in the 40's spoke of the possibility and actual existence of mathematical treatment of observations, no one spoke of stochastic modeling. Naturally the ideas of Vistelius were received with a smile and, at first, no serious attention was given to them. But time passed, and with the introduction of mathematics into geology it became increasingly clear that the matter was not simply one of making computations but of expressing geological ideas in mathematical language, that the trend established by Vistelius not only did not replace geology but led it to a higher plane, requiring more concrete and clearer concepts than were possible without the participation of mathematics. It is now clear that the proposition of Vistelius concerning the cardinal importance of stochastic modeling has been accepted and has withstood the test of time. It is receiving more and more attention, and it is never missing from the pages of geological journals.

The idea of the independence of analytical, or mathematical, geology, as it was called by then, experienced a number of vicissitudes. In 1944 it was adopted by Academician V. I. Vernadskii, and it would seem that after such authoritative support it must receive recognition. In 1947 the English journal "Nature" devoted a special article to the works of Vistelius, in which the new trend was warmly reviewed and a bright future prophesied for it. The late 40's and early 50's were unfavorable for promulgation of new ideas, and the concept of Vistelius concerning the independence of mathematical geology suffered dogmatic criticism. In 1962 a summary paper by A. B. Vistelius was published,* giving actual examples demonstrating the discipline defended by him. This paper was translated into English, French, and German. In American publication catalogs, a division on mathematical geology appeared, and a section on mathematical geology was made in the edition of a geological dictionary—the Glossary of Geology, a fundamental handbook for geologists—that was under preparation. In 1965 the Laboratory of Mathematical Geology was organized at the V. A. Steklov Mathematical Institute of the Academy of Sciences of the USSR. Thus, the second thesis of Vistelius, proposed more than 20 years previously, also proved viable and has found clear expression in present-day geology.

The actual studies of Vistelius in the field of mathematical geology may be subdivided into two branches. First of all, Vistelius introduced a number of mathematical methods into geology. This was done at least 10 years before the methods became widely accepted in geology. These were the use of the theory of random functions (random processes, 1948), one of the first applications of factor analysis (1948), and the use of discriminant functions (1950). The credit for introducing correlation analysis also belongs to Vistelius, although it was known by several geologists earlier. These methods have become operating procedures in geology.

The second branch of investigation is purely mathematical: here we find problems that were formulated by Vistelius, solved in advanced forms by mathematicians, and that have led to the appearance of specific literature. One of these problems is that of interformational erosion, solved by Academician A. N. Kolmogorov; a second refers to false correlation, recognized by Vistelius simultaneously with F. Chayes, and worked out by mathematicians in a whole series of papers. There are also a number of other problems.

In geologic relations, mathematical geology was developed by Vistelius along the following four basic branches:

1. Investigation of the effect of geologic conditions on the type of distribution function of the investigated random value. This includes stochastic modeling of geologic processes.

2. Analysis of paragenetic associations. Since 1947 Vistelius has systematically published papers devoted to methods of studying relations in the geological and geochemical sciences. These papers have dealt with evaluation of the strength of correlation, or constraint force, and determination of the form of this force on the basis of statistical and theoretical information characteristics. These investigations attracted a great number of followers, and at the present time there is apparently no geologist who does not use these methods to some extent. They have proved to be so natural that it is a source of perplexity to many that Vistelius had to publish several papers in order to convince geologists to use them.

These methods have been widely introduced in recent time into mineralogy, geochemistry, petrography, and lithology, where they have been developed by students of Vistelius.

3. Investigation of stratigraphic sections. On the basis of the theory of stochastic processes, Vistelius developed theoretical work on one of the most important problems of lithology: the problem of turbidites, posed by Kuenen. By Vistelius this problem was reduced to the Markov

---

*Problems of Mathematical Geology. Geol. i Geofiz., Sibirsk. Otd. Akad. Nauk SSSR, No. 12 (1962), pp. 3-9; No. 7 (1963), pp. 3-17; No. 12 (1963), pp. 3-10.

test, and it was thus given concrete meaning. Investigations of flysch sections have shown excellent agreement with the scheme of Kuenen, and this must be considered a great achievement of lithology. This problem is now being worked on among the mathematical projects of the students of Vistelius. In addition to the theoretical problems of general geological significance, methods of the theory of random functions in various modifications have been used by Vistelius in a number of practical projects. These methods permit one to map barren strata. This was first pointed out by Vistelius himself during his studies in Azerbaidzhan and Turkmenia. Later, these methods were used to solve a number of practical problems in Karelia, Kazakhstan, the Kuznetsk basin; and everywhere they gave positive results. These same methods of Vistelius have been used by geologists of India and the USA. On the whole, this branch of endeavor creates a strict mathematical basis for lithostratigraphy, the value of which is especially great when studying oil-bearing clastic rocks and barren metamorphic rocks of the Precambrian.

4. Questions on mapping geological features. This problem, which has developed markedly in recent years in the works of American geologists, was developed principally by students of Vistelius. Thus, in the fields of lithology and paleogeography, such work was led by M. A. Romanova, and in the field of tectonics, by M. D. Belonin. Both authors adopted the approximation surface proposed by Vistelius, which has a number of advantages in geologic interpretation over approximation of other authors, whose views have received wide recognition. In this direction, the Laboratory of Mathematical Geology, headed by Vistelius, is conducting a series of investigations aimed at deepening the interpretation of proposed models as well as increasing the strictness of application of the methods.

We have noted the wide and rapidly growing popularity of Vistelius's work. In 1958 he was chosen a member of the board of editors of the leading American journal "Journal of Geology." Now he is a member of the International Organizational Committee of Mathematical Geology. He is commissioned to report on international geological congresses having to do with the use of mathematical methods. Vistelius carries on intense consultation and organizational work on the introduction of mathematical methods into geology. He is a member of several committees and has headed the All-Union Committee of Experts on Mathematical Methods. In recent years, his works have been widely acclaimed in foreign countries. The Pergamon Press in England has published a translation of his manual on structural analysis; the Consultants Bureau in New York has published a single-volume collection of his works, and a collection of selected works is also being prepared in French translation. Vistelius acts as consultant not only to Soviet geologists but to foreign scientists as well. A number of his students work in the Soviet Union, but some also work in foreign countries (USA, China, Bulgaria, Yugoslavia, and elsewhere).

Vistelius is full of new plans and ideas. Some of the principal ones, developed in his laboratory, should be noted. One of these is devoted to Tertiary oil basins in the Caspian region. According to Vistelius's program, this theme will fill a five-volume monograph, and will cover questions that cannot be solved without using mathematical methods. It will be something of an exhibit of the application of mathematics in lithology. Three volumes of the monograph have already appeared. It is hoped that the author has the strength to finish the other two. Another theme concerns the Aptian—Cenomanian deposits of southeastern USSR, which will point out the rational use of lithologic features, chosen from mathematical considerations, for predicting oil possibilities. This subject is based on original material collected by Vistelius and his students during expeditions in the field. Lastly, a third theme is regional geochemistry. This involves the study of phosphorus distribution in granitic rocks of the world. Apparently this is the first investigation in geochemical endeavor to embrace the whole planet and to rest on precise analytical data. The significance of this theme is very great, since it involves study of the behavior of a characteristic accessory element in a most important group of rocks in all parts of the world. It is natural that, in working out such problems, questions will arise, associated with explanation of the effect of the mantle on development of the earth's crust, the origin of the granitic rocks themselves, the sources of transmagmatic solutions, and the like. Analysis of these questions requires the use of profound mathematical equipment, and this, in turn, leads to the setting up of mathematical problems. Difficulties in developing these themes are very great, but we hope that the task will prove to be within our power and that we will successfully perform it.

M. F. Dvali
D. S. Korzhinskii
Yu. V. Linnik
M. A. Romanova
O. V. Sarmanov

THE PUBLISHED WORKS OF A. B. VISTELIUS

1939

1. "The Antarctic continent," Vestn. Znaniya, No. 9, pp. 34-37.
2. "Tourmaline in carbonate veins in the vicinity of Chupa Bay (Northern Karelia)," Uch. Zap. Leningr. Gos. Univ., No. 34, pp. 60-70.
3. "Wulfenite in the Kyzyl-Kan mine (Northern Tadzhikistan)," Uch. Zap. Leningr. Gos. Univ., No. 49, pp. 56-62.

1940

4. "Geologic history of Antarctica," Vestn. Znaniya, No. 2, pp. 36-38.
5. "Mineralogy of the Andreevskii mine (Khakassiya)," Uch. Zap. Leningr. Gos. Univ., No. 45, pp. 148-157.

1941

6. "A composite dike in the region of the Gul'shad lead—zinc deposit (Balkhash region)," Dokl. Akad. Nauk SSSR, Vol. 31, No. 6, pp. 559-601.

1942

7. Mineral Assemblages from the Region of the Gul'shad Lead—Zinc Deposit (Balkhash region), Summary of Dissertation at the competition for a scientific degree among candidates of the geological and mineralogical sciences, Leningrad State University.

1943

8. "Notes on garnets from the vicinity of Lake Balkhash," Zap. Vseross. Mineral. Obshch., Pt. 72, No. 3-4, pp. 167-173.

1944

9. "Notes on analytical geology," Dokl. Akad. Nauk SSSR, Vol. 44, No. 1, pp. 27-31.
10. "Physical properties of oil-bearing strata in the Permian system," Author's abstract in the collection: Scientific Research Papers of Petroleum Specialists, Geol., No. 1, pp. 81-82.

1945

11. "Frequency distribution of porosity coefficients and energy processes in the *Spirifer* beds of the Buguruslan oil region," Dokl. Akad. Nauk SSSR, Vol. 49, No. 1, pp. 44-47.

---
* Excluding publications in abstract journals.

12. "Expression of the results of fossilization of fluctuating movements of the earth's crust by means of the series $\sum_{i=0}^{k} e^{a_i x + b_i} \cos(\omega_i x + y_i)$). Dokl. Akad. Nauk SSSR, Vol. 49, No. 7, pp. 531-535.

1946

13. "Porosity cycles and the phenomenon of phase differentiation of sedimentary strata," Dokl. Akad. Nauk SSSR, Vol. 54, No. 6, pp. 519-521.

1947

14. "Correlation of mesorhythms in the Lower Permian strata of trans-Kama Tataria and their stratigraphic significance," Dokl. Akad. Nauk SSSR, Vol. 55, No. 3, pp. 241-244.
15. "The application of the correlation coefficient during investigation of paragenetic mineral associations in clastic rocks," Dokl. Akad. Nauk SSSR, Vol. 55, No. 4, pp. 343-345.
16. "Correlation between apatite and nepheline in the Kukisvumchorr-Yukspor sphene deposit (Khibiny tundra)," Dokl. Akad. Nauk SSSR, Vol. 56, No. 2, pp. 185-188.
17. "New confirmation of Goldschmidt's observations on the place of germanium in coal," Dokl. Akad. Nauk SSSR, Vol. 58, No. 7, pp. 1455-1457.
18. "Stochastic basis of a geologically important probability distribution," Dokl. Akad. Nauk SSSR, Vol. 58, No. 4, pp. 631-634 (jointly with O. V. Sarmanov).

1948

19. "Geology of the lower Kazanian deposits of the Buguruslan region," Sov. Geol., Sb. No. 28, pp. 48-63.
20. "Some analytical methods of investigating rhythmicity," Sov. Geol., Sb. No. 28, pp. 174-182.
21. "The measure of correlation between paragenetic members and methods of studying it," Zap. Vses. Mineralog. Obshchestva, Pt. 77, No. 2, pp. 147-158.
22. "The simplest types of problems in mathematical treatment of lithologic observations," Litolog. Sb. Vses. Nauchn.-Issled. Geologorazved. Inst., No. 1, pp. 125-131.
23. "The distribution of magnesite in Paleozoic rocks on the eastern part of the Russian platform," Litolog. Sb. Vses. Nauchn.-Issled. Geologorazved. Inst., No. 2, pp. 42-49.
24. "Rounding of quartz sands of the Belinskii Bank (Volga delta)," Dokl. Akad. Nauk SSSR, Vol. 63, No. 1, pp. 69-72.

1949

25. "The mechanism of formation of sedimentary beds," Dokl. Akad. Nauk SSSR, Vol. 65, No. 2, pp. 191-194.
26. "The mechanism of correlation of stratification," Dokl. Akad. Nauk SSSR, Vol. 65, No. 4, pp. 535-538.
27. "Calcium sulfates in the Paleozoic rocks on the eastern part of the Russian platform," Geokhim. Sb. Vses. Nauchn.-Issled. Geologorazved. Inst., No. 1, pp. 142-158.

1950

28. "Paleogeographic significance of correlation between bed thicknesses (according to data on the Productive sequence of the Apsheron Peninsula)," Litolog. Sb. Vses. Nauchn.-Issled. Geologorazved. Inst., No. 3, pp. 61-73.
29. "Porosity and chemical composition of the carbonate strata on the eastern part of the Russian platform," Tr. Lab. Gidrogeol. Probl. im. F. P. Savarenskogo, Vol. 2, pp. 194-202.

30. "Distribution of enantiomorphous types of quartz," Zap. Vsosoyuzn. Mineral. Obshch., Pt. 79, No. 3, pp. 191-195.
31. "Correlation between copper content in borax waters of Azerbaidzhan and the degree of mineralization," Dokl. Akad. Nauk AzerbSSR, Vol. 6, No. 1, pp. 34-36.
32. "Mineral content of the heavy sand fraction in the lower part of the Productive sequence of the Apsheron Peninsula, the Chokrak strata of southern Dagestan, and alluvium of the Volga," Dokl. Akad. Nauk SSSR, Vol. 71, No. 2, pp. 367-370.
33. "Lower Permian pebbles and cobbles from the Productive sequence of the Apsheron Peninsula," Dokl. Akad. Nauk SSSR, Vol. 72, No. 2, pp. 369-372 (jointly with A. D. Miklukho-Maklai).

1951

34. "Porosity rhythms in the lower Kazanian deposits of southern Tataria," Tr. Leningr. Obshch. Estestvoispyt., Vol. 68, No. 2, pp. 150-167.
35. "Correlation once again (a reply to S. V. Konstantov)," Zap. Vsesoyuzn. Mineral. Obshch., Pt. 80, No. 1, pp. 79-80.
36. "The required number of grains for computation in immersions," Zap. Vsesoyuzn. Mineral. Obshch., Pt. 80, No. 3, pp. 188-190.
37. "Probability of the effect of an 'eolian field' in the diagram of L. B. Rukhin on the field of residual types of sand," Izv. Akad. Nauk SSSR, Ser. Geol., No. 1, pp. 155-156.
38. "Status of treating lithologic observations and means of improving it," Izv. Akad. Nauk SSSR, Ser. Geol., No. 3, pp. 90-104.
39. "Toward the history of disthène in the middle Miocene layer of Dagestan," Dokl. Akad. Nauk SSSR, Vol. 79, No. 1, pp. 133-136.
40. "Paleozoic pebbles and cobbles from the Productive sequence of the Apsheron Peninsula," Dokl. Akad. Nauk SSSR, Vol. 79, No. 3, pp. 499-502 (jointly with A. D. Miklukho-Maklai).

1952

41. "The Kirmakinskaya series of eastern Azerbaidzhan. I. Principal lithologic features," Dokl. Akad. Nauk AzerbSSR, Vol. 8, No. 1, pp. 17-23.
42. "Probability distributions (in reply to V. S. Dmitrievskii)," Izv. Akad. Nauk SSSR, Ser. Geol., No. 1, pp. 155-156.
43. "Mineralogy of the Miocene sand-silt strata of southern Azerbaidzhan," Dokl. Akad. Nauk SSR, Vol. 85, No. 5, pp. 1155-1158.
44. "Natural paragenetic associations of some components of oil in Azerbaidzhan," Izv. Akad. Nauk AzerbSSR, No. 2, pp. 17-31 (jointly with D. D. Zul'fugarly).
45. "Nature of changes in mineral content of concentrates during successive washing of sands," Zap. Vsesoyuzn. Mineral. Obshch., Pt. 80, No. 2, pp. 143-150 (jointly with N. N. Sarsadskii).

1953

46. Rock Salt, Bol'shaya Sov. Entsik., second edition, Vol. 19, p. 490.
47. "Treatment of microstructural diagrams," Zap. Vses. Mineralog. Obshch., Pt. 82, No. 4, pp. 271-280.
48. "Devonian limestones from the red-bed sequence of Tuarkyr," Dokl. Akad. Nauk SSSR, Vol. 90, No. 2, pp. 231-234 (jointly with A. D. Miklukho-Maklai and V. N. Ryabinin).
49. "A new discovery of the Konka horizon on the Krasnovodsk Plateau," Dokl. Akad. Nauk SSSR, Vol. 90, No. 3, pp. 445-448 (jointly with I. A. Korobkov).

1954

50. "Sands of the middle and lower Volga," in a collection of papers of the Laboratory of Aerial Methods (Labor. Aerometodov Akad. Nauk SSSR) in memory of N. G. Kellyu (1953).
51. "Basic features of the color characteristics of Cretaceous clastic sand—silt strata of the trans-Caspian region," Dokl. Akad. Nauk SSSR, Vol. 95, No. 2, pp. 367-370 (jointly with N. N. Yaroslavskaya).
52. "Mineral associations and characteristic paragenetic relations of the Aptian-Cenomanian clastic strata of the trans-Caspian region," Dokl. Akad. Nauk SSSR, Vol. 97, No. 3, pp. 503-506.
53. "Remarks on a paper by Professor P. A. Ryzhov: 'Determining the accuracy of calculating reserves of mineral deposits,'" in: Investigations of Problems of Mine Surveying, Sb. No. 29, pp. 200-201 (jointly with O. V. Sarmanov).

1955

54. "The Kirmakinskaya series of eastern Azerbaidzhan. II. Mineral associations," Dokl. Akad. Nauk SSSR, Vol. 103, No. 1, pp. 117-120.
55. "Age of the lower part of the red-bed sequence of the Cheleken Peninsula," Dokl. Akad. Nauk SSSR, Vol. 105, No. 4, pp. 786-789.

1956

56. "Problems of studying correlation in mineralogy and petrography," Zap. Vsesoyuzn. Mineral. Obshch., Pt. 85, No. 1, pp. 58-73.
57. "Subdividing the Recent deposits of the eastern Caucasus and the northern Caspian region into districts according to mineral content," Dokl. Akad. Nauk SSSR, Vol. 111, No. 5, pp. 1067-1071.
58. "The middle division of the Productive sequence of the Apsheron Peninsula and the problem of its origin," Izv. Akad. Nauk SSSR, Ser. Geol., No. 4, pp. 77-94 (jointly with A. D. Miklukho-Maklai).

1957

59. "Subdivision of unfossiliferous strata by quantitative-mineralogical, petrographic, and chemical features," Zap. Vsesoyuzn. Mineral. Obshch., Pt. 86, No. 1, pp. 99-115.
60. "The statistics of microstructural diagrams," Zap. Vsesoyuzn. Mineral. Obshch., Pt. 86, No. 6, pp. 691-703.
61. "Regional lithostratigraphy and conditions under which the Productive sequence of the southeastern Caucasus formed," Tr. Leningr. Obshch. Estestvoisp., Vol. 69, No. 2, pp. 126-150.
62. "The nature of pre-Permian vulcanism in western Turkmenia," Dokl. Akad. Nauk SSSR, Vol. 117, No. 5, pp. 867-869.

1958

63. Structural Diagrams, Izd. Akad. Nauk SSSR, Moscow-Leningrad.
64. "Spectral brightness of sand—silt rocks of Aptian, Albian, and Cenomanian ages in the trans-Caspian region," in: Geology of the Trans-Caspian, No. 1, pp. 31-67.
65. "Subdividing the alluvial deposits of the Pamirs into districts according to mineral associations," Dokl. Akad. Nauk SSSR, Vol. 118, No. 6, pp. 1158-1161.
66. "Some questions on the geology of western Turkmenia," Izv. Akad. Nauk TurkmSSR, No. 6, pp. 115-119 (jointly with I. A. Korobkov).
67. Dictionary of Petroleum Geology. The terms: autocorrelation, probability, probable error, dispersion, dispersion analysis, phase differentiation, mathematical expectation. Gostoptekhizdat, Moscow.

68. Volume-frequency analysis of sediments from thin-section data, a discussion, J. Geol., Vol. 66, No. 2, pp. 224-226.
69. "Paragenesis of sodium, potassium, and uranium in volcanic rocks of Lassen Volcanic National Park, California," Geochim. Cosmochim. Acta, Vol. 14, No. 1-2, pp. 29-34.

1959

70. "Geology at the University of Chicago (translation of a report of E. K. Olson)," Vestn. Leningr. Gos. Univ., Ser. Geol. i Geogr., Issue 3, No. 18, pp. 137-138.
71. Address at the session of the Technical Council of the Ministry of the Petroleum Industry and the Academy of Sciences of Turkmenia at Ashkhabad, 1956. "Objectives and potentials of prospecting and exploratory work on oil and gas in the western parts of Central Asia," pp. 352-355, Izd. Akad. Nauk TurkmSSR, Ashkhabad.
72. "Origin of the red-bed sequence on the Cheleken Peninsula. Experience in using absolute age of clastic minerals for solving problems of lithology and paleogeography," Dokl. Akad. Nauk SSSR, Vol. 125, No. 6, pp. 1307-1310.
73. "Correlation between percentage values," Dokl. Akad. Nauk SSSR, Vol. 126, No. 1, pp. 22-25 (jointly with O. V. Sarmanov).

1960

74. "The morphometry of clastic particles," Tr. Labor. Aerometodov Akad. Nauk SSSR, Vol. 9, pp. 135-202.
75. "Peculiarities of germanium concentration in coal (in reference to the review of V. M. Ershov)," Izv. Akad. Nauk SSSR, Ser. Geol., No. 8, p. 100.
76. "Skew frequency distributions and the fundamental law of the geochemical processes," J. Geol., Vol. 68, No. 1, pp. 1-22.
77. "Age of the red-bed sequence on the northwestern Krasnovodsk Plateau," Izv. Akad. Nauk TurkmSSR, Ser. Fiz.-Tekhn., Khimich. i Geol. Nauk, No. 3, pp. 108-111 (jointly with I. A. Korobkov and M. A. Romanova).

1961

78. Data on the Lithostratigraphy of the Productive Sequence of Azerbaidzhan, Izd. Akad. Nauk SSSR, Moscow-Leningrad.
79. "The middle Caspian land mass (in reference to papers by Tamrazyan and Rikhter)," Byul. Mosk. Obshchestva Ispytatelei Prirody, Ser. Geol., Vol. 36, Issue 1, pp. 148-151.
80. "Sedimentation time trend functions and their application for correlation of sedimentary deposits," J. Geol., Vol. 69, No. 6, pp. 703-728.
81. Discussion of paper by F. Chayes "On correlation between variables of constant sum," J. Geophys. Res., Vol. 66, No. 5, p. 1601.
82. "The absolute age of the clastic part of sand—silt deposits in southwestern Central Asia," Dokl. Akad. Nauk SSSR, Vol. 138, No. 2, pp. 422-425 (jointly with A. Ya. Krylov).
83. "On the correlation between percentage values: major component correlation in ferromagnesium micas," J. Geol., Vol. 69, No. 2, pp. 145-153 (jointly with O. V. Sarmanov).

1962

84. Red Beds of the Cheleken Peninsula (Lithostratigraphy and Geologic Structure), Izd. Akad. Nauk SSSR, Moscow-Leningrad (jointly with M. A. Romanova).
85. "Phosphorus in the granitoidal rocks of Tien Shan," Geokhimiya, No. 2, pp. 116-135.
86. "Problems of mathematical geology. I. On the history of the question," Geol. i. Geofiz., Sibirsk. Otd. Akad. Nauk SSSR, No. 12, pp. 3-9.

1963

87. Phase Differentiation of Paleozoic Deposits of the Middle Volga and Trans-Volga Regions, Izd. Akad. Nauk SSSR, Moscow-Leningrad.
88. "Functions of probability distributions of accessory element concentrations in rocks and minerals," Teor. Veroyatnostei i ee Primeneniya, Vol. 8, No. 2, pp. 232-233.
89. "Problems of mathematical geology. II. Models of processes and paragenetic analysis," Geol. i Geofiz., Sibirsk. Otd. Akad. Nauk SSSR, No. 7, pp. 3-17.
90. "Problems of mathematical geology. III. The random process," Geol. i Geofiz., Sibirsk. Otd. Akad. Nauk SSSR, No. 12, pp. 3-10.
91. Foreword to the book of F. Chayes: Quantitative Mineral Analysis of Rocks in Thin Section under the Microscope [Russian translation], IL, Moscow. [In English: Petrographic Modal Analysis, Wiley, New York (1956).]
92. "Programming problems of geology and geochemistry for use with all-purpose electronic computers," Geol. Rudn. Mestorozhd., No. 3, pp. 34-48 (jointly with T. B. Yanovskaya).
93. "Dispersion of clastic material in the Aptian-Cenomanian basin of southeastern USSR," Dokl. Akad. Nauk SSSR, Vol. 150, No. 6, pp. 1319-1322 (jointly with M. E. Demina).
94. "Rounding of quartz grains," in: Geochemistry, Petrography, and Mineralogy of Sedimentary Rocks. A collection in honor of the 60th birthday of L. V. Pustovalov, pp. 233-253 (jointly with M. E. Demina).
95. "Functions of probability distributions of phosphorus concentrations in granitoidal rocks of Switzerland, the Guianas, and equatorial Africa," Dokl. Akad. Nauk SSSR, Vol. 152, No. 6, pp. 1449-1452.

1964

96. "Principal types of mathematical solutions of problems in modern geology," Razvedka i Okhrana Nedr., No. 6, pp. 18-25.
97. "Problems of geochemistry and information measures," Sov. Geol., No. 12, pp. 5-26.
98. "Mathematical methods in geology," Sov. Geol., No. 12, pp. 148-149.
99. "The problem of formation of sedimentary beds (author's abstract of a report)," Byul. Mosk. Obshchestva Ispytatelei Prirody, Ser. Geol., Vol. 39, No. 3, p. 148.
100. "Probability and statistical problems of geology," Transactions of the 4th Mathematical Conference [in Russian], 1961, Vol. 2, Reports of Sections, pp. 329-335.
101. "Paleogeographic reconstruction by absolute age determinations of sand particles," J. Geol., Vol. 72, No. 4, pp. 483-486.
102. "Informational characteristics of frequency distribution in geochemistry," Nature, Vol. 202, No. 4938, p. 1206.
103. "Stochastic model for the generation of the bedding of red-beds from the Cheleken Peninsula (Caspian Sea)," Bull. Geol. Soc. Am., Program of Annual Meetings, p. 213.
104. "A discussion on the statistical analysis of fabric diagrams," J. Geol., Vol. 4, Pt. 1, pp. 224-228.
105. "Distribution of the heavy fraction in sands from deposits of the central Karakum," Dokl. Akad. Nauk SSSR, Vol. 158, No. 4, pp. 860-863 (jointly with M. A. Romanova).
106. "Phosphorus in granitic rocks of North America," Bull. Geol. Soc. Am., Vol. 75, pp. 1055-1092 (jointly with V. J. Hurst).

1965

107. "Theory of formation of sedimentary beds," Dokl. Akad. Nauk SSSR, Vol. 164, No. 1, pp. 158-160 (jointly with T. S. Feigel'son).
108. "Nature of the sequence of beds in sections of some sedimentary sequences," Dokl. Akad. Nauk SSSR, Vol. 164, No. 3, pp. 629-632 (jointly with A. V. Faas).

109. "Variations in thickness of beds in a section of Paleozoic flysch in the southern Urals," Dokl. Akad. Nauk SSSR, Vol. 164, No. 5, pp. 1115-1118 (jointly with A. V. Faas).
110. "Probleme der mathematischen Geologie. I. Zur Geschichte," Z. Angew. Geol., Vol. II, No. 5, pp. 265-268.
111. "Probleme der mathematischen Geologie. II. Modelle von Vorgängen und die Analyse von Paragenesen," Z. Angew. Geol., Vol. II, No. 6, pp. 306-313.
112. "Probleme der mathematischen Geologie. III. Der Zufallsprozess," Z. Angew. Geol., Vol. II, No. 7, pp. 356-359.

1966

113. Red-Beds of the Cheleken Peninsula. Lithology. Experiment in Stochastic Modeling of Processes Leading to the Formation of Sedimentary Beds, Izd. Nauka, Moscow-Leningrad.
114. "Principaux types de solutions mathématiques des problèmes géologiques actuels," Bureau de Recherches Géologiques et Minières. Serv. d'Inform. Géol., Centre Sci. et Techn., D'Orleans-La Source, No. 4701.
115. "Problèmes de géologie mathématiques," Idem, No. 4703.
116. "Programmation des problèmes géologiques et géochimiques pour l'emploi des calculatrices électroniques universelles," Idem, No. 4702.
117. Structural Diagrams, Pergamon Press, Oxford.
118. "Formation of the Belaya granodiorites on Kamchatka (an experiment in stochastic modeling)," Dokl. Akad. Nauk SSSR, Vol. 167, No. 5, pp. 1115-1118.
119. "A stochastic model of crystallization of alaskites and the corresponding transient probabilities," Dokl. Akad. Nauk SSSR, Vol. 170, No. 3, pp. 653-655.
120. Review of the book of V. F. Morkovkina: Chemical Analyses of Volcanic Rocks and Rock-Forming Minerals, Geokhimiya, No. 5, pp. 617-620.
121. "Trend surfaces," J. S. African Inst. Mining Met., Symposium on Mat. Stat. and its Comp. Appl. in Ore Valuation, Johannesburg, pp. 66-72.

1967

122. "Crystallization of alaskites from the Karakul'dzhur River (central Tien Shan)," Dokl. Akad. Nauk SSSR, Vol. 172, No. 1, pp. 165-168.
123. "Phosphorus and some trace elements in granitic rocks of Brazil," J. Geol., Vol. 75, No. 6 (jointly with D. Guimaraes and V. A. Galibin).
124. Studies in Mathematical Geology, Consultants Bureau Inc., New York.
125. "Mathematical techniques in making geological interpretation," 7th World Petroleum Congress P. D. 5, pp. 13-25, Mexico.
126. "Stochastic matrix of quasieutectic granites (as exemplified by granites from the vicinity of Darvar in northern Croatia)," Dokl. Akad. Nauk SSSR, Vol. 175, No. 6, pp. 1363-1367.

I. Geologic Hypotheses and Probability Distributions

GEOCHEMICAL BEHAVIOR OF ELEMENTS IN THE LITHOSPHERE*

B. A. Choubert

National Research Center
Paris, France

An original conception of the evolution of the crustal composition is discussed. It is based on results of studies of Precambrian rocks in central Africa and the Guianas, and also studies of the chemical compositions of igneous rocks from other parts of the world. It is shown that the primary material of igneous rocks was similar to peridotite but changed gradually by introduction of leucocratic elements. Further development of the problem by means of mathematical methods is recommended.

The geologist who works in regions where Precambrian rocks are widespread (such as central Africa, South America, Canada, and the like) is constantly faced with great difficulties because of the absence of fossils in the sedimentary strata and because of changes in the crystalline rocks, which have commonly been recrystallized and metamorphosed to such an extent that it is difficult to classify them.

The Precambrian shields consist of the deep parts of ancient mountain systems, where only the roots of former fold zones are preserved. These roots are generally disposed around huge massifs consisting chiefly of granitoids of various ages but of common mineral composition. The area occupied by mafic rocks is negligible in comparison.

It is known that petrographers are interested chiefly in rocks of rare mineralogy, and these are almost never encountered in ancient platforms. Classic petrographic study and description of granitoids is not very interesting and it requires a different approach. The predominance of granitoids of ordinary mineral composition leads one to think that there are fundamental laws according to which precrystalline material gradually acquires with time a common chemical composition, and, on cooling, yields granites that are independent of their geographical position.

---

*This report was read at a colloquium of the German Geological Society at Strasbourg (France), which met March 3-10, 1965. It is published in the present collection with additions and some changes made for the Seminar on Mathematical Geology in Leningrad, April 12, 1965.

We still know very little about the lithosphere. It is taken for granted that the elements in the earth's crust are zonally arranged, concentrated in shells, forming a geosphere with diminishing specific gravity from the base to the top. The heavy metals are concentrated in the center of the earth, and there is an increase in Si content toward the periphery, and dominant Mg is progressively displaced by Al. The lithosphere corresponds approximately to the silicon shell, but detailed subdivisions of this shell differ with different authors. We have not advanced far from ideas of the 19th century, when Edward Suess suggested that the earth is divided into three zones: NiFe, SiMg, and SiAl. Scientists are trying to find refinements of these subdivisions by means of geophysics or by study of meteorites.

The history of the lithosphere is a history of the evolution of a siliceous solution, which may be represented as a mixture of atoms in a state of motion, existing in flexible frameworks of quantitative equilibrium in like manner and with known constancy. Magmatic rocks are in this case unique material, the study of which will allow us to recognize the laws controlling the precrystalline material.

The composition of the precrystalline material, that is, the primary material from which magmatic rocks originated, may be studied on the basis of chemical analyses that permit determination of the loss taking place prior to the cooling of this material and to its crystallization, and also the loss arising during the various changes caused by metamorphism.

However, as it turns out, this introduces a new difficulty, associated with the impossibility of obtaining a single composition from even a single rock, taking samples from very closely spaced localities, no matter what analytical method we use, spectrographic or chemical. Fluctuations in chemical composition prove to be rather large, and this makes it impossible to speak of the nature or type of any particular massif. Deviations, arising from a number of different causes, are combined with facies variations [Shaw, 1964]. A certain number of analyses are therefore necessary in order to obtain a proper view of the composition of any single rock. In general, the many thousands of chemical analyses of rocks to be found in the literature cannot be studied without using statistical methods. These methods are sufficiently well known so that it is hardly necessary to emphasize the importance of their use in this paper.

Let us decide now what it is we are going to subject to statistical analysis. We are interested in the global material before its crystallization. In view of this, we shall not consider crystals. We may speak of molecules, neutral atoms, or positive and negative ions. A molecule is an unacceptable concept to apply to minerals, since the boundary between molecules and crystals is not clear, although both sometimes have their own special properties. The purpose of our study is better served by the concept of atoms and ions, without regard for whether they have electrical charges or valence that may affect their chemical behavior. The size of an atom or ion is measured by its radius or diameter, which may alter under the influence of such factors as ionization and coordination number. These properties of atoms and ions are very important, and we shall try to discover whether the composition of the atomic mixture making up the lithosphere corresponds to general laws, depending on the chemical properties and physical peculiarities (filling space) of the atoms, or whether it simply depends on the quantitative relations among the different atoms or groups of atoms, i.e., on their concentrations in the mixture or solution.

It is well known that analyses of magmatic rocks are given by weight percentages of oxides of the ten principal elements and of water, though some secondary components are occasionally included: S, Cl, Ba, Sr, Cr, and others.

We will first study the eight components: Al, Fe + Mn, Mg, Ca, Na, K, Ti, and P. The primary material is considered to be a solution of all components in silicon in unoxidized and precrystalline form. Atoms of silicon and oxygen are therefore not used in the computations.

In the mixture of these eight elements, the relations of which are highly variable, some common statistical patterns are found (a statistical constancy of ratios and correlation) that apply to almost all magmatic rocks in combination. These results have been described in detail in several publications, to which we will not now turn our attention.

The statistical patterns that were obtained may be expressed in the following form.

Statistical Constancy among Total Charges

$$\text{I. } 3Al + 2Fe + 2Mg + nTi \cong 3(2Ca + Alc + n'Ti + 5P + \varepsilon),$$

where Mg, Fe, and the other symbols represent concentrations of the various elements: Alc = K + Na; n + n' = 4; ε represents elements not present in constant amounts.

Statistical Constancy among Total Atoms

$$\text{II. } Al + Mg \cong Fe + Ca + Na + K + Ti + P + \varepsilon.*$$
$$\text{III. } Al \cong 2Ca' + Alc + \varepsilon \text{ (the most frequent case).}$$

For alkalic rocks we find $Al \cong Alc + \varepsilon$ or $Al \cong Na' + K + \varepsilon$, where Ca' = Ca of feldspars; Ca' + Ca" = Ca; Na' = Na of feldspars; Na' + Na" = Na.

Correlation between Total Charges and Summed Diameters

$$\text{IV. } \frac{EB_{val}}{EN_{val}} = \frac{EB_{diam}}{EN_{diam}}$$

where EB represents the "light-colored" elements Al + Ca' + Na + K + Ti + P; EN represents the "dark-colored" elements: Ca" + Mg + Fe.

Correlation between Mg and Total "Dark-Colored" Elements when Al + Mg = 50% (cf. Statistical Constancy II)

$$\text{V. } Mg = A(EN) \pm b,$$

where A is the angular coefficient of regression; b is a segment along the ordinate axis.

VI. From (I) and (II), by multiplying all members of (II) by 3 and subtracting them from (I), we obtain

$$5Fe \cong 3Ca + Mg + Ti + 12P$$

[if Ti is found in identical quantities on the left and right sides of equation (I)].

This result is very interesting. It shows that alkali feldspars are independent of other elements and may therefore form and develop in rocks without disturbing the overall equilibrium (feldspathization).

Let us recall that the indicated equations have a statistical significance, and their treatment is therefore different from the treatment of algebraic equations in the determinate sense.

---

*Recent research (1968) on more important material tends to show that a significant part of the "nondosed" elements (factor ε in the above formula) could be the hydrogen contained in micas as well as in amphiboles. The importance of this element, which is generally overlooked and is almost wholly unsusceptible to detection by ordinary techniques, will be emphasized in a future paper.

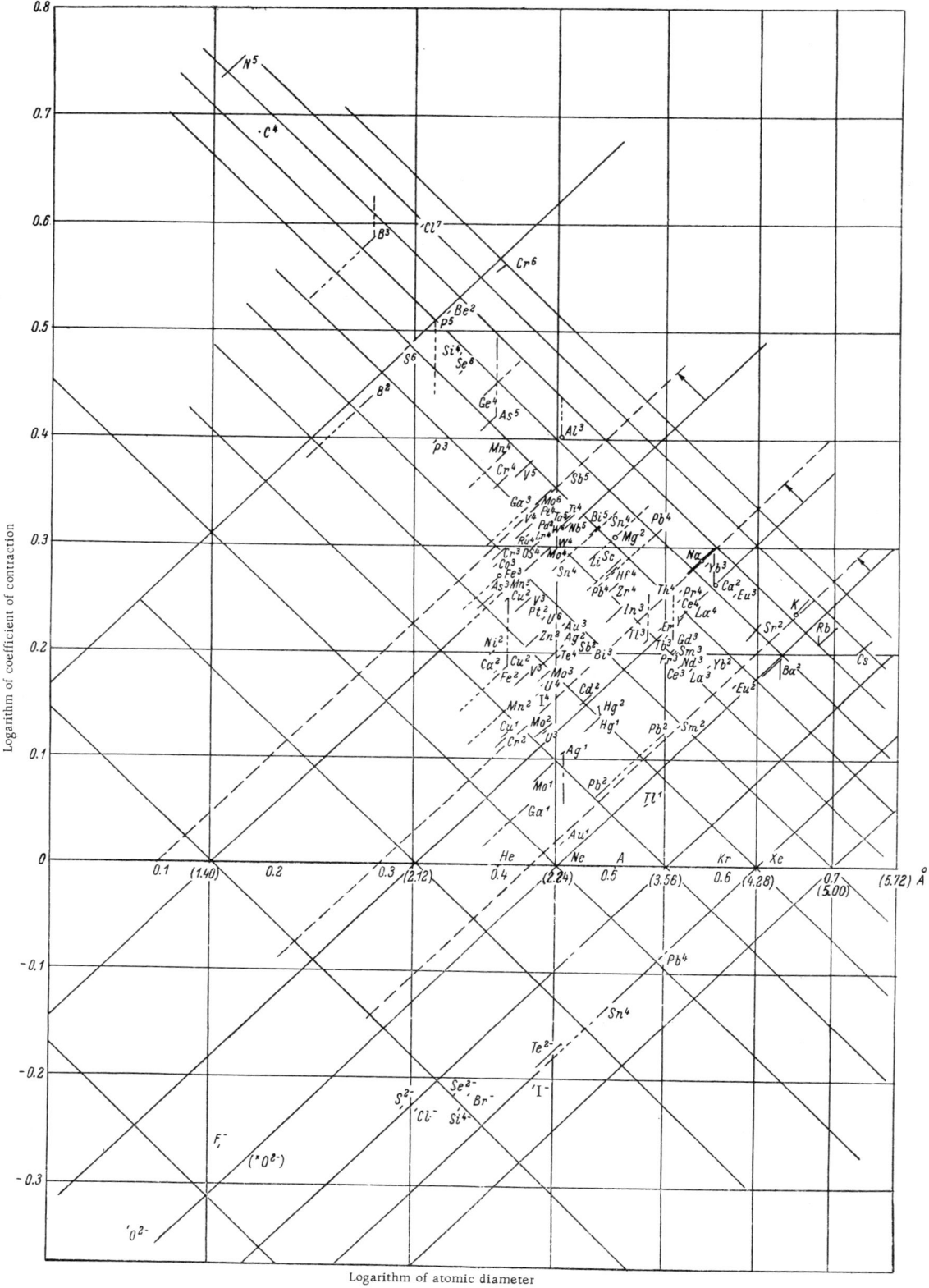

Fig. 1. Classification of elements by atomic (ionic) diameter. Vertical lines represent boundaries for classes of atomic diameters with an interval of 0.72 Å; 45° lines represent boundaries for classes of ionic diameters. In this figure all possible variations of diameters occupy areas limited by the greatest and least values of the ratio D_{at}/D_{ion} (areas are not illustrated in this figure).

Table 1. Distribution of Elements by Atomic Diameter (interval of 0.72 Å)

f'_4 0.68–1.40	f'_3 1.40–2.12	f'_2 2.12–2.84	f_1 2.84–3.56	f''_2 3.56–4.28	f''_3 4.28–5.00	f''_4 5.00–5.72
H F (?)	N C B S Cl P	(P), Be Se, Mn Br, Ga **Fe**, Ni Co, Cr As, Cu V, Zn Ru, Rh Os, Ir I, Ce Mo, W Pd, Pt U, Sn He	(Sn), Te Ta, Nb **Al**, Ag Au, Sb **Ti**, Sc Cd, Hg Li, Bi Zr, **Mg** Ne, Ar In, Tl Pb	Th Nd Y Ce La **Na**	Sr Ba Xe **K** Rb	(Rb) Co

Note. Chemical analyses borrowed from the works of Washington (166), MacDonald (12), Buddington (4), Johannsen (11), Hall (8), and Choubert (1). Analyses made before 1906 are not included.

They indicate the existence of average equilibbrium conditions, which may be lacking in particular circumstances. We assume that these equilibrium conditions apply to the precrystalline state of the material from which magmatic rocks have formed.

Classification of Atoms by Size

In addition to definite relations among the atoms and a balance between charges (valences), the packing of atoms is also very important, i.e., the role played by the size of particles when they occupy space under great pressures without forming crystalline structures.

According to calculations of A. F. Kapustinskii, pressures in the lower layers of the lithosphere may even change the shape of the electron orbits, and this, of course, also changes the chemical properties of the elements.

The radius (or diameter) is one of the most important characteristics of atoms. Diameters of neutral atoms measured in metallic structures change through ionization. Cations are smaller, anions larger. The diameters also change because of external factors such as coordination number.

It is known that at high pressures even hydrogen has metallic properties. In view of this, we ascribe chief significance, in our studies, to atomic diameters in metallic structures.

Figure 1 shows the distribution of atomic and ionic diameters depending on the D_{at}/D_{ion} ratio, which may be called the "coefficient of contraction." The logarithm of the coefficient of contraction is plotted on the ordinate, the logarithm of atomic diameter on the abscissa. This graph shows the variations of two systems: ionic and atomic. In the atomic system the points are displaced vertically, in the ionic, at an angle of 45°.

If all elements in the Mendeleev table are classified according to their diameters, then, by choosing the size of a class to be 0.72 Å (Table 1), we obtain seven classes [Choubert, 1963].

We have already noted the value of such a classification of atoms [Choubert, 1963]. It permits one to recognize the alkali metals (Na and K) as a different kind of material, added to the elements that make up peridotites. If there were no alkalies, then peridotites would exhibit symmetrical distribution of elements according to size, at least for the frequencies f'_2 and f''_2 (the frequencies f_3 and f_4 have been but poorly studied thus far).

Computations, based on more than 3000 analyses, confirm the view that, on the average, Fe = Ca (at.%). This "statistical regularity" supports the hypothesis of a peridotite nucleus having a normal distribution of atoms according to size, made up of Mg (+Al), Fe, Ca, and the minor elements in formula (VI). This nucleus is preserved during the entire evolutionary history of the mixture, decreasing with increase in content of alkali metals from peridotites to granites. This equilibrium is statistical, however; there are always differences in the ratio of iron to calcium in the different rocks. The interpretation of this balance is flexible; we have to do with an unstable evolutionary equilibrium, i.e., with an average state, in which iron is sometimes less abundant, sometimes more abundant, than calcium [Choubert, 1963].

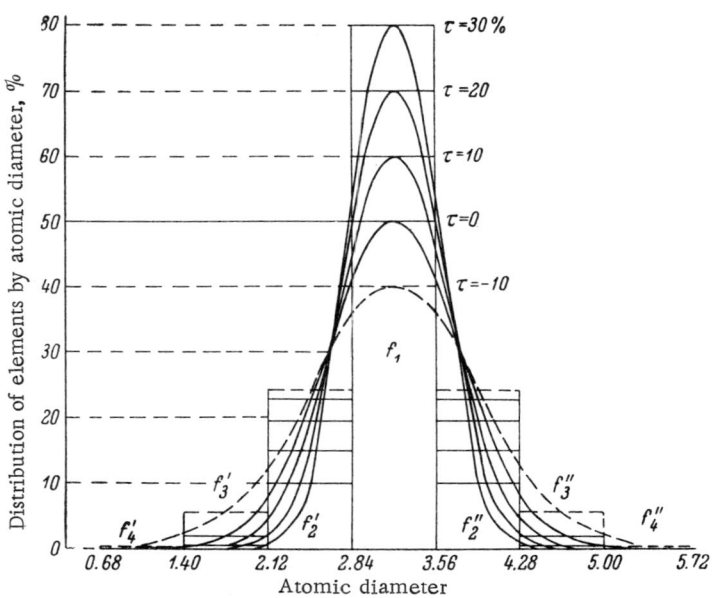

Fig. 2. Different forms of elemental distribution, corresponding to a peridotitic nucleus in case of normal distribution of diameters in period I. $\tau = f_1 - 50$ at.%; $f_3' + f_4' = \delta$; $f_3'' + f_4'' = \Delta$.

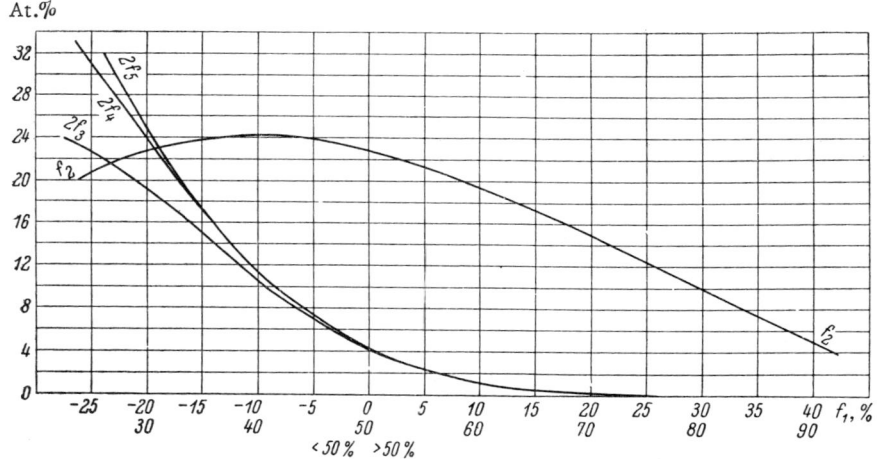

Fig. 3. Dependence of f_2, $2f_3$, $2f_4$ on τ for normal frequency distribution.

The formulas that we obtained on the basis of general calculations refer to mixtures that are nearly symmetrical, in which the proportions of Fe and Ca are approximately the same. They cannot apply to rocks that are entirely unstable, such as end members of the iron series (collobrièrite, nelsonite, ferrohypersthenite) or the calcium series (carbonatite, okaite, etc.) [Choubert, 1963]. Another aspect of classifying atoms by size is interesting if we recognize the validity of the hypothesis that it is possible to obtain values of f_3 and f_4 from known values of f_1 and f_2 from the normal distribution of elements in the peridotitic nucleus. In other words, when the distribution is symmetrical, it is sufficient to know the value of Fe (or Ca) and τ (see below) in order to determine f_3 and f_4. The frequencies f_3' and f_4' consist of gases, and we may thus form som judgment of their quantities, something that was previously impossible (Fig. 2).

Equation (II) permits us to consider the sum Al + Mg as C = 50%. The symbol τ designates the combination of elements $f_1 = 50\%$. For simplicity, we use δ to designate the total elements $f_3' + f_4'$, and $\Delta = f_3'' + f_4''$.

In Fig. 3, the evolution of $\delta + \Delta$ and f_2 in relation to τ is shown.

Evolution of the Composition of the Precrystalline Mixture with Time

A study of magmatic intrusions, chiefly those that follow one after the other within ancient Precambrian shields, has shown that the chemical compositions of these rocks change from older to younger. Backlund [1943] and Marmo [1962] have noted that in rocks of the Scandinavian shield more or less basic diorites and quartz diorites existed; then more acidic granodiorites appeared, and still more alkalic granites. The guianan shield follows the same rule in its relations of intrusive rocks, and we have devoted several years to refinement of the character of this evolution. Basing this work on a large number of chemical analyses [Choubert, 1960], we have been able to establish the fact that the diminution in magnesium content follows an exponential law, on the average, yielding a curve that is similar to the curve for radioactive decay and being compensated by a symmetrical increase in aluminum content.

The concentrations of Mg affect other components. In general, the "dark-colored elements" (Mg + Fe + Ca") also decrease exponentially in favor of the "light-colored elements." As one follows the curve upward he goes from granitic rock to diorite, then to gabbro (basalt), and, finally, to pyroxenite and peridotite (Fig. 4).

These last rocks, especially the "unsaturated," contain less than 60% of the primary constituent atoms. Recomputation of the composition (on the assumption of Al + Mg = 50%) shows that peridotite and pyroxenite occur at the beginning of the curve, not far from the origin. The missing part of the components may be restored if the hypothesis of normal atomic distribution is followed [Choubert, 1963].

If evolution takes place on a continental scale, as in the Guianan shield, great periods of evolution notably coincide with the absolute ages of important geological events. Stages marked by principal episodes of granitization in the Precambrian have the following average ages (in millions of years): 500 ± 100, 1200 ± 100, 1900 ± 120, 2600 ± 120, 3300 ± 130, 4000 ± 150 with a rather constant interval, about 700 million years. This simple coincidence (?) of duration was indicated by the study of U^{235} decay as early as 1958; new age determinations in Guiana confirm the hypothesis [Choubert, 1960, 1965].*

Of course, these figures express but general average data. In a single mass, ultramafic rock is not always older than gabbro, although it is richer in magnesium. These facies associations form because of the well-known process of differentiation, which may be considered a segregation of dark-colored elements within the gabbroic rock.

If we add material to the composition of peridotite and pyroxene on the basis of the relation Al + Mg = 50% we obtain "preperidotite," which is very similar to some basalts rich in Fe, Mg, and Ca at the same time. These are rather rare on the surface of the earth. This may explain why peridotite and pyroxenite are commonly found together; our "preperidotite" corresponds to a mixture of proportions that are unstable in the upper parts of the earth's crust.

---

*In a recent work we have defined more precisely the rhythmicity of maximal magmatic phases through geologic time as well as the value of the period (approximately 720 million years): B. Choubert, "Réflexions sur la finalité des mesures géostatistiques," Bull. Soc. Géol. France (1967) (in press).

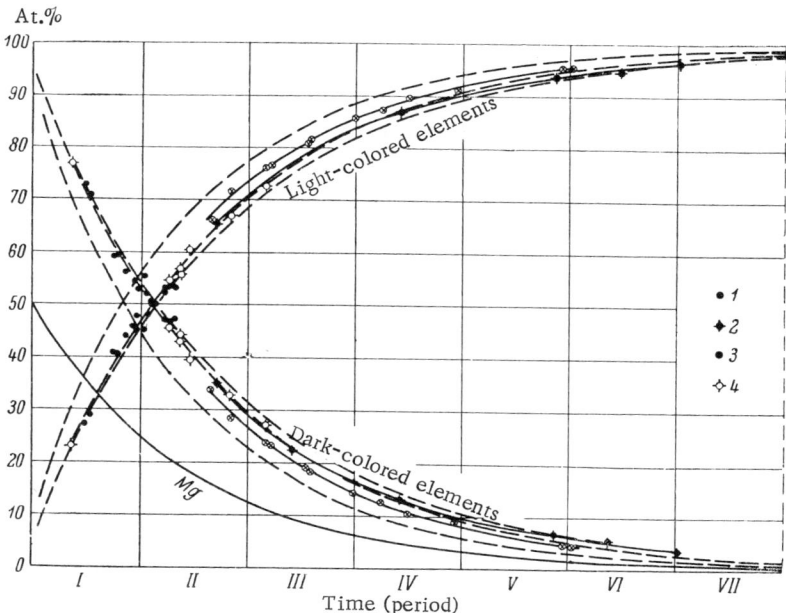

Fig. 4. Evolutionary curves of Mg (with Mg + Al = 50 at.%). Prototypes: 1) Flamanville granite (France); 2) gabbro—tonalite—granodiorite—granite of the Sierra Nevada (California); 3) Mauna Loa basalt (Hawaiian Islands); 4) basalt—andesite of Mauna Kea (Hawaiian Islands).

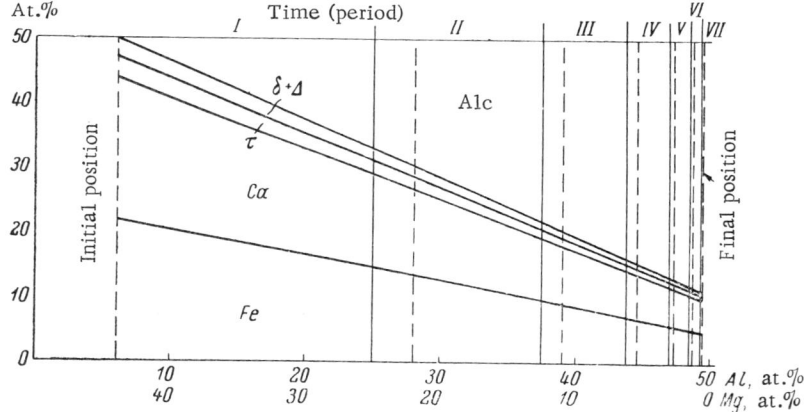

Fig. 5. Rectilinear evolution of a symmetrical mixture for Mg = 44, $\tau = 3$, $\delta + \alpha = 3.02$ at.% (in the initial position, period I) and for Fe = Ca = 5 at.% (in the final position, period VII).

During formation of the sial, basic rocks appear repeatedly, and these, in turn, are involved in the process of evolution. We thus obtain the "progeny": peridotites, gabbro, diorites, granites, all progressively younger and coexisting with the parents. It is not always possible to distinguish one from the other. Metamorphism sometimes makes it impossible to distinguish them in the field. Analyses of gabbro, diorite, and other rocks therefore prove to be much the same for older and younger masses. The average chemical composition is not strongly changed in the processes, but the classification according to time of formation becomes subject to error.

Stages of decreasing magnesium content with time are also traced into younger massifs, such as in the Sierra Nevada batholith of California, where Larsen described the sequence of appearance as gabbro, tonalite, granodiorite, and granite. This is a repetition on a small scale (in time and amount of material) of the same general phenomenon, and it raises the very interesting problem that we have noted.

Calculation of Prototypes

We have now assembled a number of statistical relations, illustrated by formulas, forming flexible frameworks, into which changes in the chemical composition of magmatic rocks may be introduced.

In their solid forms, crystalline aggregates — plutonic rocks and volcanic glasses forming at the surface — give only an approximate idea of the composition of the mixture from which they arose. All rocks are thus incomplete in the sense that they have lost most of the volatile elements during their evolution, a small part of which is sometimes preserved in bubbles in crystals.

Most of the metals were deposited in joints of the surrounding rocks during the formation of veins and mineralized zones. Others formed disseminated zones of mineralization within the host rocks. Lastly, a whole series of such processes as magmatic differentiation, crystallization differentiation, migmatization, and feldspathization shifted the final product from its initial composition.

The formulas that relate elements or groups of component elements allow us to calculate what we may consider to be the initial composition of the rock. We thus obtain a prototype, a "full" mixture of atoms having a symmetrical peridotitic nucleus (in the sense of atomic distribution by size) and including not only the major elements but also the accessory elements, normally not determined in analyses.

Such a calculation is possible under the following conditions:

1. The rock must be "symmetrical," i.e., correspond to the equation

$$Fe = Ca = \frac{Fe + Ca}{2} = f_2.$$

This is possible when the difference between Fe and Ca is not especially large (otherwise an asymmetrical distribution would be obtained).

2. The evolution must be rectilinear, i.e., depending on the Mg content, a linear change must take place among the other elements of the peridotitic nucleus from peridotite to granite (Fig. 5).

Use of the Formulas in Case of a Symmetrical Nucleus

If $f_2' = f_2''$ (and $f_3' = f_3''$, $f_4' = f_4''$) is assumed for our initial hypothesis, we may then obtain a generalization of formula (VI)

$$2Fe_0 - Mg_0 = [3(y-1) - x]\tau_0 + 3(m-1)(\delta_0 + \Delta_0), \qquad (1)$$

where Fe_0, Mg_0, etc., represent concentrations of these elements at the initial point (origin), period I. At this same point, $\tau_0 = f_1 - 50$ at.% (τ replaces Ti); τ is the combination of elements of the central class except for Al + Mg; $\delta_0 + \Delta_0 = (f_3' + f_4') + (f_3'' + f_4'')$ at the same point; $\delta + \Delta$ replaces P in the formulas given above; m is the average valence of $\delta + \Delta$; $x + y = n$ is the average valence of τ.

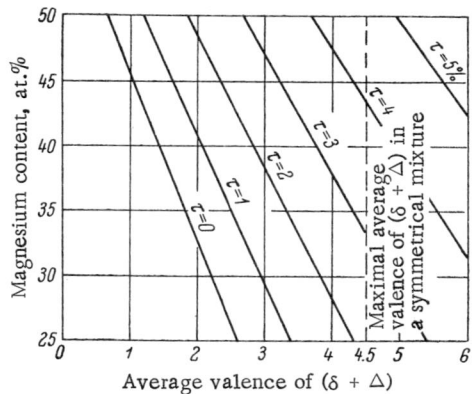

Fig. 6. Magnesium contents at the different periods. Values of τ depend on Mg concentrations and average valences $\delta + \Delta$ in the interval corresponding to equilibrium of τ with iron [left side of equation (VI)].

This formula in period VII is converted to

$$2\text{Fe} - \text{Mg} = 3(n-1)\tau + 3(m-1)(\delta + \Delta), \quad (2)$$

since all elements of the sum of τ are found on the right side of formula (I). In (1), y may be zero when τ does not exceed 5% ($f_1 = 55\%$) and the percentage of Mg is rather high (Figs. 6, 7, 8). In this case we have

$$2\text{Fe} - \text{Mg} = -(n+3)\tau + 3(m-1)(\delta + \Delta).$$

In the different periods (I–VII), the magnesium content has the following values (in at.%):

I	II	III	IV
50—25	25—12.5	12.5—6.25	6.25—3.125

V	VI	VII
3.125—1.5625	1.5625—0.78	0.78—0.39

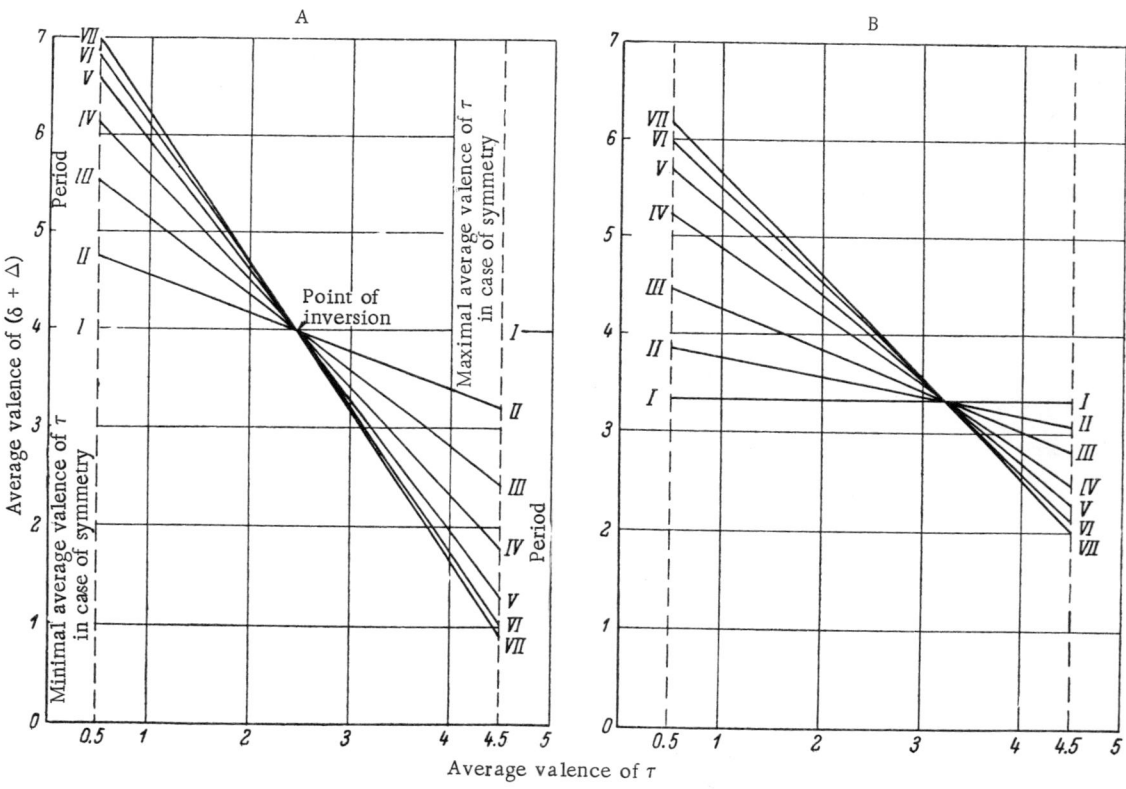

Fig. 7. Average valence of τ and $\delta + \Delta$ for different periods of evolution. A) Initial position, $\tau = 4$ at.%; average valence $\delta + \Delta = 4$; final position Fe = Ca = 7 at.% (period VII). B) Initial position $\tau = 3$ at.%; average valence $\delta + \Delta = 3.316$; final position Fe = Ca = 5 at.% (period VII).

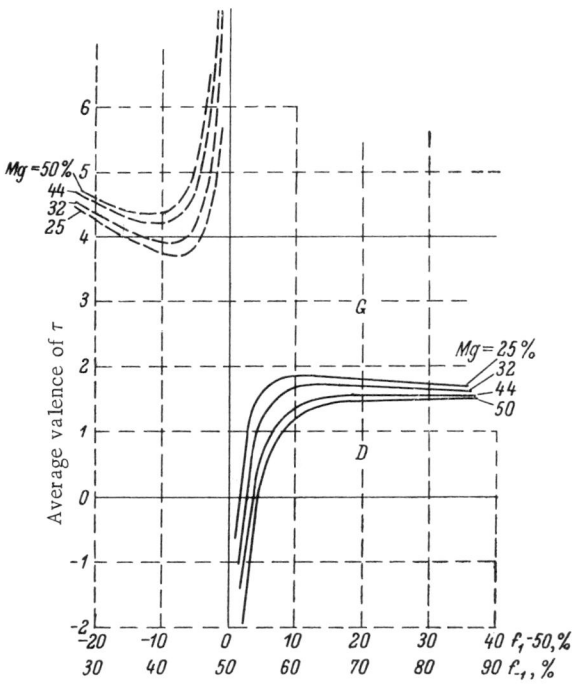

Fig. 8. Curves limiting sectors G and D (period I). G) Proportion of valence of τ on the left side of equation (II); D) the same for the right side (for average valences of $\tau = 4$ and $\delta + \Delta = 4$ and different concentrations of Mg).

Lastly, what is very interesting, Ca is separated into Ca' and Ca" according to the formulas (when symmetry obtains):

$$2Ca' = Al - (Na + K) \text{ (the particle } \varepsilon \text{ is discarded)}$$
$$2Ca'' = Mg - [\tau + (\delta + \Delta)]. \tag{3}$$

The calculation reduces to solution of the equation

$$(Al + Mg) - (Ca + Fe + Alc) = \tau + (\delta + \Delta),$$

where Ca = Fe; and Al, Mg, and the others are known.

This problem, apparently very simple, is complicated by the fact that each element may be present in amounts differing from the average, and this requires appropriate corrections.

Search for Genetic Criteria

Until this point we have examined the character of individual rocks and have shown a method that permits us to reconstruct the chemical composition of the precrystalline mix from which magmatic rocks — volcanic and plutonic — derived. This method is based on the hypothesis that there exists a peridotitic nucleus with symmetrical distribution of atoms according to size.

The concept of a peridotitic nucleus is an *a priori* assumption for analyzing questions concerning the origin of the mix. We have seen that the exponential dependence of the composition on time, referring to the Mg content, is the backbone of the whole process of evolution and (among other things) of the gradual displacement of the two principal groups of elements that form the melanocratic and leucocratic parts of the rock, one by the other.

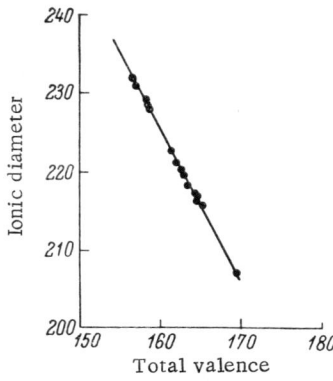

Fig. 9. Correlation between totals of ionic diameters and the valence of different samples of prototypes of the Flamanville granite (France).

This genetic problem has still another aspect: what common features are found among samples taken from a single massif (regardless of size), or, in different words, what are the criteria, when we study a series of samples selected at random, that might show us if these samples have a common origin? From the preceding discussion we may see that, if the magnesium curve has any connection with the remote past, the question confronting us has to do with the time of formation of the massif.

It may be assumed that the hypothetical initial mix, existing under certain temperature and pressure conditions, had some composition that will be reflected in its final solidified form. We studied this question in one of our previous works [Choubert, 1963].

In this study, it was shown that a dependence of the following form exists:

$$N_{at} = A \Sigma_{diam} + b, \qquad (4)$$

or, by replacing the number of atoms (for 100 valences) by the number of valences in 100 atoms,

$$\Sigma_{diam} \cong -A \Sigma_{val} + b, \qquad (4')$$

where A is the slope of the line and b is the segment on the ordinate axis.

In combining this correlation with that connecting valence with the diameters of the light-colored and dark-colored elements, we are able to correct the number obtained for alkalies of the Nancy standard [Choubert, 1963].

In calculating the prototype we obtain:

$$\frac{EB_{at}}{\Sigma_{at}} \cong \frac{EB_{val}}{\Sigma_{val}} \cong \frac{EB_{diam.\ ion}}{\Sigma_{diam.\ ion}} \qquad (5)$$

and

$$\frac{EN_{at}}{\Sigma_{at}} \cong \frac{EN_{val}}{\Sigma_{val}} \cong \frac{EN_{diam.\ ion}}{\Sigma_{diam.\ ion}},$$

where EB represents total atoms in the composition of the "light-colored" part, $Al + Ca + Na + K + \tau + (\delta + \Delta)$; EN represents total atoms in the "dark-colored" part, $Fe^{2+} + Fe^{3+} + Mg + Ca"$.

Below are given the percentage relations among contents of atoms, valence, and summed atomic diameters:

Atoms	77.185	84.435	79.19	76.5	75.155	73.355	88.26
Valence	78.333	83.89	79.335	77.1	74.993	73.695	88.754
Diameter	77.64	84.45	78.995	76.38	75.434	73.13	88.394
Atoms	81.05	89.68	90.835	89.73	96.42	86.08	
Valence	81.00	90.05	91.15	89.53	96.41	86.40	
Diameter	80.96	90.00	90.907	89.82	96.38	86.60	

The figures illustrate the results of our computations for several analyses of the Flamanville granite (analyzed by J. Goni), France.

The values of EB and EN_{val} and also of EB and $EN_{diam.\ ion}$ are obtained by multiplying the percentage values of atoms of EB and EN by the corresponding numbers of valences and ionic diameters and then calculating the sum. The regression of one to the other is rectilinear for prototypes of rocks originating from a single mass or a single flow of lava (Fig. 9).

As study of magmatism in the Guianan Precambrian has shown [Choubert, 1960], EB and EN are exponential, depending on the time of evolution. In a system of semilog coordinates (log of Mg concentration as the ordinate, time as the abscissa), the evolution of Mg is rectilinear (negative). The same is true for EN, with a slope that depends on the final value of Ca = Fe in period VII. The regression of Mg to ΣEN may therefore be defined in the following way:

$$Mg \cong A\,(\Sigma EN) \pm b, \tag{6}$$

where $\Sigma EN = Mg + Fe + Ca''$ (at.%).

The genetic features of rocks of a single mass are expressed by equations (4)-(6). The subdivision into EB and EN or, more precisely, into two groups of elements that take part chiefly in the formation of the leucocratic and melanocratic parts of the rock (EB and EN) during crystallization corresponds to a very precise separation of different categories of atoms, their valences, and their diameters.

Rocks with Normal and Disturbed Evolution

The proposed method of computing the composition of prototypes leads to the subdivision of rocks into two categories: 1) rocks with an excess of Al + Mg (frequencies $f_2 - f_3 - f_4$, deficient relative to 50 at.%), and, 2) rocks with insufficient Al + Mg (frequencies $f_2 - f_3 - f_4$ in excess).

The existence of these two categories poses an interesting problem touching the origin and evolution of the initial mix. It is explained by the fact that, from one period to the next, material that existed at first is gradually lost, and in this way EB replaces EN.

Apart from development of concentrations of Fe, Ti, and other elements (magmatic segregations), which are the consequence of these replacements, material which should have been "removed" or "lost" remains in place because of introduction of alkalies, and the normal course of the process is disturbed. As a result, one may observe enrichment in places of those elements that should have been removed from the mix: Fe, Ca, and Ti, and also parts of δ and Δ. On the other hand, alkali (Na) may also form excessive concentrations.

These local, secondary concentrations are generally distributed throughout the mass, where they are found together with normal types.

Hybrid rocks are dominant in some massifs. These are characteristic of syenites, as a rule (shonkinite, nepheline syenite, ijolite, etc.). Despite the wide geographic distribution of such rocks, according to Fersman they constitute only 1% of the total magmatic rocks.

Pegmatites

As is well known, granitoidal rocks are accompanied by a network of distinctive bodies called pegmatites. A. Lacroix spoke of "residual solutions"; other authors have used the term "granitic juice."

One peculiar feature of pegmatites is the high content of rare elements in them: Li, Be, Nb, Ta, Sn, W, rare earths, and others. Long ago a relation was established between the presence of trace elements and pneumatolysis (the action of "mineralizing agents" according to Michel Levy), since many minerals in these rocks contain F, Cl, B, and related elements. Rosenbusch called these "fumarole facies."

There is no doubt that the material in all these pegmatite veins and lenses before their emplacement was part of the granitic mix, with which they are now genetically related. The segregation of residual solutions takes place unevenly, sometimes in large quantity, sometimes in negligible amounts, since some granites are not accompanied by pegmatites. In this case, the combination of elements characteristic of these bodies remains within the granite mass, but it is highly improbable that this is ever completely true.

The preservation of "residual solutions" in rocks is manifested in enrichment of the initial mix in the "light-colored" elements, but we know already that an increase in the amount of alkalic feldspars takes place without disturbing overall equilibrium. The calculation therefore shows that, after calculating the amount of anorthite, the appearance of a pegmatitic residue for Al = Alc proves impossible.

But, as calculation of the Nancy standard has shown, and also the American standard G, almost always Al $\gtreqless$ Alc. An excess of one or the other may indicate the presence of a residual solution. In this case, the formation of feldspar is impossible. These excesses at the time of crystallization involve other minerals (micas, amphiboles, alkalic tourmalines, etc.). The suggested problem deserves more intensive study, since precise computation of prototypes depends on it.

In conclusion, we may raise the question of whether these investigations, in addition to scientific interest, have any practical use. There is no doubt that until the fundamental geochemical processes become known, the behavior of each element in particular remains a mystery. The problem may therefore be examined from the practical point of view, since theoretical reconstruction of the composition of the initial material may permit us to find new deposits of useful minerals, arising from negligible quantities of the corresponding elements included in the rocks. It has been noted already, in using methods based on geochemical data, that deposits are definitely related to the distribution of metals in magmatic masses. But, regardless, geochemistry is related to petrography and historical geology.

Views on the Metallogeny of Trace Elements

The correlations we have discussed are very important, since they permit us to compute τ when the Mg content is insufficient, and also, therefore, to compute $\delta + \Delta$ and Fe = Ca. As we have noted, τ and $\delta + \Delta$ consist exclusively of secondary elements, whereas f_2 includes many metals. Thus, precise computation of prototypes may have some influence on prospecting for mineral deposits.

Investigation of concentrations of minor elements has heretofore been a study associated with mineral deposits. However, prospecting cannot be fruitful if it is divorced from the basic problems of geochemistry. This fact was confirmed by P. Routhier [1963] in his book on mineral deposits.

Determination of the relations among different types of mineralization and different families of rocks has not yet passed the stage of direct observation, and summaries in this field still continue to be a collection of all possible examples.

Except for deposits associated with segregation (or protocrystallization) or concentrations in the immediate vicinity of intrusive masses (peribatholithic deposits) or with pegmatites, all groups of accumulations of metals have no clear connection with rocks, because of the distance they are removed from the source. In view of this, we commonly do not know to which mass or which type of rock a metallic deposit belongs. Of course, we do not have the slightest conception of the amount of metals initially present in gabbros or granites. We do not speak of "traces" of metals that are now readily measured, but we wish to determine the total quantity having "potentially" existed in some intrusive mass, the source of mineralization.

The exact content, let us say, of niobium and tantalum in a pegmatite vein becomes known when all the veins have been worked out. It is therefore perfectly understandable that considerable interest should be found in indirect methods that permit a better understanding of the formation of mineral deposits. The potential of rocks relative to mineral deposits and also to other elements remains an enigma, and numerous observations in the field merely confirm the inadequacy of the old theories, supplying little that is new.

Specialists in different disciplines have tried to solve some of these problems by means of mathematical methods. Of these, we should note first, in the USSR, F. Y. Loewinson-Lessing, who was a pioneer in this field, then A. B. Vistelius, O. V. Sarmanov, and N. K. Razumovskii; in the USA, F. Chayes; in Canada, D. M. Shaw; in Switzerland, P. Niggli; in France, F. Blondel, G. Materon (in connection with mineral deposits), the author of these pages, and others.* At present the path is more or less marked out, but investigation of the possibilities along this path have only begun, so broad is the field of endeavor.

In order to determine the contents of elements from τ and $(\delta + \Delta)$, the average valence values of which we know, the problem lies in using atoms, in the size classes, that have similar diameters and valences that range between known limits. This problem is soluble, since characteristic associations of elements and minerals exist (paragenetic combinations) in well-studied deposits. Thus, a foreseeable number of possibilities is obtained, and the final selection is backed by the considerations of chemical, mineralogical (paragenesis), geological (environment), and mathematical (probability) disciplines.

Determining Loss during the Evolutionary Process

The initial contents of a number of elements decline from one period to the next. In this process a considerable part of the Fe, Ca, τ, and $(\delta + \Delta)$ are lost, and, of course, Mg, which is replaced by Al.

The "loss" generally follows the same exponential law as the magnesium content, the amount decreasing by half from one period to the next.

In Table 2 an example of computing loss (in at.%) is given. Mg = 44%, Ti = 4%, $(\delta + \Delta)$ = 2.66% (period I), Fe = Ca = 7% (period VII).

By knowing that, in addition to Fe and Ca, classes f_2^l and f_2^{ll} contain minor elements, particularly most of the metals, it becomes clear that a large number of them will depart with Fe and Ca, which naturally constitute the chief part of the loss.

We know that iron forms the main concentrations between stages I and II and between II and III in ultramafic rocks, gabbros, and diorites, and this corresponds to field observations. During the same interstages anorthosites and other rocks rich in calcium also appear. Deposits of titaniferous iron, chromium, nickel, and platinum are also associated with basic and ultra-

---

*We do not pretend that this list is complete; the number of individuals using statistical methods is constantly growing.

Table 2. Balance of Elements by Periods (I-VII)

Elements	I	Loss	II	Loss	III	Loss	IV	Loss	V	Loss	VI	Loss	VII
Fe	21.67		14.220		10.494		8.631		7.699		7.233		7
Ca	21.67		14.220		10.494		8.631		7.699		7.233		7
Ca		7.4512		3.7256		1.8628		0.9314		0.4657		0.23285	
Fe		7.4512		3.7256		1.8628		0.9314		0.4657		0.23285	
τ	4		2.625		1.937		1.593		1.421		1.335		1.292
τ		1.376		0.688		0.344		0.172		0.086		0.043	
$\delta + \Delta$	2.66		1.744		1.288		1.059		0.945		0.888		0.8595
$\delta + \Delta$		0.921		0.456		0.228		0.114		0.057		0.0285	
Σ		17.1994		8.5952		4.2976		2.1488		1.0744		0.5372	

basic rocks. Copper is associated with andesites; pyrite—sphalerite—galena accumulations with granodiorites, pegmatites with granites, and so forth.

Concentrations may also form from elements that were previously extracted from the mass but are again drawn in during the evolutionary process. This may be the origin of carbonatites and iron in connection with hybrid rocks (syenites, shonkinites, ijolites). A change in the external conditions (decrease in pressure and temperature) causes the removal of elements in a definite sequence (staggered), beginning with the most "refractory" and ending with those that crystallize at temperatures equal to or lower than that of granitoids. These problems have been studied by many investigators, particularly Fersman. The crystallization temperature of rocks ranges from 1500 down to 600°C. The crystallization of different minerals begins at different temperatures, some beginning a little earlier and terminating much later: pneumatolytic minerals form earlier (1000-400°) than hydrothermal (310-100°).

The classification of some elements that separated from solution during evolution, depending on composition of the rock and its relation to the different periods, is shown schematically in Table 3.

The superposition of external factors, such as depth, pressure, temperature, and geographic position relative to the focus, leads to complication of the chemical reactions, which are more or less interrelated. The paragenetic associations are not very characteristic for each stage of the evolution. The boundaries between different associations is imprecise, and the causes for concentration of any particular element (either principal or secondary) differ (initial concentration, chemical composition of the host rocks, or simply the necessity of maintaining internal equilibrium).

The nonmetallic elements S, As, Sb, F, Cl, B, and others are segregated during the entire process of evolution.

Dynamics of Mineralization

Equilibrium between valences permits one to explain the scheme of mineralization in the following way.

The first case, the normal course of evolution:

a. Increase in amount of Al^{3+} at the expense of Mg^{2+} during the first period shifts the equilibrium of valences, except for those greater than two, in f_2 classes, and leads to removal of Cr, V, Pt (along with Fe), and also Th, the rare earths (along with Ca), and others.

b. The appearance of Na^+ (and K) leads to removal of Cu^+.

Table 3. Classification of Elements by Periods and Rocks (I–VII)

Elements	P. I	P. II	P. III	P. IV	P. V	P. VI	P. VII
	peridotites	gabbros	diorites	granodiorites, monzonitic granites			
	pyroxenites	norites	quartz., diorites	akeritic granites		alkali granites	
				(syenites, nepheline syenites) (monzonites)			
	basalts sensu lato		andesites	dacites		rhyolites (trachytes, phonolites)	
C							
Pt							
Pd, Os, Ir, Cr							
Ni, Co							
V							
Cu							
Au							
Ag							
Pb							
Zn							
U							
(TR, Zr, Th)							
Ta, Nb, Li							
Sn							
W, Mo							
Ba							
Cs, Rb							

Note. Elements in parentheses belong chiefly to rocks indicated in parentheses. Intensity of line indicates quantitative values.

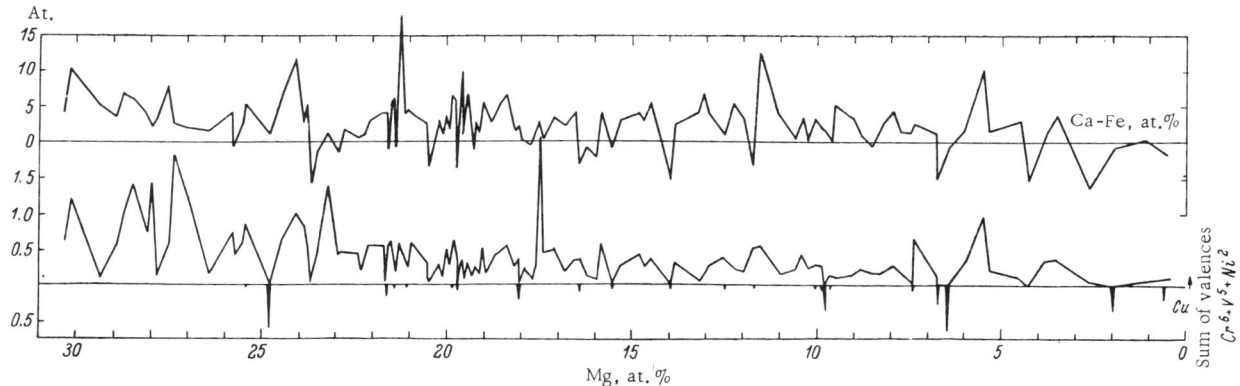

Fig. 10. Relations between the difference Ca − Fe (at.%) and the sum of valences $Cr^{6+} + V^{5+} + Ni^{2+}$. The negative black peaks show correlation between a particular difference Ca − Fe and the valence of Cu^{2+}. All values have been calculated for Al + Mg = 50 at.%.

c. Lastly, removal of Sn, W, Bi, and other such elements, the high valences of which are no longer compensated by the low valences of the alkalies (from the time that Al is sufficiently abundant to "saturate" the alkalies).

From the point of view of atomic diameters, losses take place from small diameters to large, from the class f_2^1 to class f_1, taking into account changes caused by transition to the ionic system, by which ions of the metals Mo, W, and Sn now prove to be in the same class as

Mg and Fe (Fig. 1). This sequence corresponds to the general evolution of the mean diameter, which increases slowly at first, after the first period, and then more and more rapidly [Choubert, 1963].

Extraction of the principal elements follows the same tendency: Fe comes out of solution more rapidly than Ca. It becomes understandable why hybridization begins with enrichment of calcium (predominantly calcium types in periods II, III, and IV, and types rich in iron in periods V, VI, and VII).

The second case, disturbed evolution (syenite, agpaite): here, concentrations of Zr, Ti, rare earths, V, Th, and Nb are especially characteristic, and also Fe and Ca. The rocks are commonly rich in F, Cl, and P.

According to normal distribution, the sequence of events will be the following:

a. Intense introduction of alkalies (Na) in the prototype, as a consequence of which $2Ca + Alc > Al + Mg$, and as a result of this

b. Increase in τ with removal of Ti, Zr, Nb, and, sometimes, part of Al;

c. Simultaneous increase of $\delta + \Delta$ in the prototype; relative abundance of P, Cl, and F (apatite, sodalite, villiaumite), and also Sr, Ba, and K;

d. Restoration of equilibrium requires the segregation, in part, of Ca with rare earths, Th, and related elements as well as Fe, Mo, V, and others.

The third case, pneumatolytic: in going from peridotites to granites, pneumatolytic segregations become increasingly abundant, forming pegmatites. With introduction of alkali metals, the central class (τ) increases, while all the others decrease. The decrease of $\delta + \Delta$ progressively frees gases and elements of larger diameter (Rb, Cs, Ba, etc.), which may thus be segregated under appropriate circumstances, taking different metals with them.

Geochemical Behavior of Cr, Ni, (Co), and V at Low Concentrations (Parts of a Percent and Traces)

From a preliminary study of the behavior of these metals on the basis of analyses published in the literature, it may be shown that in 80% of the examples this group appears in amounts determined by the ordinary methods of classical chemical procedures when Ca is more abundant than Fe (in at.%).

Rocks in which Fe is more abundant than Ca commonly contain Cu. This means that excesses of valence of Cr—Ni—V compensate Fe in f_2' when it is too low; on the other hand, the presence of Cu^+ equalizes Fe when it is too high relative to f_2'' (Fig. 10).

Analyses of rocks in which the contents of elements with Cr, Ni, V > 0.01% (by weight) without copper or but traces of copper represent 78.31% of the total where Ca > Fe; with Cr, Ni, V, Cu > 0.01%, the value is 50.0% where Ca ≷ Fe; without Cr, Ni, or V (or only traces) and Cu > 0.01%, the value is 60.0% where Ca < Fe.

Similar results were obtained with rocks in which the content of Cr, Ni, and V was considerably lower (Fig. 11).

For example, in the analyses made by Nockolds and Allen [1953], traces of elements were determined spectrographically. Of these, 83% that contain Cr, V, Ni, and Co are for rocks with Ca > Fe. Only 19% show Ca < Fe, but unfortunately, Cu was not determined.

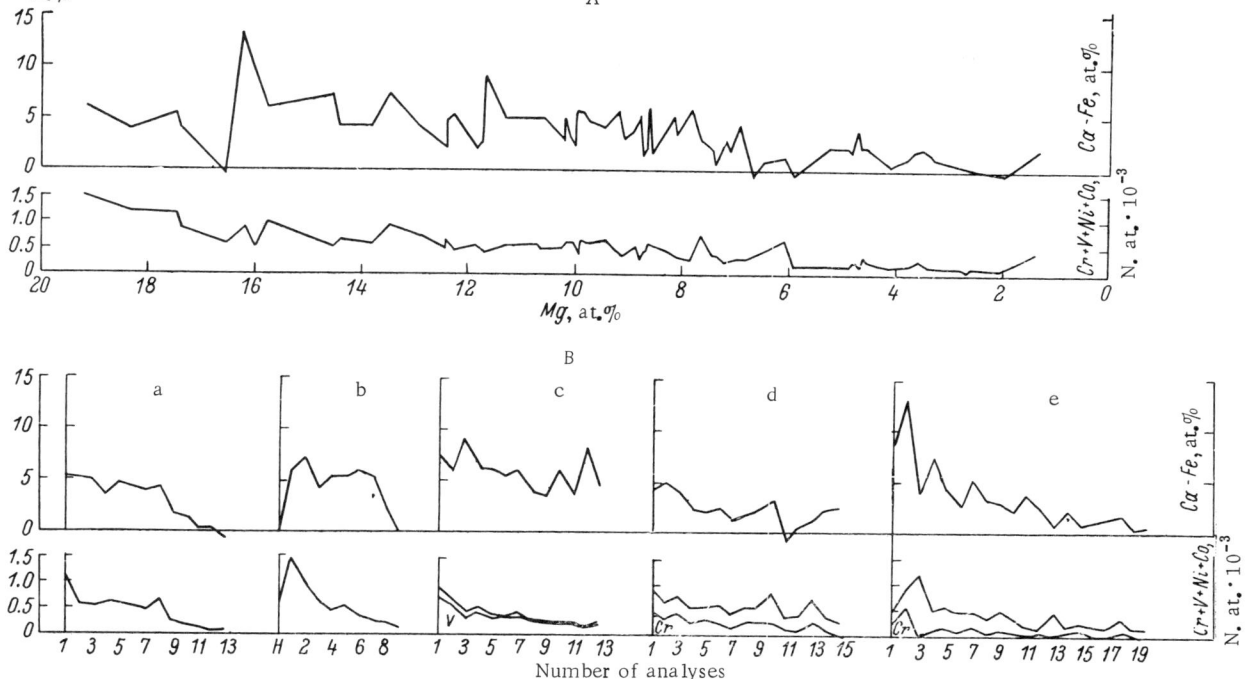

Fig. 11. Contents of elements in dark rocks. A) Relations between Ca—Fe at.% and Cr + V + Ni + Co (atoms · 10^6 — N at. · 10^3) for a series of analyses of a number of geographic associations, studied by Nockolds and Allen. B) The same for a number of geographic associations studied by the same authors: a) Crater Lake (Oregon); b) Lassan Volcanic National Park; c) Lesser Antilles; d) Caledonian rocks of Scotland; e) Southern California Batholith.

A correlation is noted between the contents of these metals and excess of Ca, the variations of which take place more or less concordantly.

But even when Fe is more abundant than Ca, it is possible to find a notable similarity between changes in Cr + Ni on the one hand and Ca − Fe on the other (chiefly in the amphibolites of Connemara, where 20 or 30 analyses show Ca < Fe) [Evans, 1960].

How do we explain why deposits of Cr, Ni, V, and other elements that are economically important are found in association with ultramafic rocks, whereas, in small concentrations, these elements are associated with rocks with abundant Ca? We think that this confirms the mechanism we mentioned above: first, some of the metals are removed because they became excessive during evolution; second, those same metals are preserved in the rock since they are necessary for overall equilibrium of the mass of atoms.

Elements in Unstable Positions

Some elements occur at the boundary between two classes (Fig. 1). For example, P, S, and Cl belong to class f_3' and f_2'; Sn, W, Fe, and Ni are found between f_2' and f_1; Pb, Th, and, in part, the rare earths are between f_1 and f_2'', and, lastly, Sr is between f_2'' and f_3''.

Furthermore, the classification by atomic diameter undergoes changes during ionization of the atoms. If we draw the boundary between ionic classes through the same points on the abscissa axis where boundaries of the atomic classes occur, the lines are inclined at 45°.

Furthermore, with time and increase in amount of Al, the system is shifted to the left. The line passing through the center of equilibrium of the new system will lie between Mg^{2+} and Fe^{2+}. As a result, a whole series of metallic ions ($Mo^{3,4+}$, W^{6+}, Sn^{4+}, $V^{2,3,4+}$, U^{4+}, and others) occur together with Mg and Fe; others (Ge^{4+}, Ga^{3+}, P^{3+}, B^{2+}, and others) will be grouped with Al; Ag, Au, Pb^{2+}, $Hg^{1,2+}$, and U^{3+} will be with Na, K, Ca^{2+}, and Sr^{2+} (all these will be combined in a single class); and, lastly, Ba^{2+} and Rb will be with Cs.

These shifts, together with the chemical characteristics (similar or incompatible), may explain the geochemical behavior of many elements, such as the division of Ca into Ca' and Ca", the separation of elements into EB and EN, the correlation between Ga^{3+} and Al^{3+}, of Pb^{2+} and Zn^{2+} with Fe^{2+}, and others. They also mark paragenetic relations among the elements occurring in belts perpendicular to the boundaries of the ionic classes. In this connection W^6, Ta^5, Nb^5, Ti^4, Bi^5, Sn^4, Li, Zr^4, and TR^3 (rare earths) are found in pegmatitic associations, whereas Fe^3, Mn^3, Cu^2, (V^3), Zn^2, Ag^2, (Au^3), and Pb^2 occur in hydrothermal associations.

The shift of some metals to a new "central class" of ionic diameters leads to their association with Fe and Mg (classic ferromagnesian group), and this new geochemical position transfers them to the period of granitoids, from which they are split off at the end of the evolutionary process, during the formation of pegmatites.

The unstable position of some elements in bordering zones permits them to appear in two neighboring classes. In order to restore equilibrium, it becomes necessary for a separation to take place between the right and left sides of the equation [formula (I)]. The relative number of elements may vary, depending on circumstances (for example, the group τ, or Ca' and Ca").

Radioactive Elements

The effect of radioactive elements is observed chiefly in the zone of their clarkes. Everyone knows that K^{40} splits into A^{40} and Ca^{40}. The proportion of the two daughter elements must increase during the course of evolution of the lithosphere.

The numerous transformations taking place through radioactivity supply a certain amount of hydrogen and helium in the vicinity of the new element, so that the clarkes of these gases are highly variable. They should increase with time. The various consequences of radioactivity probably have considerable significance, but we do not yet know the extent of this [Choubert, 1960].

Summary

As we have hoped, study of a large number of analyses has permitted us to discover several laws that control the composition of the atomic solution of the lithosphere. An exponential curve defining the relation between number of elements and time indicates a mechanism of sequential replacement of "dark-colored" elements (EN) by "light-colored" elements (EB). A peridotitic nucleus is the inherited part, which is transferred from one period to the next, gradually diminishing.

The chronology of events is reflected in the relations of the principal elements which, playing the role of leaders, determine the behavior of secondary elements. Evolution of the shape of the Gaussian curve by growth of the central ordinates takes place in parallel with the introduction of alkali metals, reflects these changes, and explains the gradual development of pneumatolysis, which reaches a maximum in the granitoidal rocks.

In comparing our results with the data of classic petrography one may convince himself that some of the concepts revealed by classical methods of investigation (microscopy, modal calculations, and the like) correspond to systematic relations of atoms or their physical properties. The use of statistics permits one thus to refine and explain the phenomena delineated by petrographers and geochemists.

REFERENCES

Backlund, H. G., "Einblicke in das geologische Geschehen des Präkambriums," Geol. Rundschau, Vol. 34 (1943).

Choubert, B., "Corrélation entre le nombre et l'encombrement spatial des atomes dans une venue magmatique," Soc. Géol. France, Mém. 79 (1943).

Choubert, B., "Les granites précambriens des Guyanes et leur origine probable," Mém. Carte Géol. France, Paris, 176 (1960).

Choubert, B., "Principes évolutifs de la formation des roches ignées," Soc. Géol. France, Mém. 98 (1963).

Choubert, B., "État actuel de nos connaissances sur le géologie de la Guyane Francaise," Comm. Soc. Géol. France, No. 2, pp. 65-66 (1965).

Evans, B. W., and Leake, B. E., "The composition and origin of the striped amphibolites of Connemara, Ireland," J. Petrol., Vol. 1, No. 3, Oxford (1960).

Macheras, G., "Détermination d'un critère pratique des différenciations magmatiques," Bull. Soc. Géol. France, Sér. 7, Vol. 2, No. 1, pp. 78-86 (1960).

Marmo, V., "On granites," Bull. Comm. Geol. Finland, No. 20, Helsinki (1962).

Nockolds, S. R., and Allen, R., "The geochemistry of some igneous rock series," Geochim. Cosmochim. Acta, Vol. 4, No. 3, pp. 105-142 (1953).

Routhier, P., Les Gisements Métallifères: Géologie et Principes de leur Recherche, Paris (1963).

Shaw, D. M., Interprétation Géochimique des Éléments en Traces dans les Roches Cristallines, Paris (1964).

Vistelius, A. B., "Problems of mathematical geology. I. History of the question," Geol. i Geofiz., Sibirsk. Otd. Akad. Nauk SSSR, No. 12, pp. 3-9 (1962).

Vistelius, A. B., "Problems of geochemistry and information measures," Sov. Geol., No. 12, pp. 5-26 (1964).

GENERATION OF THE LOG-NORMAL FREQUENCY DISTRIBUTION IN SEDIMENTS

G. V. Middleton

McMaster University
Hamilton, Ontario, Canada

Following a discussion of several examples, an "adequate" stochastic model is defined as one where the probability assumptions are implicit in the corresponding physical model. A model for the generation of log-normal size distributions in sands is proposed, based upon the concept of a repeated "sorting event" and the application of the central limit theorem. This model appears to be somewhat less than adequate, by the above definition, but may be improved by further investigation.

Introduction

The experience of many years of work on the size frequency distribution observed in samples of natural sediments, has led sedimentary petrologists to analyze sediment size distributions in terms of an ideal model, the log-normal distribution. It has been found that, although few samples show a very close approximation to the log-normal size distribution [Friedman, 1962], many samples of sands approximate this distribution more closely than other mathematical models, and in fact the approximation is quite close, particularly in the central part of the size range. Some other samples may be regarded as mixtures of two or more different log-normal size distributions.

Thus, it has been found convenient to plot size distributions as cumulative curves on logarithmic probability paper (where a log-normal distribution will plot as a straight line), and to derive graphical or numerical measures of sorting, skewness, and kurtosis which permit comparison of sampled distributions with an ideal log-normal model.

In spite of this established practice, there have been few attempts to explain why sediments tend to approach an ideal log-normal size distribution. It is the purpose of this paper to make a few preliminary observations about the type of stochastic processes which must be involved in the generation of log-normal size distributions.

It is a pleasure to dedicate this paper to Dr. Andrei B. Vistelius who has pioneered in the application of stochastic models and statistical methods in geology. I much regret that, owing to my own restricted linguistic ability and the scarcity of translations, this paper will give inadequate credit to work published in the USSR. I offer it, nevertheless, in the spirit of international cooperation in science.

Requirements for an "Adequate" Stochastic Model of Size-Frequency Distributions

The geologist does not directly observe natural size-frequency distributions which obey any mathematical law [Middleton, 1962]. In all observations, the methods of sampling and measurement play an important role in determining the nature of the observed size distribution. The correct procedure is for the geologist to formulate a hypothesis about the nature and significance of the size distribution which he will observe if he follows a certain sampling and analytical procedure, and then to test the hypothesis by observation. An "adequate" model of the observed size distribution should take into account the generation of size distributions from parent rock materials by weathering, the way in which these distributions are modified by transportation and deposition, and the way in which the observed size distribution is generated by the particular methods of sampling and measurement which are used.

The nature of what constitutes an "adequate model" needs further discussion. For example, consider Kolmogoroff's theory for the generation of log-normal size distributions by crushing. It is not necessary to assume any specific initial size distribution nor any very definite physical properties of the material which is being crushed. It is assumed merely that the probabilities of particles being crushed into smaller ones in unit time are independent of each other, and that the crushing process continues for a long time. This is sufficient to generate the log-normal distribution, and (perhaps) to satisfy the mathematician. The scientist would, however, wish to understand why it is possible to make the basic assumption. Can this probability statement be derived from more primitive physical statements about the process involved in the crushing itself?

Another example is suggested by a recent paper of Leopold and Langbein [1962], in which the concept of random walk is applied to a consideration of stream profiles. Modifying their argument slightly, it can easily be shown that if the probability of an increment of downward erosion (dH) is proportional to the height above base level (H), the stream will erode an exponential long profile, since, on the average,

$$dH = -kH\,dx, \qquad (1)$$

and integrating yields

$$H = H_0 e^{-kx}, \qquad (2)$$

where H_0 is the height of the source, k is a constant, and x is the distance from the source. But why is the probability of unit downward erosion proportional to the height above base level?

An adequate model may, therefore, be defined as one which assumes only a set of physical conditions. The probability assumptions should be implicit in the physical assumptions, or should be of the simplest kind (e.g., certain alternatives considered to be equally probable). They should not involve arbitrary assumptions of complex distributions unless these are themselves well rooted in theory or observation.

An example where such complex distribution may reasonably be assumed is provided in a paper by Kottler [1950] on the log-normal size distribution of particles in photographic emulsions. Kottler assumes that the times at which particles start to grow in the emulsion are normally distributed. By assuming that the rate of growth is proportional to the size of the crystal (a reasonable assumption on chemical grounds), it is shown that the resulting size distribution is log-normal.

Let x be the size of the crystal at time t

$$dx = kxdt, \tag{3}$$

$$x = x_0 e^{kt}, \tag{4}$$

or

$$t = a + b \ln x, \tag{5}$$

where a and b are constants. Thus, if t is normally distributed, x will be log-normally distributed. Kottler presents various arguments to support his assumption of a normal distribution of time of beginning of growth, which although not entirely convincing, do make the model plausible.

It may be noted that a similar argument should apply to crystallization from a magma, so long as there is no strong interaction of crystals. It may be expected, therefore, that phenocrysts and certain early-crystallizing accessory minerals will show log-normal size distributions.

An example of an "adequate" model for the generation of an observed size distribution is provided by the derivation by Bennett [1936] of the size distribution observed in crushed coal. The observed size distribution is found to approximate closely, in many cases, the Rosin—Rammler distribution, which states that the distribution function (cumulative curve) is given by

$$R = 100 e^{-bx^n}, \tag{6}$$

where R is the wt.% retained on a sieve of opening x, and b and n are constants.

Bennett [1936] gives the following derivation of this law from probability theory. Suppose that a fragment of coal breaks by brittle fracture at points of weakness, irrespective of position. Set up three coordinate axes and consider the intercept of a line Θ_z, parallel to one of the axes, with the fragment. The intercept, of length L_z, will intercept future fracture planes at N_z points. The average distance between the fracture planes (i.e., the average thickness of the particles) will be $\bar{a}_z = L_z/N_z$. The probability that p fracture planes will be cut by a line of length a_z is given by Poisson's series (since N_z is large). The probability that no fracture planes will be cut is given by

$$P_{(0)} = e^{-\frac{a_z}{\bar{a}_z}}. \tag{7}$$

This is the same as the probability that the thickness of a fragment will be greater than a_z. Hence, the probable number of fragments whose thickness in the z direction is greater than a_z is

$$R_{(a_z)} = N_z e^{-\frac{a_z}{\bar{a}_z}}. \tag{8}$$

Similar arguments hold for the x and y coordinates. The total number of fragments produced by fracturing the original fragment is

$$N = N_x N_y N_z. \tag{9}$$

Equation (9) can therefore be rewritten as a percentage

$$R_z = 100 e^{-\frac{a_z}{\bar{a}_z}}. \tag{10}$$

It is not immediately apparent that the three ratios of the form $a_z/\bar{a}_z$ can be replaced by a single average $a/\bar{a}$, but this can be shown by writing an expression for the volume of particles having a volume exceeding $a_x a_y a_z$.

$$R_z = 100 e^{-\frac{a_z}{\bar{a}_z}}. \tag{11}$$

This equation, reduced to a percentage, proves to have a form similar to equation (11), namely,

$$R = 100 e^{-\frac{a}{\bar{a}}}. \tag{12}$$

In order to take account of the interaction of many particles in the breakage process, which leads to departures from the ideal law for breakage of a single particle, we write

$$a = \lambda x^n,$$

where λ and n are constants, a is the "ideal diameter," and x is the "actual diameter." It follows that

$$R = 100 e^{-\left(\frac{x}{\bar{x}}\right)^n}. \tag{13}$$

In general, n is close to unity, being larger or smaller than unity if there are proportionately fewer or more fines produced by breakage of many particles than by breakage of only one particle.

This derivation of the Rosin—Rammler law appears to be an excellent example of an "adequate" mathematicophysical model for breakage. The basic assumptions are few and are carefully compared with the physical reality (see Bennett's paper for a much fuller discussion than is given above). The assumptions do not include such unrealistic ones as isotropy or homogeneity of the material. The law has been thoroughly investigated empirically and found to hold for many different types of crushing machinery.

The law has also been found to hold for many in situ or little-transported rock materials [Krumbein and Tisdel, 1941; Kittleman, 1964]. As no physical assumptions were made by Bennett which are inconsistent with breakage mechanisms other than crushing, the model may be considered an adequate explanation of broken, unsorted rock materials, such as weathered rock, tuff, and explosion breccia, and evaporite solution—collapse breccia [Middleton, 1961].

Derivation of the Log-Normal Size Distribution.
General Discussion

There are two main mathematical models for generating the log-normal distribution: (1) by transformation of a normal distribution, and (2) by use of the Central Limit Theorem. The first model is illustrated by the theory of Kottler, given above. The second is well known from its exposition in a book by Cramer [1946] and its use by Miller and Goldberg [1955] in a paper on the log-normal distribution of elements.

Use of the central limit theorem permits generation of either a normal or a log-normal distribution, by a slight change in the basic assumptions. There are three different possible approaches:

1. **The Addition Model.** It is assumed that the quantity which is measured is the joint effect (or sum) of a large number of independent random variables. For example, the total error of measurement may be considered to be the sum of a large number of small errors

due to uncontrolled variations in temperature, pressure, etc., uncontrolled human factors, and so on. If the random variables are $x_1, x_2, \ldots, x_n$, then it follows from the central limit theorem that

$$X = \sum_{i=1}^{n} x_i. \tag{14}$$

X is asymptotically normally distributed, i.e., $f(x)$ becomes very close to normal as n tends to infinity. The assumptions involved in the application of the central limit theorem are very broad, and are satisfied by practically any real physical variables. The only critical assumption is the assumption of independence for the random variables.

   2. **The Multiplication Model.** If in the above model it is assumed that the quantity is equal to the product of the effects of the random variables, then

$$X = \prod_{i=1}^{n} x_i, \tag{15}$$

or

$$\log X = \sum_{i=1}^{n} \log x_i, \tag{16}$$

from which it follows that $\log X$ is normally distributed or that X is log-normally distributed.

   3. **The Proportionate Effect Model.** In this model, we assumed that the effect of the n-th random variable is a random proportion of the sum of the (n − 1) preceding effects of the random variables $x_1, x_2, \ldots, x_{(n-1)}$. Hence,

$$x_n - x_{n-1} = k_n x_{n-1}. \tag{17}$$

It follows that

$$\sum_{i=1}^{n} \frac{x_i - x_{i-1}}{x_{i-1}} = \sum_{i=1}^{n} k_i. \tag{18}$$

and if $x_i - x_{i-1}$ increases little as $n \to \infty$, then

$$\sum_{i=1}^{n} \frac{x_i - x_{i-1}}{x_{i-1}} \to \int_{x_0}^{x_n} \frac{dx}{x} = \log x_n - \log x_0. \tag{19}$$

Hence,

$$\log x_n = \log x_0 + k_1 + k_2 + \ldots + k_n \tag{20}$$

and it follows that x_n is log-normally distributed (Aitchison and Brown, 1957, pp. 22-23).

   Aitchison and Brown [1957, p. 25] have pointed out that it is not necessary to assume any time-sequence of effects in this (or in any other) model. Note that

$$x_n - x_{n-1} = k_n x_{n-1} ; \tag{21}$$

therefore

$$x_n = x_0 (1 + k_1)(1 + k_2) \ldots (1 + k_n); \tag{22}$$

so that the proportionate effect model may easily be reduced to the "multiplication model."

Epstein [1947] derived the log-normal law for sizes produced by breakage, by using a multiplication model. The physical assumptions are that breakage proceeds by a large number of discrete independent random events (breakages). The probability of breakage of a particle is independent of the size of the particle, and the distribution of the sizes $f(d)$ which result from breakage of particles with diameter d_0 is dependent only on the ratio (d/d_0). It follows that the distribution of sizes after n breakages is the distribution of the product of $(n + 1)$ independent random variables, and that it is asymptotically log-normal (for details, see Epstein's paper, or Aitchison and Brown [1957, pp. 26-27]). It may be noted that the physical assumptions are not entirely realistic. In some crushing processes the larger particles are more likely to be broken than the smaller ones, and many materials are not homogeneous (a homogeneous material is implied by the second assumption). It is not surprising, therefore, that many crushed materials do not show a log-normal size distribution.

The problem of explanation of the natural occurrence of the log-normal size distribution in sediments is therefore resolved into an attempt to find a physical model which is consistent with the known processes of sediment formation, and which leads (with a minimum of further assumptions) either to the multiplication model or to the proportionate effects model. To be convincing, the physical model should also suggest deviations from the mathematic assumptions of the mathematical model which will give rise to the types of deviations from log-normality of size distributions which are observed in nature.

Rogers and others [1963] have suggested that the observed log-normality arises from breakage. If this were so, then the "breakage theories" of Kolmogoroff [1941] and Epstein [1947] might be considered to provide a satisfactory explanation of the log-normality of sands. It seems to the writer, however, that the size distributions observed in sands cannot be explained satisfactorily in terms of breakage, but must be explained in terms of sedimentary sorting processes. Whatever the total size distribution produced by weathering and breakage processes during transportation, it does not determine the local size distribution of the sand-sized particles, as finally deposited.

Inman [1949] explained, in a qualitative way, the way in which fluid mechanical sorting processes tended to give rise to log-normal distributions in the fine sand sizes. What is required is a stochastic model that follows the general approach suggested by Inman. In the following section, the writer presents a first attempt at such a model.

The Model

We consider the final size distribution $f_n(x)$ to arise from an initial size distribution $g(x)$, by a series of discrete "sorting events." The clearest physical example is that of a long straight beach with wave action resulting in longshore drifting of sand along the beach, from a "source" at one end.

Each wave takes sand into suspension, redeposits some of it on the beach, and removes some of it to the offshore region. This is the "sorting event."

We consider first a sand of uniform size. Then a given layer of sand on the beach at a given point is transformed by the sorting event into a layer which is either thicker or thinner, i.e., the weight of sand per unit area (y) is either increased or decreased. The weight of sand in the "moving layer" is a function of the type and magnitude of the wave, i.e., over a period of time it defines a frequency distribution $f(y)$ which depends upon various hydrodynamic factors.

In reality, the situation is complicated by the fact that the size is not uniform. However, for each size x there will be a conditional distribution function $f(y/x)$, and the size distribution in the "moving layer" will therefore be given by $f(x,y)$. Now if $f(x)$ and $f(y)$ are independent we can write $f(x,y) = f(x)f(y)$, or in terms of random variables

$$z = xy. \tag{23}$$

Since x is a size and y is a weight, $f(z)$ will be a size—weight frequency distribution of the usual type. After the n-th sorting event,

$$z_n = x_0 y_1 y_2, \ldots, y_n \tag{24}$$

and it follows that if x_0 and the y_i are independent random variables, z_n will be asymptotically log-normally distributed.

The assumptions are: *

1. The $f(y)$ are functions of independent random variables. This will not be true over the short run because of the autocorrelation of successive waves, but it will be true over the long run, because the hydrodynamic factors may be considered to vary randomly from month to month, or at least from year to year.

2. $f(x)$ and $f(y)$ are functions of independent random variables. $f(x)$ may be considered to be the size distribution in the source of the sediment on the beach. It is therefore a primary datum. $f(y)$ is defined by the interaction of the hydrodynamic properties of the waves approaching the beach and the beach materials. Since the beach slope and type depend upon the coarseness of the source materials, it may be objected that y is not independent of x. However, we may reasonably assume that $f(y/x)$ depends only on the ratio y/x, i.e., the weight of any given size remaining on the beach after a sorting event depends only on the proportion of that size which was present in the source. Aitchison and Brown [1957, p. 26] show that this statement is equivalent to a statement of proportionate effect, and leads to the log-normal distribution.

It is interesting to consider one way in which such a model may break down. Suppose that the hydrodynamic factors are such that y = 0 for all $x \leq d_0$, i.e., no sand at all is deposited when the diameter falls below a certain value. Then $z_n = 0$ for all $x \leq d_0$, and thus gives rise to the so-called "three parameter log-normal distribution," which may appear, after the usual phi-transformation, to be a normal distribution skewed toward the larger grain sizes. This would appear to be a reasonable explanation for the commonly observed slight negative skewness of beach sands. The model developed above will now be applied to sand in other physical situations.

At first, it may not appear that **fluvial transport** has much in common with wave transport. Nevertheless, the essential aspects of the model are preserved, since the sand moves by a series of steps down the river (from point-bar to point-bar, for example). Thus, once again, there are a series of discrete "sorting events," acting on an initial size distribution. The principal modifications are:

---

*From the mathematical point of view, the model may be formulated in the following way. Let x_0 be the "initial" size of the sand particle, conforming to the fundamental distribution law with a density $g(x)$, in which x_0 is nearly a constant, and let y_i be the value of the effect on this size by the i-th sorting event. The effect amounts to multiplying the size by a factor equal to y_i (since grinding or granulation of the sand particles takes place, $y_i \leq 1$). The factors do not depend on each other or on the size the particle has already attained. Then, after n sorting events, the size of the particles is determined by the formula

$$x_n = x_0 y_1 y_2, \ldots, y_n; \tag{24}$$

then z_n at a large n is asymptotically lognormal (Editor's footnote).

1. The number of events will not be as large as in beach transport. It should, therefore, be easier to detect a downstream modification of the size distribution. The rate of convergence of the distribution to log-normal is most rapid at first and proceeds more slowly after the number of events is already large. In wave action, the number of events becomes large very rapidly, so that approximately log-normal distributions should be found even very close to the sediment source. In rivers, we expect to see a more gradual development of the log-normal distribution downstream.

2. The model breaks down in an important way, because the $f(y)$ are no longer functions of independent random variables. The magnitude of the "sorting events" decreases systematically downstream, i.e., the competency of the stream decreases. Since the range of magnitude of the "sorting events" is very large for any section of the river (due to floods and seasonal variation in discharge), this is not a major defect of the model, but will eventually result in a non-log-normal distribution. The effect will be just the opposite of that noted for wave action, i.e., the distribution will be truncated at the coarse end, resulting in a positive skewness.

It is thought that both of these effects, predicted from the model proposed here, are consistent with observations on actual river sands. Inman [1949] noted a change downstream from positive to negative skewness in the Mississippi River sands. The positive skewness of river sands is consistent with the proposed model but it is to be expected that some river sediments may show instead a change from strong negative to slight negative skewness, without passing through a symmetrical stage. The development of strong negative skewness in sands with average diameter less than 0.18 mm is correctly explained by Inman [1949] and cannot be predicted from this model. A breakdown of the model is to be expected when the physical laws, which relate hydrodynamic factors to size, change for any substantial proportion of the "sorting events." Exactly such a change takes place at low values of friction (shear) velocity and at average grain sizes which are small enough to permit the development of a laminar boundary layer at the bed of the stream.

Conclusions

It is concluded from this study that an adequate stochastic model for the generation of size-frequency distributions is one in which the principal assumptions are physical in nature; the validity of the mathematical assumptions is judged by reference to the physical model, and the validity of the physical model is judged by the usual scientific criteria of observation and hypothesis testing.

The model for the generation of the log-normal size distribution of sand-sized sediments which is proposed in this paper is based upon the concept of an original (unspecified) size distribution which is modified by a series of "sorting events." This concept leads to a "multiplication model" for the generation of the log-normal distribution, by application of the central limit theorem.

Applying the criterion for an "adequate model," stated above, it can be seen that the proposed model cannot be considered to be fully adequate, because of a number of probability assumptions whose validity is not fully apparent from the nature of the physical model which has been proposed. It is hoped, nevertheless, that this paper may stimulate work leading to the proposal of a more satisfactory model.

I wish to acknowledge a discussion with Dr. Igor Chernenko, which helped me in the development of the model proposed in this paper.

REFERENCES

Aitchison, J., and Brown, J. A. C., The Lognormal Distribution, Cambridge University Press, Cambridge, 176 pp. (1957).

Bennett, J. G., "Broken coal," J. Inst. Fuel, Vol. 10, pp. 22-39 (1936).

Cramer, H., Mathematical Methods of Statistics, Princeton University Press, Princeton, N. J., 575 pp. (1946).

Epstein, B., "The mathematical description of certain breakage mechanisms leading to the logarithmico-normal distribution," J. Franklin Inst., Vol. 244, pp. 471-477 (1947).

Friedman, G. M., "On sorting, sorting coefficients, and the log-normality of the grain-size distribution of sandstones," J. Geol., Vol. 70, pp. 737-753 (1962).

Inman, D. L., "Sorting of sediments in the light of fluid mechanics," J. Sediment. Petrol., Vol. 19, pp. 51-70 (1949).

Kittleman, L. R., Jr., "Application of Rosin's distribution in size-frequency analysis of clastic rocks," J. Sediment. Petrol., Vol. 34, pp. 483-502 (1964).

Kolmogoroff, A. N., "Uber das logarithmisch normale Verteilungegesetz der Dimensionen der Teilchen bei Zerstückelung," Dokl. Akad. Nauk SSSR, Vol. 31, p. 99 (1941) [in Russian and German].

Kottler, F., "The distribution of particle sizes," J. Franklin Inst., Vol. 250, pp. 339-356, 419-441 (1950).

Krumbein, W. C., and Tisdel, F. W., "Size distribution of source rocks of sediments," Am. J. Sci., Vol. 238, pp. 296-305 (1940).

Leopold, L. B., and Langbein, W. B., Concept of Entropy in Landscape Evolution, US Geol. Survey Prof. Paper 500-A, 20 pp. (1962).

Middleton, G. V., "Evaporite solution breccias from the Mississippian of Southwest Montana," J. Sediment. Petrol., Vol. 31, pp. 189-195 (1961).

Middleton, G. V., "On sorting, sorting coefficients, and the lognormality of the grain-size distribution of sandstones: a discussion," J. Geol., Vol. 70, pp. 754-756 (1962).

Miller, R. L., and Goldberg, E. D., "The normal distribution in geochemistry," Geochim. Cosmochim. Acta, Vol. 8, pp. 53-62 (1955).

Rogers, J. J. W., Krueger, W. C., and Krog, M., "Sizes of naturally abraded materials," J. Sediment. Petrol., Vol. 33, pp. 628-632 (1963).

CORRELATION OF JOINT TRENDS WITH THE ELEMENTS OF TECTONIC STRUCTURES

L. D. Knoring

All-Union Petroleum Scientific Research Institute for Geological Exploration
Leningrad, USSR

It is shown that all joint systems developed on local uplifts have a strict spatial correlation with the orientation of bedding planes. In order to make a valid comparison of joint trends, the probability distribution of joint trends of a single system has been examined. This distribution agrees closely with the Fisher distribution at the initial time of joint development. Through plastic deformation of the material in the bed, this distribution is systematically destroyed with time.

At present a number of points of view exist concerning the nature of the correlation between the direction of joints developed in sedimentary strata in tectonically deformed regions and the elements of tectonic structures.

1. Joints are not related either to fold structures or to any elements of these structures [Parker, 1942; Novikova, 1951]. They preserve trends that are general for great regions of the earth's crust [Shat-skii, 1945; Dorofeeva, 1963; Shul'ts, 1964].

2. Joints are related to the fold in its entirety: to its form, the strike of its axis, the dip of the axial plane, and so forth [Permyakov, 1949; Azhgirei, 1956; de Sitter, 1960; Éz, 1962; Chiang Chu-ch'i, 1963, et al].

3. Joints are related to bedding planes. The orientation changes systematically from point to point as the attitude of the bed changes [Kirillova, 1949; Belousov, 1952; Mikhailov, 1956; Kazimirov and Kuznetsova, 1960; Kalacheva and Knoring, 1963; Gromov, 1963; et al.].

The purpose of the present paper is investigation of the validity of the indicated points of view. To accomplish this, jointing of six tectonic uplifts in regions differing in history of their tectonic development were studied, and the orientations of 46,935 joints were measured.

The measurements were treated mathematically. This treatment consisted of the simplest test of the hypothesis that a correlation exists between joint trends and elements of tectonic structures.

Brief Description of the Investigated Uplifts

As indicated, six structures were studied in detail: the Khristoforova and Southern Markovo uplifts (Siberian platform), the Shugurovo uplift (Russian platform), the Kurtun folds (the Baikal marginal depression), and the Chil'-Dara and Western Aruk-Tau anticlines (Tadzhik depression).

The Shugurovo and Khristoforova uplifts are typical platform structures with dips no greater than 1-2°. Both structures, according to the strata investigated, are asymmetrical uplifts with northeasterly trend, steeper southeastern limb, and quaquaversal southwestern termination.

The Shugurovo uplift is on the southwestern slope of the Tatar arch within the Sok-Sheshma dislocation. In the Kazanian strata, in which the investigated joints occur, the structure is 30 × 15 km. In the underlying strata, the crest of this uplift is divided into two domes resting on a common base. The sag between the domes passes into the zone of greatest uplift in the Upper Permian strata.

Jointing was studied chiefly in sandstones and limestones of the *Conchifera* horizon of the upper Kazanian strata and the upper *Spirifer* beds of the lower Kazanian strata. These latter are the most widespread within the Shugurovo uplift, and they furnish the greatest number of outcrops, rather uniformly distributed throughout the region. Jointing was studied in 104 exposures that represent all the principal elements of the uplift, and 6200 measurements of joint orientation were made.

The Khristoforova uplift is in the Angara-Lena monocline of the Angara lower Paleozoic platform basin (syneclise). On this structure, jointing in a sand—clay sequence of the middle subdivision of the Upper Cambrian upper Lena series was studied. The uplift is outlined by this series as indicated above. Jointing was studied in 84 outcrops irregularly distributed over the area of the structure, most of the exposures being grouped in the southeastern half. In the northwestern part and the northeastern domal part, exposures are almost completely absent. Altogether, 6967 measurements were made on joint attitudes on this uplift.

The Southern Markovo uplift, according to the upper horizons, is a platform structure complicated by salt tectonics, as a result of which the dips on the limbs reach high values for platform uplifts (4-8°).

This uplift lies within the Markovo arch, which has a trend that is nearly northerly. Along the top of the lower horizon of the Lower Ordovician Ust'-Kut series, the uplift is considerably elongated, with dimensions of 16 × 7 km. The structure is asymmetrical, the western limb being steeper (dips of 8-10°), the western, gentler (dips of 4°). Jointing was studied in 21 outcrops in which limestones in the lower horizon of the Ust'-Kut series are exposed. Most of the outcrops are on the western limb of the uplift and on the crest. About two outcrops were found in the nose and on the eastern limb. In this uplift the orientations of 2250 joints were measured.

The group of Kurtun folds are found in the narrow Angara-Lena zone of linear folds in the Baikal marginal depression, separating the southern part of the Siberian platform from the Baikal fold zone.

Three structures that were studied are steep, almost isoclinal folds, trending to the northeast. Of these, the southern two are asymmetrical, dips on the southeastern limbs ranging from 30-50 to 70-75°. On the northwestern limbs, the beds are nearly vertical or are even overturned. Where overturned, the beds dip 80-85°. The folds are separated from each other by a narrow synclinal trough. The southernmost of the three folds passes gradually into a broad syncline. The crest and southwestern limb of the second anticline is complicated by faults. The third, northernmost fold is separated from its neighbors by a narrow synclinal trough. Structurally it is nearly a symmetrical fold with gently dipping limbs (as compared with its neighbors), chiefly about 30-40°, locally increasing to 60°.

Joints in the clastic sequence of the Sinian Kachergat series were studied in the above folds. Of 61 studied exposures, most occur in the central part of the northern two folds and in the synclinal troughs that separate them from neighboring folds. On the northwestern nose of the southern fold and in the northern and southern synclines, jointing is characterized by solitary outcrops. In all, 11,118 measurements on joint directions were made on the three Kurtun folds.

The Chil'-Dara and Western Aruk-Tau anticlines are, respectively, in the Kulyab synclinorium, which is characterized by well-defined ridge-like anticlines and numerous minor dislocations, and the Kafirnigan anticlinorium, a distinguishing feature of which is the development of broad and widely spaced box folds.

The Western Aruk-Tau anticline is a box fold with a trend nearly to the north, 35 km long and up to 5 km wide. In the central part the fold axis plunges and the single fold splits into three anticlines, which may be traced throughout the northern half of the structure. These three anticlines, together with the intervening synclines, form the broad arch of the Western Aruk-Tau anticline in the zone where the crest is highest. All the folds have ridge-like forms, considerable linear extension, and, as a rule, are sharply asymmetrical. The western fold is overturned to the west, the eastern to the east. Dips on the limbs range from 10 to 75°.

In the southern half, the Western Aruk-Tau anticline is a single fold with a steeper western limb (dips here reaching 25°, whereas they do not exceed 15° on the eastern limb) and gently inclined beds on the crest.

The Western Aruk-Tau anticline consists of lower Paleogene carbonate rocks almost throughout its entire extent, chiefly limestones and dolomites. In the uppermost parts of the fold, Upper Cretaceous rocks are exposed, consisting of limestone, clay, sandstone, and gypsum. Good exposures on the anticline permitted detailed study of joints in the constituent rocks. Joints were traced along 19 profiles in 157 outcrops, where a total of 10,800 measurements were made on attitudes of joints.

The Chil-Dara anticline is a narrow, sharply defined ridge-like fold with a northeasterly trend and dimensions of 10 × 3 km. It is made of Paleogene and Neogene rocks. The fold is asymmetrical, with the northwestern limb steeper (dips of 60-80°) than the southeastern, in which the beds dip 45-60°.

Against this background, secondary disharmonic folds with crests parallel to the general dip direction of the beds are developed in medium- and fine-bedded carbonate and sand—clay beds of the Middle Eocene Alai strata to the Lower Oligocene Isfara-Khanabad strata. The northwestern limb is complicated by a fault with displacement up to 350 m, extending along the entire limb almost parallel to the fold axis.

Jointing in the Chil'-Dara anticline was studied in 104 outcrops, which are rather uniformly distributed over the entire area of the fold; in all, 9600 measurements of joint attitudes were made. Joints in secondary folds in the above-indicated horizons were also studied.

Method of Analyzing Observations

In order to test the validity of the previously noted points of view, the following method was employed. The poles of the joint planes were plotted on a Kavraiskii net and were analyzed for the pertinent structural element. In this analysis, agreement of the observed direction distribution in a certain association of joints with the Fisher distribution appeared. If the observed distribution relative to the pertinent structural element corresponded to the Fisher distribution,

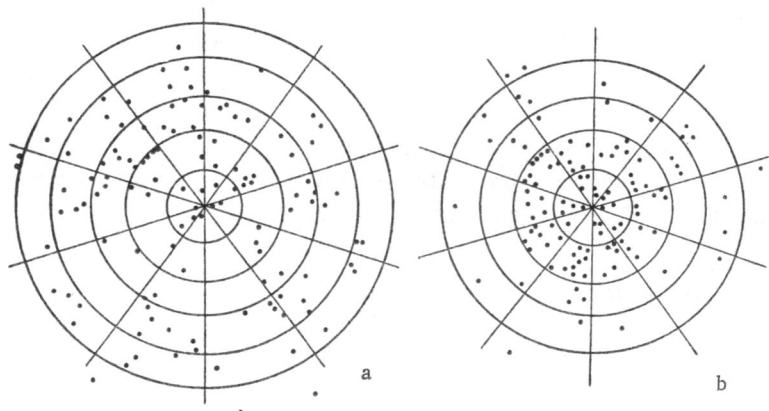

Fig. 1. Distribution of unit vectors of joints around the summed vector placed at the center of the net, indistinguishable from the Fisher distribution. a) Chil'-Dara anticline, N = 133; b) Shugurovo uplift, limestone, N = 114. Measurements were made at a single locality; circles were drawn at 5° intervals, azimuthal lines at 36° intervals.

all joints of the given association were considered to belong to a single set. Inspection of the position of these systems relative to the structural elements permitted us to interpret the causes responsible for development of the joints.

The Fisher distribution as a test to indicate which system joints may belong to was chosen on the basis that variations in joint trends are determined, from the geological point of view, by conditions of the central limit theorem.

A Fisher distribution with the parameter k greater than 4 corresponds to normal distribution according to this theorem. As we know [Vistelius, 1958], the Fisher distribution density has the form:

$$\frac{k}{2shk} e^{-k\cos\theta} \sin\theta d\theta,$$

where k is a parameter serving as a measure of direction concentration.

Thus, when the investigated joints formed a single system, we should have observed a good agreement between their densities and the Fisher distribution. When the observed frequencies did not correspond to the Fisher distribution, it could be assumed that either the material, reduced to a single series of variations, is heterogeneous, i.e., contains different joint systems, or individual joint systems have been noninvariantly disturbed by plastic deformation that affected the material of the bed after formation of the joints.

In the method we have used, in addition to careful geological analysis of the data, it was important to find the optimal method, in the given situation, for the goodness of fit test for the Fisher distribution with the observed frequencies. Since this question, except for rare exceptions [Vistelius, 1958], has not been discussed in the literature on structural analysis, despite its importance, we shall dwell on this in more detail.

Table 1. Distribution of Unit Vectors around the Summed Vector for Angles ψ

No. of sector	Value of sector, deg	Frequency of vectors in sector, m_i	Math. expect. of number of sector, N_{pi}
1	0—36	15	11.4
2	36—72	11	11.4
3	72—108	10	11.4
4	108—144	13	11.4
5	144—180	16	11.4
6	180—216	10	11.4
7	216—252	13	11.4
8	252—288	10	11.4
9	288—324	9	11.4
10	324—360	9	11.4
		114	

Note. $\chi^2 = 3.72$.

The Goodness-of-Fit Test of the Direction Distribution of Joint Systems with the Fisher Distribution by the Chi-Square Method

For testing the correspondence between the observed distribution and the Fisher distribution, all measurements of joint orientations for a system obtained in a single bed at a single outcrop were plotted on an equatorial net. Joints with poles that clearly gave isolated accumulations on the net were studied. The summed vector of such an accumulation was superimposed on the center of the net, and the unit vectors representing the given accumulation were moved along the corresponding parallels.* The entire zone occupied by the vectors was then subdivided into equal sectors and circular zones (Fig. 1). The size of class interval was chosen so that the observed unit vectors should be more or less uniformly distributed in each one and should number at least five in each, better ten. Fulfillment of this condition requires a large number of observations. The chi-square method was therefore used only when the number of joints belonging to a single system in a given bed and an actual point on the structure were large (> 100).

Testing of the H_0 hypothesis for goodness of fit between the observed distribution of joint-system trends and the Fisher distribution reduces to the following [Watson and Irving, 1957].

For the azimuthal angles ψ between the summed vector and each unit vector and for the angles θ of deviation of the unit vectors from the summed vector, values of χ^2 were computed by the formula

$$\chi^2 = \sum_{i=1}^{l} \frac{m_i^2}{N p_i} - N,$$

Table 2. Distribution of Unit Vectors around the Summed Vector for Angles θ

No. of circular zone	Interval occupied by ring zone, deg	Frequency of vectors in circular zone, m_i	θ_i	$\Phi(\theta_i)$	Prob. of vector falling in circular zone, p_i	Math. expect. of vectors in circular zone, N_{pi}
1	0—5	19	0	1.00000	0.19808	22.58112
2	5—10	63	5	0.80192	0.57097	65.09058
3	10—15	13	10	0.23095	0.09334	10.64076
4	15—20	15	15	0.13761	0.10763	12.26982
5	>20	4	20	0.02998	0.02998	3.41772
					1.00000	114.00000

Note. $\chi^2 = 1.864$.

---
* Each orientation measurement was considered a unit vector.

where m_i is the number of unit vectors falling in a given class (a sector or a circular zone, respectively); p_i is the probability of falling within the class $p_i = 1/l$ for a sector (l is the number of sectors), $p_i = e^{-k(1-\cos\theta_i)} - e^{-k(1-\cos\theta_{i+1})}$ for a circular section; and N is the total number of unit vectors in the system.

If the value of χ^2 obtained in both cases is less than the tabular value [Vistelius, 1958, Table IV) at the specific number of degrees of freedom and the 5% level of significance, the H_0 hypothesis is not disproved by the existing material.

The number of degrees of freedom when testing for goodness of fit is $l - 1$ with the first condition, $\nu - 3$ with the second, where l is the number of sectors and ν is the number of circular zones.

Tables 1 and 2, prepared on the basis of Fig. 1b, serve as illustrations of calculations by the indicated method.

The first three columns of Table 1 were completed after grouping the data by sectors. The mathematical expectation given in this table was computed by the formula $Np_i = N/l$.

The value of χ^2 obtained (3.72) was less than the tabular value $\chi^2_{0.05} = 16.9$ for 9 (10 − 1 = 9) degrees of freedom and a 5% level of significance.

Consequently, the hypothesis of uniform distribution of vectors for the angles ψ is not disproved by observations.

The first four columns of Table 2 were also filled from data in Fib. 1b. Values of the function $\Phi(\theta_i)$ in column 5 were found by the equation $\Phi(\theta_i) = e^{-k(1-\cos\theta_i)}$, where calculation of k was made by $k = (N-1)/(N-R)$. Determination of the probabilities that vectors would fall in a particular circular zone (column 6) was made by means of the expression $p_i = e^{-k(1-\cos\theta_i)} - e^{-k(1-\cos\theta_{i+1})}$. For example, the probability that a vector will fall within the circle of radius 5° is equal to $1.00000 - 0.80192 = 0.19808$. In column 7, the mathematical expectation of vectors in the circular zones is given. The total for column 6 should be unity, but for column 7 it should be the number of unit vectors.

The computed value of χ^2 (1.864) is less than the tabular value $\chi^2_{0.05} = 5.99$ for a 5% level of significance and 2 (5 − 3 = 2) degrees of freedom. Consequently, the distribution of angular deviations of vectors from the summed vector according to the law dictated by the Fisher distribution is not contradicted by the observations.

The Goodness-of-Fit Test for Trends of Joint Systems with the Fisher Distribution for a Large Number of Small Samples

The method described above was used only when it was possible to find a rather large number of joints belonging to a single system in a single bed at a single locality. In a number of places, however, because of inadequate exposures or other reasons, it was not possible to collect sufficient data. Furthermore, even where exposures were good, it was not always possible to have confidence in referring a large number of joints to a single local distribution. Under these conditions, it was necessary to test the possible relation of all systems to the association having Fisher distribution, assuming that each system might have its own distribution parameters. Since the Fisher distribution is symmetrical, and we could advance no other symmetrical distributions as rival hypotheses, everything reduces in this case to testing the lack of contradiction of the observed symmetrical distribution. For this purpose, the method of small samples was used, a technique already used in geology [Vistelius, 1960].

Table 3. Calculation of the Required Parameters for the Method of Small Samples

Seq. No.	Azimuth of dip direction of joint					Standard dev., σ	Dev. from mean $x_k - \bar{x}$	Index τ_k
	x_1	x_2	x_3	x_4	Mean $\bar{x}$			
1	36	38	35	35	36	1.414	0.0	0.000
2							+2.0	1.414
3							−1.0	−0.707
4							−1.0	−0.707
5	202	200	200	200	200.5	1.0	+1.5	+1.500
6							−0.5	−0.500
7							−0.5	−0.500
8							−0.5	−0.500
9	338	344	348	328	339.5	8.7	−1.5	−0.172
10							+4.5	+0.517
11							+8.5	+0.976
12							−11.5	−1.321
13	330	332	330	332	331	1.155	−1.0	−0.864
14							+1.0	+0.864
15							−1.0	−0.864
16							+1.0	+0.864
17	110	100	104	100	106	4.91	+4	+0.815
18							+4	+0.815
19							−2	−0.407
20							−6	−1.222
21	284	284	284	290	285.5	3.0	−1.5	−0.500
22							−1.5	−0.500
23							−1.5	−0.500
24							+4.5	+1.500
25	104	106	102	102	103.5	1.919	+0.5	+0.265
26							+2.5	+1.302
27							−1.5	−0.782
28							−1.5	−0.782
29	80	90	86	86	85.5	3.97	−5.5	−1.385
30							+4.5	+1.134
31							+0.5	+0.125
32							+0.5	+0.125
33	336	336	343	330	336.25	5.11	−0.25	−0.049
34							−0.25	−0.049
35							+6.75	+1.321
36							−6.25	−1.223
37	335	342	344	340	340.25	3.86	−5.25	−1.360
38							+1.75	+0.453
39							+3.75	+0.971
40							−0.25	−0.065
41	332	318	318	320	319.5	1.92	+2.5	+1.302
42							−1.5	−0.781
43							−1.5	−0.781
44							+0.5	+0.259
45	322	322	318	318	320	2.37	+2.0	+0.843
46							+2.0	+0.843
47							−2.0	−0.843
48							−2.0	−0.843
49	150	142	144	150	146.5	4.11	+3.5	+0.852
50							−4.5	−1.095
51							−2.5	−0.608
52							+3.5	+0.852
53	112	112	112	92	107	10.0	+5.0	+0.500
54							+5.0	+0.500
55							+5.0	+0.500
56							−15.0	−1.500
57	263	270	263	264	265	3.47	−2.0	−0.576
58							+5.0	+0.441
59							−2.0	−0.576
60							−1.0	−0.288

Table 3 (continued)

Seq. No.	Azimuth of dip direction of joint					Standard dev., σ	Dev. from mean, $x_k - \bar{x}$	Index τ_k
	x_1	x_2	x_3	x_4	Mean, $\bar{x}$			
61	293	296	302	300	297.75	4.08	−4.75	−1.139
62							−1.75	−0.426
63							+4.25	+1.532
64							+2.25	+0.551
65	286	295	285	295	290.25	5.5	−4.25	−0.772
66							+4.75	+0.863
67							−5.25	−0.955
68							+4.75	+0.863
69	105	105	105	108	105.75	1.505	−0.75	−0.498
70							−0.75	−0.498
71							−0.75	−0.498
72							+2.25	+1.495
73	73	76	76	76	75.25	1.505	−2.25	−1.495
74							+0.75	+0.498
75							+0.75	+0.498
76							+0.75	+0.498
77	36	36	43	30	36.25	5.11	−0.25	−0.049
78							−0.25	−0.049
79							+6.75	+1.321
80							−6.25	−1.223
81	28	28	28	30	28.5	1.0	−0.5	−0.500
82							−0.5	−0.500
83							−0.5	−0.500
84							+1.75	+1.500
85	42	42	34	42	40	4.0	+2.0	+0.500
86							+2.0	+0.500
87							−6.0	−1.500
88							+2.0	+0.500
89	143	146	145	144	144.5	1.29	−1.5	−1.162
90							+1.5	+1.162
91							+0.5	+0.387
92							−0.5	−0.387
93	232	220	225	228	226.25	5.055	+5.75	+1.137
94							−6.25	−1.238
95							−1.25	−0.227
96							+1.75	+0.346
97	290	282	284	290	286.5	4.11	+3.5	+0.852
98							−4.5	−1.095
99							−2.5	−0.608
100							+3.5	+0.852
101	193	196	195	194	194.5	1.29	−1.5	−1.162
102							+1.5	+1.162
103							+0.5	+0.387
104							−0.5	−0.387
105	47	45	36	38	41.5	3.76	+5.5	+1.463
106							+3.5	+0.930
107							−5.5	−1.463
108							−3.5	−0.930
109	172	176	173	175	174	1.83	−2.0	−1.093
110							+2.0	+1.093
111							−1.0	−0.546
112							+1.0	+0.546
113	285	285	287	286	286	1.155	−1.0	−0.864
114							−1.0	−0.864
115							+1.0	+0.864
116							+1.0	+0.864
117	35	40	40	35	37.5	2.89	−2.5	−0.865
118							+2.5	+0.865
119							+2.5	+0.865
120							−2.5	−0.865

Table 3 (continued)

Seq. No.	Azimuth of dip direction of joint					Standard dev., σ	Dev. from mean $x_k - \bar{x}$	Index τ_k
	x_1	x_2	x_3	x_4	Mean, $\bar{x}$			
121	210	216	220	210	214	4.9	−4.0	−0.816
122							+2.0	+0.408
123							+6.0	+1.224
124							−4.0	−0.816
125	196	202	198	206	200.5	4.44	−4.5	−1.012
126							+1.5	+0.338
127							−2.5	−0.563
128							+5.5	+1.238
129	250	248	250	250	249.5	1.0	+0.5	+0.500
130							−1.5	+1.500
131							+0.500	+0.500
132							+0.5	+0.500
133	210	210	208	206	208.5	1.53	+1.5	+0.980
134							+1.5	+0.980
135							−0.5	−0.327
136							−2.5	−1.634
137	205	205	212	205	206.75	4.46	−1.75	−0.392
138							−1.75	−0.392
139							+5.25	+1.132
140							−1.75	−0.392
141	170	170	168	166	168.5	1.53	+1.5	+0.98
142							+1.5	+0.98
143							−0.5	−0.327
144							−2.5	−1.634
145	38	37	40	37	38	1.414	0.0	0.000
146							−1.0	−0.707
147							+2.0	+1.414
148							−1.0	−0.707
149	7	2	5	0	3.5	3.1	+3.5	+1.127
150							−1.5	−0.484
151							+1.5	+0.484
152							−3.5	−1.127
153	340	332	334	340	336.5	4.11	+3.5	+0.852
154							−4.5	−1.095
155							−2.5	−0.608
156							+3.5	+0.852
157	355	356	350	352	353.25	2.755	+1.75	+0.632
158							+2.75	+0.998
159							−3.25	−1.177
160							−1.25	−0.454
161	255	255	257	260	256.75	2.36	−1.75	−0.741
162							−1.75	−0.741
163							+0.25	+0.106
164							+3.25	+1.384
165	353	353	352	352	352.75	0.5	+0.25	+0.500
166							+0.25	+0.500
167							+0.25	+0.500
168							−0.75	−1.500
169	12	12	15	14	13.25	1.5	−1.25	−0.833
170							−1.25	−0.833
171							+1.75	+1.167
172							+0.75	+0.500
173	80	80	78	88	81.5	4.43	−1.5	−0.340
174							−1.5	−0.340
175							−3.5	−0.790
176							+6.5	+1.467
177	342	344	348	348	345.5	3.0	−3.5	−1.167
178							−1.5	−0.500
179							+2.5	+0.833
180							+2.5	+0.833

Table 3 (continued)

Seq. No.	Azimuth of dip direction of joint				Standard dev., σ	Dev. from mean $x_k - \bar{x}$	Index τ_k
	X_1	X_2	X_3	X_4	Mean, $\bar{X}$			
181	82	86	83	85	84	1.83	−2.0	−1.093
182							+2.0	+1.093
183							−1.0	−0.546
184							+1.0	+0.546
185	84	84	83	82	83.25	0.957	+0.75	+0.783
186							+0.75	+0.783
187							−0.25	−0.261
188							−1.25	−1.310
189	2	4	8	8	5.5	3.0	−3.5	−1.167
190							−1.5	−0.500
191							+2.5	+0.833
192							+2.5	+0.833
193	342	352	342	340	344	5.45	−2.0	−0.367
194							+8.0	+1.468
195							−2.0	−0.367
196							−4.00	−0.734
197	128	127	128	128	127.75	0.5	+0.25	+0.500
198							−0.75	−1.500
199							+0.25	+0.500
200							+0.25	+0.500

Note. Computations of means and standard deviations for the azimuths of the joints were made by ordinary methods, since the azimuths within each small sample differed little from each other, and in no samples were observations from the first and fourth quadrants used simultaneously.

The test consisted of separating joints by angles and directions of dip. In the first case, joints were chosen at all outcrops for identity of dips, and in the second for identity of dip direction (azimuth).

Below we have described our use of the method of small samples for studying the distribution law for joint trends in sandstones of the Sinian Kachergat series on the Kurtun fold. The test was made only for the azimuth of the dip. In each outcrop, four joints with identical angles of dip were selected, where there was no doubt that all belonged to the same system. Such sampling was carried out at 50 outcrops on a single member of the sandstones.

Thus, in all, 200 joints were selected. An essential fact in this was that all 200 joints did not have to belong to a single system. It was important merely that four joints within a single outcrop actually represent a single system.

The data were treated in the following manner. The measured azimuths of dip directions were tabulated (Table 3). The following values were then calculated from these data for each outcrop.

1. The mean arithmetic azimuth of dip direction, $\bar{X}$.

2. The standard deviation (σ), characterizing the measure of scatter, determined from the equation

$$\sigma = \sqrt{\frac{\sum_{i=1}^{4}(X_i - \bar{X})^2}{3}}.$$

Table 4. Values of Distribution Functions

k	τ_k	$\tilde{F}_k = \dfrac{k - \frac{1}{2}}{n}$	$\tau_k + 1.732$	F_k	$\lvert \tilde{F}_k - F_k \rvert$
1	−1.634	0.0025	0.098	0.0283	0.0258
2	−1.634	0.0075	0.098	0.0283	0.0208
3	−1.500	0.0125	0.232	0.0670	0.0545
4	−1.500	0.0175	0.232	0.0670	0.0495
5	−1.500	0.0225	0.232	0.0670	0.0445
6	−1.500	0.0275	0.232	0.0670	0.0395
7	−1.500	0.0325	0.232	0.0670	0.0345
8	−1.495	0.0375	0.237	0.0685	0.0310
9	−1.463	0.0425	0.239	0.0691	0.0266
10	−1.385	0.0475	0.347	0.1003	0.0528
11	−1.360	0.0525	0.372	0.1075	0.0550
12	−1.321	0.0575	0.411	0.1188	0.0613
13	−1.310	0.0625	0.422	0.1220	0.0595
14	−1.238	0.0675	0.494	0.1428	0.0853
15	−1.223	0.0725	0.509	0.1471	0.0746
16	−1.223	0.0775	0.509	0.1471	0.0696
17	−1.222	0.0825	0.510	0.1474	0.0649
18	−1.177	0.0875	0.555	0.1604	0.0729
19	−1.167	0.0925	0.565	0.1633	0.0708
20	−1.167	0.0975	0.565	0.1633	0.0658
21	−1.162	0.1025	0.570	0.1647	0.0622
22	−1.162	0.1075	0.570	0.1647	0.0572
23	−1.139	0.1125	0.597	0.1714	0.0589
24	−1.127	0.1175	0.605	0.1748	0.0573
25	−1.095	0.1225	0.637	0.1841	0.0616
26	−1.095	0.1275	0.637	0.1841	0.0566
27	−1.095	0.1325	0.637	0.1841	0.0516
28	−1.093	0.1375	0.639	0.1847	0.0472
29	−1.093	0.1425	0.639	0.1847	0.0422
30	−1.012	0.1475	0.720	0.2081	0.0606
31	−0.955	0.1525	0.777	0.2246	0.0721
32	−0.930	0.1575	0.802	0.2318	0.0743
33	−0.865	0.1625	0.867	0.2507	|0.0882|
34	−0.865	0.1675	0.867	0.2507	0.0832
35	−0.864	0.1725	0.868	0.2509	0.0784
36	−0.864	0.1775	0.868	0.2509	0.0734
37	−0.864	0.1825	0.868	0.2509	0.0684
38	−0.864	0.1875	0.868	0.2509	0.0634
39	−0.843	0.1925	0.889	0.2569	0.0644
40	−0.843	0.1975	0.889	0.2569	0.0594
41	−0.843	0.2025	0.889	0.2569	0.0544
42	−0.833	0.2075	0.899	0.2598	0.0523
43	−0.833	0.2125	0.899	0.2598	0.0473
44	−0.816	0.2175	0.916	0.2647	0.0472
45	−0.816	0.2225	0.916	0.2647	0.0422
46	−0.790	0.2275	0.942	0.2722	0.0447
47	−0.782	0.2325	0.950	0.2746	0.0421
48	−0.782	0.2375	0.950	0.2746	0.0371
49	−0.781	0.2425	0.951	0.2748	0.0323
50	−0.781	0.2475	0.951	0.2748	0.0273
51	−0.772	0.2525	0.960	0.2774	0.0249
52	−0.741	0.2575	0.991	0.2864	0.0289
53	−0.741	0.2625	0.991	0.2864	0.0239
54	−0.734	0.2675	0.998	0.2884	0.0209
55	−0.707	0.2725	1.025	0.2962	0.0237
56	−0.707	0.2775	1.025	0.2962	0.0187
57	−0.707	0.2825	1.025	0.2962	0.0137
58	−0.707	0.2875	1.025	0.2962	0.0087
59	−0.608	0.2925	1.124	0.3248	0.0323
60	−0.608	0.2975	1.124	0.3248	0.0273
61	−0.608	0.3025	1.124	0.3248	0.0223
62	−0.576	0.3075	1.156	0.3341	0.0266
63	−0.576	0.3125	1.156	0.3341	0.0216
64	−0.563	0.3175	1.169	0.3378	0.0203
65	−0.546	0.3225	1.206	0.3485	0.0260
66	−0.546	0.3275	1.206	0.3485	0.0210
67	−0.500	0.3325	1.232	0.3560	0.0235
68	−0.500	0.3375	1.232	0.3560	0.0185
69	−0.500	0.3425	1.232	0.3560	0.0135
70	−0.500	0.3475	1.232	0.3560	0.0085
71	−0.500	0.3525	1.232	0.3560	0.0035
72	−0.500	0.3575	1.232	0.3560	0.0015
73	−0.500	0.3625	1.232	0.3560	0.0065

Table 4 (continued)

h	τ_k	$\tilde{F}_k = \dfrac{h - \frac{1}{2}}{n}$	$\tau_k + 1.732$	F_k	$|\tilde{F}_k - F_k|$
74	−0.500	0.3675	1.232	0.3560	0.0115
75	−0.500	0.3725	1.232	0.3560	0.0165
76	−0.500	0.3775	1.232	0.3560	0.0215
77	−0.500	0.3825	1.232	0.3560	0.0265
78	−0.498	0.3875	1.234	0.3566	0.0309
79	−0.498	0.3925	1.234	0.3566	0.0359
80	−0.498	0.3975	1.234	0.3566	0.0409
81	−0.484	0.4025	1.248	0.3607	0.0418
82	−0.454	0.4075	1.278	0.3693	0.0382
83	−0.426	0.4125	1.306	0.3774	0.0351
84	−0.407	0.4175	1.325	0.3829	0.0346
85	−0.392	0.4225	1.340	0.3873	0.0352
86	−0.392	0.4275	1.340	0.3873	0.0402
87	−0.392	0.4325	1.340	0.3875	0.0450
88	−0.387	0.4375	1.345	0.3887	0.0488
89	−0.387	0.4425	1.345	0.3887	0.0538
90	−0.367	0.4475	1.365	0.3945	0.0530
91	−0.367	0.4525	1.365	0.3945	0.0580
92	−0.340	0.4575	1.392	0.4023	0.0562
93	−0.327	0.4625	1.405	0.4060	0.0565
94	−0.327	9.4675	1.405	0.4060	0.0615
95	−0.288	0.4725	1.444	0.4173	0.0552
96	−0.261	0.4775	1.471	0.4251	0.0524
97	−0.227	0.4825	1.505	0.4349	0.0476
98	−0.172	0.4875	1.560	0.4508	0.0367
99	−0.065	0.4925	1.667	0.4818	0.0107
100	−0.049	0.4975	1.683	0.4864	0.0111
101	−0.049	0.5025	1.683	0.4864	0.0167
102	−0.049	0.5075	1.683	0.4864	0.0211
103	−0.049	0.5125	1.683	0.4864	0.0261
104	0	0.5175	1.732	0.5005	0.0170
105	0	0.5225	1.732	0.5005	0.0220
106	0.106	0.5275	1.838	0.5294	0.0019
107	0.125	0.5325	1.857	0.5367	0.0042
108	0.125	0.5375	1.857	0.5367	0.0008
109	0.259	0.5425	1.991	0.5754	0.0329
110	0.265	0.5475	1.997	0.5771	0.0296
111	0.338	0.5525	2.070	0.5982	0.0457
112	0.340	0.5575	2.072	0.5988	0.0413
113	0.346	0.5625	2.078	0.6005	0.0380
114	0.387	0.5675	2.119	0.6124	0.0449
115	0.387	0.5725	2.119	0.6124	0.0399
116	0.408	0.5775	2.140	0.6185	0.0410
117	0.453	0.5825	2.185	0.6315	0.0490
118	0.484	0.5875	2.216	0.6404	0.0529
119	0.498	0.5925	2.230	0.6445	0.0520
120	0.498	0.5975	2.230	0.6445	0.0480
121	0.498	0.6025	2.230	0.6445	0.0430
122	0.500	0.6075	2.232	0.6450	0.0375
123	0.500	0.6125	2.232	0.6450	0.0325
124	0.500	0.6175	2.232	0.6450	0.0275
125	0.500	0.6225	2.232	0.6450	0.0225
126	0.500	0.6275	2.232	0.6450	0.0175
127	0.500	0.6325	2.232	0.6450	0.0125
128	0.500	0.6375	2.232	0.6450	0.0075
129	0.500	0.6425	2.232	0.6450	0.0025
130	0.500	0.6475	2.232	0.6450	0.0025
131	0.500	0.6525	2.232	0.6450	0.0075
132	0.500	0.6575	2.232	0.6450	0.0125
133	0.500	0.6625	2.232	0.6450	0.0175
134	0.500	0.6675	2.232	0.6450	0.0225
135	0.500	0.6725	2.232	0.6450	0.0275
136	0.500	0.6775	2.232	0.6450	0.0325
137	0.500	0.6825	2.232	0.6450	0.0375
138	0.517	0.6875	2.249	0.6500	0.0375
139	0.546	0.6925	2.278	0.6583	0.0342
140	0.546	0.6975	2.278	0.6583	0.0392
141	0.632	0.7025	2.364	0.6832	0.0193
142	0.783	0.7075	2.515	0.7268	0.0193
143	0.783	0.7125	2.515	0.7268	0.0143
144	0.815	0.7175	2.547	0.7361	0.0186
145	0.815	0.7225	2.547	0.7361	0.0136
146	0.833	0.7275	2.565	0.7413	0.0138
147	0.833	0.7325	2.565	0.7413	0.0088
148	0.833	0.7375	2.565	0.7413	0.0038
149	0.833	0.7425	2.565	0.7413	0.0012

Table 4 (continued)

k	τ_k	$\widetilde{F}_k = \dfrac{k - \frac{1}{2}}{n}$	$\tau_k + 1.732$	F_k	$\lvert \widetilde{F}_k - F_k \rvert$
150	0.843	0.7475	2.575	0.7442	0.0033
151	0.843	0.7525	2.575	0.7442	0.0083
152	0.852	0.7575	2.584	0.7468	0.0107
153	0.852	0.7625	2.584	0.7468	0.0157
154	0.852	0.7675	2.584	0.7468	0.0207
155	0.852	0.7725	2.584	0.7468	0.0257
156	0.852	0.7775	2.584	0.7468	0.0307
157	0.852	0.7825	2.584	0.7468	0.0357
158	0.863	0.7875	2.595	0.7500	0.0375
159	0.863	0.7925	2.595	0.7500	0.0425
160	0.864	0.7975	2.596	0.7502	0.0473
161	0.864	0.8025	2.596	0.7502	0.0523
162	0.864	0.8075	2.596	0.7502	0.0573
163	0.864	0.8125	2.596	0.7502	0.0623
164	0.865	0.8175	2.597	0.7505	0.0670
165	0.865	0.8225	2.597	0.7505	0.0720
166	0.930	0.8275	2.662	0.7693	0.0582
167	0.971	0.8325	2.703	0.7812	0.0513
168	0.976	0.8375	2.708	0.7826	0.0549
169	0.980	0.8425	2.712	0.7838	0.0587
170	0.980	0.8475	2.712	0.7838	0.0637
171	0.980	0.8525	2.712	0.7838	0.0687
172	0.980	0.8575	2.712	0.7838	0.0737
173	0.998	0.8625	2.730	0.7890	0.0735
174	1.093	0.8675	2.825	0.8164	0.0511
175	1.093	0.8725	2.825	0.8164	0.0561
176	1.127	0.8775	2.859	0.8263	0.0512
177	1.132	0.8825	2.864	0.8277	0.0548
178	1.134	0.8875	2.866	0.8283	0.0592
179	1.137	0.8925	2.869	0.8291	0.0634
180	1.162	0.8975	2.894	0.8364	0.0611
181	1.162	0.9025	2.894	0.8364	0.0661
182	1.167	0.9075	2.899	0.8378	0.0697
183	1.224	0.9125	2.956	0.8543	0.0582
184	1.238	0.9175	2.970	0.8583	0.0592
185	1.302	0.9225	3.034	0.8768	0.0457
186	1.302	0.9275	3.034	0.8768	0.0507
187	1.321	0.9325	3.053	0.8853	0.0472
188	1.321	0.9375	3.053	0.8853	0.0522
189	1.384	0.9425	3.116	0.9005	0.0420
190	1.414	0.9475	3.146	0.9092	0.0383
191	1.414	0.9525	3.146	0.9092	0.0433
192	1.441	0.9575	3.173	0.9170	0.0405
193	1.463	0.9625	3.195	0.9234	0.0391
194	1.467	0.9675	3.199	0.9245	0.0430
195	1.468	0.9725	3.200	0.9248	0.0477
196	1.495	0.9775	3.227	0.9326	0.0449
197	1.500	0.9825	3.232	0.9340	0.0485
198	1.500	0.9875	3.232	0.9340	0.0535
199	1.500	0.9925	3.232	0.9340	0.0585
200	1.532	0.9975	3.264	0.9433	0.0542

3. The difference between X_k and $\overline{X}$, where any of the four values may be used for X_k if the sample is representative. In the present example, because of the small size of the sample, in order to use all information to the fullest and to improve the reliability of the result, all four values of azimuth were used.

4. The dimensionless value τ was determined from the formula

$$\tau_k = \frac{X_k - \overline{X}}{\sigma}.$$

The values of τ_k obtained were arranged in a variational series, and at each point the empirical distribution function $\widetilde{F}_k$ was calculated, $\widetilde{F}_k = (k - \frac{1}{2})/n$ (n is the number of observations, 200 in the present example), and also the theoretical function $F_k = (\tau_k + 1.732) \cdot 0.289$.

All these data have been given in Table 4. The absolute difference between F_k and $\widetilde{F}_k$ was determined and is shown in the last column of Table 4.

The test of the initial hypothesis reduces to the following. If the greatest value D_{200} of the absolute difference $|F_k - \widetilde{F}_k|$ for the 5% level of significance is smaller than the value of $\lambda_5/\sqrt{n}$, the hypothesis of symmetrically distributed azimuths of dip directions of the joints is not contradicted by observational data.

For the present example, when n = 200, using the value $\lambda_5 = 1.358$,* we find

$$\frac{\lambda_5}{\sqrt{n}} = \frac{1.358}{\sqrt{200}} = 0.0960.$$

The value $D_{200} = 0.0882$, obtained by us in our observations, lies within the range of permissible values (0.0882 < 0.0960).

It follows, therefore, that the hypothesis that the distribution of azimuth deviations for the dip directions of joints is symmetrical (i.e., a Fisher distribution in the present case) for a particular value of dip of the joints is not contradicted by observational data.

Analysis of Observations

In all, we measured 46,935 joint orientations, and these were plotted on 834 structure diagrams. All the material was selected by means of the indicated methods. It was found that some distributions of joint trends conformed to the Fisher distribution (Table 5, Fig. 1) and some did not (Table 5, Fig. 2). In connection with the fact that the lack of agreement appeared most clearly in clays and clayey rocks, trends of joint systems in limestones and sandstones only were used for comparison of orientations. The test is as follows.

If the first point of view is valid, i.e., if there are joints that preserve their orientation over large areas, the joints must belong to a single structure, even more to a group of folds in the region, since the conclusion that world-wide joints exist is generally drawn on the basis of analyzing composite rose diagrams. On such diagrams particular directions will predominate only if the joints of a given orientation prove to be sufficiently numerous within each individual local structure of the region.

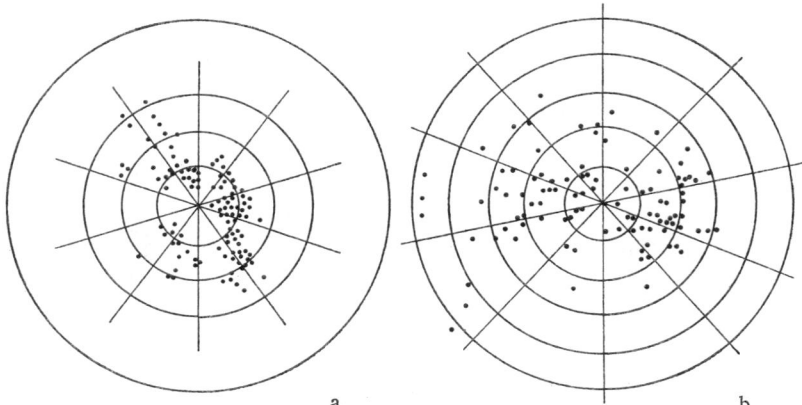

Fig. 2. Irregular girdle distribution of unit vectors for azimuth angles around the summed vector placed at the center of the grid. a) Chil'-Dara anticline, clay, N = 120; b) Shugurovo uplift, limestone, N = 102.

---
* See Dunin-Barkovskii and Smirnov, 1955, Table 4.4, p. 290.

Table 5. Goodness-of-Fit Test for Trend Distribution of Joint Systems with the Fisher Distribution

Chi-Square Method

Structure	Geotectonic zone	Lithology of bed	No. of observ.	Value of k	Ser. θ		Ser. φ	
					χ^2	$\chi^2_{0.05}$	χ^2	$\chi^2_{0.05}$
Chil'-Dara anticline	Tadzhik depression	Limestone...	133	28.634	7.89	16.9	6.27	7.81
		Clay.......	110	280.00	34.00	16.9	3.70	3.84
Shugurovo uplift	Russian platform	Limestone...	114	58.160	3.72	16.9	1.864	5.99
		Limestone...	102	16.631	12.36	16.9	1.93	3.84
Khristoforova uplift	Siberian platform	Sandstone...	100	32.293	5.5	16.9	1.9	3.84
		Clay.......	96	57.23	27.31	16.9	1.57	3.84

Small Sample Method

Structure	Geotectonic zone	Lithology of bed	No. of observ.	Distribution of			
				azimuths of dip directions		dip angles	
				D_n	$\frac{\lambda_5}{\sqrt{n}}$	D_n	$\frac{\lambda_5}{\sqrt{n}}$
Western Aruk-Tau anticline	Tadzhik depression	Limestone..	200	0.0410	0.0960	0.0257	0.0960
Kurtun folds	Zone of linear Baikal folds	Sandstone..	200	0.0882	0.0960	0.0897	0.0960
Khristoforova uplift	Siberian platform	Sandstone..	50	0.153	0.192	0.172	0.192

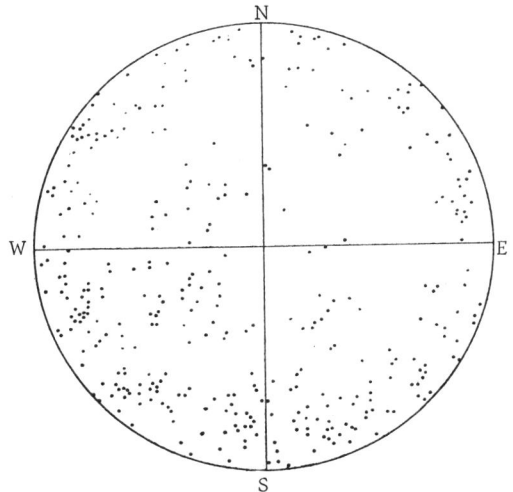

Fig. 3. Distribution of summed vectors of joint systems on the Kurtun folds.

On diagrams of summed vectors of systems plotted for individual structures, joints that persist in trend over the entire structure must give clear unimodal concentrations. As a first approximation it may be stated that the distribution of unit vectors within such accumulations do not differ from the Fisher distribution. An example of such a diagram is given in Fig. 3. Each point on the diagram represents the orientation of a system (its summed vector) computed from a large number of measurements (up to 130) of joints in the given system (unit vectors). Such diagrams were plotted for all the indicated structures. Clear concentrations could not be found on any of the diagrams. In order to avoid any erroneous conclusion, we combined similar orientations on the joint diagrams, assuming that they belonged to a single system. This hypothesis was tested on the basis of the Fisher distribution. In all cases it gave negative results.

Table 6. Results of Testing the Identity of Orientations of Joint Sets in Homogeneous Parts of the Structure

Information about the structure	No. of outcrops treated	No. of degrees of freedom columns k_1	rows k_2	$F_{0.05}$ at k_1 and k_2 degrees of freedom	Observed values of F at k_1 and k_2 degrees of freedom
Tadzhik depression, Chil'-Dara anticline, Kamalinskii member (Neogene, N_1 km), sandstone, northwestern limb	6	10	124	1.91	19.25
Tadzhik depression, Chil'-Dara anticline, Alai beds (Paleogene, Pg_2 a), sandstone, southeastern limb	4	6	74	2.22	2.41
Tadzhik depression, Western Aruk-Tau anticline, Bukhara beds (Paleogene, Pg_1 bh), limestone, eastern limb	11	20	416	1.60	6.41
Russian platform, Shugurovo uplift, Kamyshla beds (Permian, P_2 kaz), limestone, southeastern limb	8	14	120	1.77	5.43
Zone of linear Baikal folds, Kurtun folds, Kachergat series (Sinian, S_n katsch), sandstone, southeastern limb	4	6	96	2.19	2.20
Siberian platform, Southern Markovo uplift, Ust'-Kut series (Ordovician, O_1 ust), limestone, western limb	5	8	62	2.09	7.77

If we start on the basis that the second point of view is valid (that the joints are related to the fold as a whole), it is natural to expect a system of joints in rocks of a particular composition to be of the same orientation within a limb in those segments in which the axial plane has maintained its direction. Tests of the sameness of orientations in joint systems were made by means of the Fisher distribution [Vistelius, 1958]. Results of the test, shown in Table 6, do not allow us to state that this point of view is valid.*

The validity of the next point of view (that the joints are related to bedding planes) was tested by examination of structure diagrams plotted on Kavraiskii nets [Vistelius, 1958]. Orientations of joints measured at a single outcrop within a single bed were plotted on each diagram. Measurements of joints in different beds or different parts of the structure were not plotted on the same diagram. In all, 834 structure diagrams were analyzed. Positions of the joint systems on the diagrams indicate that the diversity of joints, in both folded and platform zones, is characterized by a definite orientation in the following system of coordinates: the X axis lies in the bedding plane and points in the dip direction; the Y axis coincides with the strike of the bed and is directed to the right of the positive direction of the X axis; and the Z axis is perpendicular to the bedding plane and is directed upward.

The orientations of all joint systems were changed correspondingly with change in orientation of the coordinate axes (Fig. 4). The following systems of joints were thus distinguished.

1. A set of tension joints parallel to the XZ plane, set I.

2. A set of tension joints parallel to the YZ plane, set II.

3. Two pairs of conjugate sets of shear joints perpendicular to the XY plane, passing through the Z axis and mutually symmetrical relative to the X and Y axes — sets III-IV and III^a-IV^a.

---

*The hypothesis that the summed vectors are identical is not contradicted by observations if $F < F_{0.05}$.

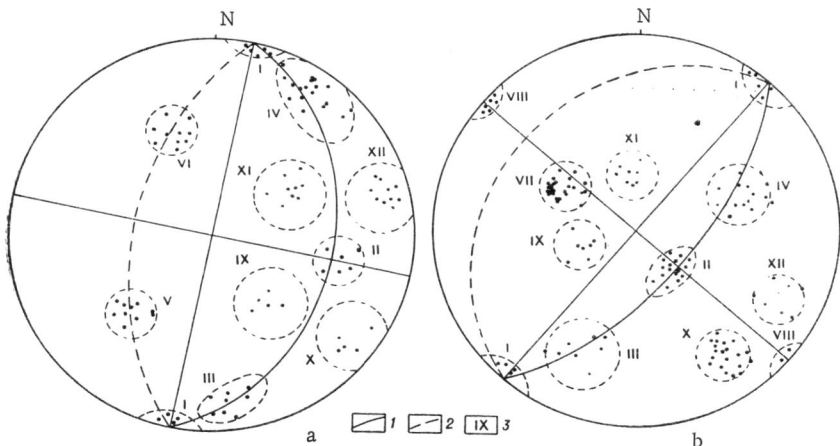

Fig. 4. Change of joint directions in the Western Aruk-Tau anticline in connection with change in attitude of the bed. Symmetry of the corresponding joint sets may be seen. a) Limestone, dipping 35° to the west (azimuth of 285°); sandstone, dipping 65° to the northwest (azimuth of 312°). 1) Great circle representing the intersection of the surface of the sphere and the bedding plane; joints perpendicular to the bedding plane lie on this arc; 2) great circle representing intersection of the surface of the sphere and a plane perpendicular to the bedding plane, extending parallel to the strike of the bed; joints with strikes coinciding with the direction of dip of the bed lie on this arc; the diameter joining the two ends of the arc represents the projection of the strike line of the bed; joints with strikes parallel to the strike of the bed appear on the diagrams as points on this diameter; 3) symbol of joint set.

4. A pair of conjugate sets of shear joints perpendicular to the YZ plane, passing through the X axis and symmetrically inclined to the Z axis — sets V-VI.

5. A pair of conjugate sets of shear joints perpendicular to the XZ plane, passing through the Y axis, the Z angle bisecting the angle between them — sets VII-VIII.

6. Four sets of joints equally inclined to the XY plane; traces of their intersections with the indicated plane form rhombs, the diagonals of which are the X and Y axes; in intersecting, these sets form the side faces of rhombic pyramids, the height of which is equal to the Z axis — sets IX, X, XI, and XII. The mechanical type these joints belong to is not clear. It is possible that they developed when the two main normal stresses were of equal value. The rhombic pyramid formed by them may apparently be considered an approximation of a cone, which should be the shape of the joint in this situation.

7. Joints parallel to the XY plane (along the bedding). We did not study these, and we shall not therefore consider them.

On the structure diagram, all joint sets occupy a definite position, and they systematically lie on appropriate lines representing projections of the indicated coordinate planes (Fig. 4). The picture observed on the diagrams is characteristic for most of the diagrams studied, and it indicates a correlation between the trend of all joints and the position of the bed.

In those rare cases in which deviation of joint orientation from the systematic position in the introduced coordinate system occurs, it is still difficult to deny that such correlation exists.

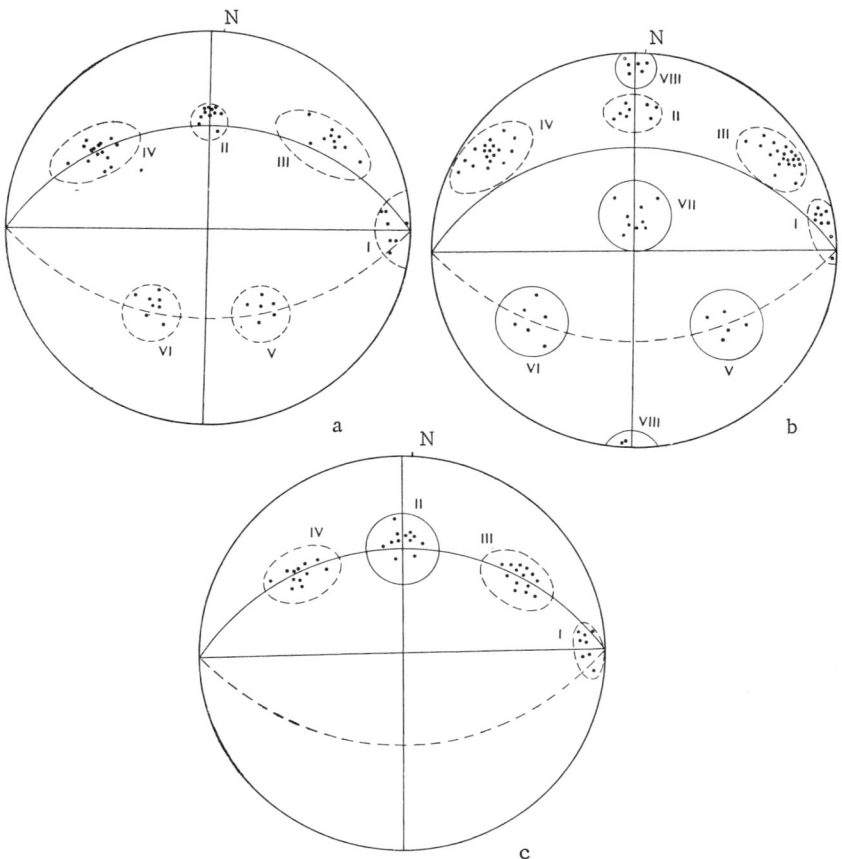

Fig. 5. Distribution of joint orientations in the Chil'-Dara anticline in beds of different composition within a single outcrop. a) Sandstone with basal clay cement (basal = floating grains); b) clay; c) sandstone with contacting grains and interstitial cement. Beds dip 41° to the south (azimuth of 173°). Symbols mean the same as in Fig. 4.

For example, Fig. 5 shows the distribution of joint orientations in a layer of clay and the underlying and overlying sandstones (of different composition) in a single exposure on the Chil'-Dara anticline. The nature of the joint distribution on the indicated diagrams leads one to believe that the deviation of joint directions from the corresponding lines is a secondary feature, that it must be due to deforming processes that developed in the rocks after the joints formed, causing displacement and disturbance of the initial orientation. These processes have been manifested most strongly in the more plastic rocks — clays (Fig. 5b) — and are almost completely absent in the sandstones (Fig. 5a).

Similar phenomena are noted wherever the systematic position of joints has been disturbed in the introduced system of coordinates. It is characteristic that all the joint sets do not display appreciable displacement, only those completely defined.*

The assumption of plastic deformation appears probable, since development of the destructive process jointly with plastic and elastic deformation of rocks may now be considered a proved fact [Gzovskii, 1963].

---

* These are joints of earlier formation than those that exhibit no appreciable displacement.

Subsequent development of plastic deformation leads not only and not so much to change in the systematic position of joints in coordinates of the bed as to change in the probability distribution law of joint directions belonging to a single set. We have noted above that general views concerning formation of a single set of joints lead one to assume that the orientation distribution of such joints must conform to the Fisher distribution with the parameter k greater than 4. The existence of such sets confirms the assumption. But sets were also observed that do not conform to the Fisher distribution. As a rule, in such sets the uniformity of azimuth distribution of the unit vectors about the summed vector has been disturbed, whereas the distribution law of radial deviations according to the Fisher distribution has been preserved (Table 5). In this case, the vectors are not regularly distributed, but are strung out in a belt of definite trend, chiefly along the strike of the bed. Such an irregular girdle distribution of vectors is especially marked in the more plastic rocks, clays (Fig. 2a); it is less noticeable in limestones (Fig. 2b). The distribution here described attests to the effect of some group (or a few groups) of causes, leading to deviation of joint attitudes from the ideal position in a single direction at some finite value. Plastic deformation developing in rocks after formation of the joints must clearly be considered such a cause.

The indicated distribution was observed in all sets displaced from their regular position relative to the bed and also in sets in which such displacement was not noted (Fig. 5c).

Thus, study of the direction distribution of joint sets and their relation to attitudes of the bed in all probability not only leads to confirmation of the fact that the rock experienced finite plastic deformation after the joints were formed, but also, in some measure, to determination of the magnitude of this deformation (according to the degree of irregularity of azimuth distribution of the unit vectors around the summed vector of the set).

The results we have obtained permit us to make the following definition of a joint set, differing from that generally accepted. By a joint set we mean a group of joints with orientations at a given point at the moment they form belonging to a particular Fisher distribution with the parameter k greater than 4.

Conclusions

As a result of our investigations, we may consider the following established.

1. A set is a combination of joints whose trends at any given point at the time of formation belong to a particular Fisher distribution with the parameter k greater than 4.

2. In all, as many as 14 sets of joints have been distinguished on local structures, the orientations systematically related to the orientation of the bedding plane.

3. Observed deviations of joint sets from the indicated pattern are apparently due to the effect of plastic deformation in the rock after formation of the joints. These sets, in turn, deviate from the corresponding coordinate planes in a single direction and approximately at a single angle.

4. Plastic deformation leads also to a change in the type of probability distribution law for the directions of joints belonging to a single set. Fundamentally, this change is a disturbance of the uniformity of distribution of joint orientations for azimuths about their mean values, i.e., a failure to fulfill one of the conditions of the Fisher distribution.

The author is happy to use this occasion to express his thanks to A. B. Vistelius for his constant assistance in this work.

REFERENCES

Azhgirei, G. D., Structural Geology, Izd. Mosk. Gos. Univ. (1956).

Belousov, V. V., "Tectonic fractures, their types and mechanism of formation," Tr. Geofiz. Inst., No. 17 (144), Moscow (1952).

Chiang Chu-Ch'i, "Mechanism of formation of tectonic fractures (as exemplified in Dagestan)," Vestn. Mosk. Gos. Univ., Ser. IV, No. 5, pp. 21-28 (1963).

De Sitter, L. U., Structural Geology, McGraw-Hill (1959).

Dorofeeva, T. V., "Experience in the use of methods of investigating jointed reservoir rocks, Southern Minusinsk basin," in: Joint Investigations of Jointed Reservoir Rocks and Experience in Computing their Petroleum Reserves, Tr. Vses. Nauchn.-Issled. Geologorazved. Inst., No. 214, Gostoptekhizdat, Leningrad (1963).

Dunin-Barkovskii, I. V., and Smirnov, N. V., Theory of Probability and Mathematical Statistics in Technology (General Part), Gostekhizdat, Moscow (1955).

Éz, V. V., "The relation of jointing in coals of the Donets Basin to fold structures," in: Folds of the Earth's Crust, Their Types and Mechanism of Formation, Izd. Akad. Nauk SSSR, Moscow (1962).

Gromov, V. K., "Experience in the use of methods of investigating jointed reservoir rocks, Bashkirian ASSR," in: Joint Investigations of Jointed Reservoir Rocks and Experience in Computing their Petroleum Reserves, Tr. Vses. Nauchn.-Issled. Geologorazved. Inst., No. 214, Gostoptekhizdat, Leningrad (1963).

Gzovskii, M. V., Fundamental Questions of Tectonophysics and Tectonics of the Baidzhansai Anticlinorium, Parts III and IV, Izd. Akad. Nauk SSSR, Moscow (1963).

Kalacheva, V. N., and Knoring, L. D., "Jointed reservoir rocks for oil and gas in promising Lower Cambrian strata of the Irkutsk amphitheater," Byul. Nauchn.-Tekhn. Inf., No. 1 (45) (1963).

Kazimirov, D. A., and Kuznetsova, K. I., "Experience in studying jointed sedimentary strata in folds of southwestern Fergana," in: Problems of Tectonophysics, Gosgeoltekhizdat, Moscow (1960).

Kirillova, I. V., Some Questions on the Mechanism of Folding (in Connection with a Study of the Internal Structure of Folded Strata), Tr. Geofiz. Inst., No. 6 (133), Izd. Akad. Nauk SSSR, Moscow (1949).

Mikhailov, E. A., Field Methods of Studying Joints in Rocks, Gosgeoltekhizdat, Moscow (1956).

Novikov, A. S., "Jointing in sedimentary rocks on the eastern part of the Russian platform," Izv. Akad. Nauk SSSR, Ser. Geol., No. 5, pp. 68-85 (1951).

Parker, Y. M., "Regional systematic jointing in slightly deformed sedimentary rocks," Bull. Geol. Soc. Am., 53:381 (1942).

Permyakov, E. N., Tectonic Fracturing of the Russian Platform, Izd. Mosk. Obshchestva Ispytatelei Prirody, Moscow (1949).

Shat-skii, N. S., Tectonic Sketches of the Volga–Ural Oil Region and Adjoining Parts of the Western Slope of the Southern Urals, Mat. k Posnaniyu Geol. Stroeniya SSSR (Material for understanding the Geologic Structure of the USSR), No. 2 (6), Izd. Mosk Obshchestva Ispytatelei Prirody (1945).

Shul'ts, S. S., "Study of world-wide jointing," in: Rock Deformation and Tectonics, Report of Soviet Geologists, Problem 4, International Geological Congress, 22nd session, Izd. Nauka, Moscow (1964).

Vistelius, A. B., Structure Diagrams, Izd. Akad. Nauk SSSR, Moscow-Leningrad (1958).

Vistelius, A. B., "The skew frequency distributions and the fundamental law of geochemical processes," J. Geol., Vol. 68, No. 1, pp. 1-22 (1960).

Watson, G. S., and Irving, E., "Statistical methods in rock magnetism," Monthly Not. Roy. Astron. Soc., Geophys. Suppl., Vol. 7, No. 6, pp. 289-299 (1957).

II. Use of the Specific Properties of Multidimensional Space in Solving Geological Problems

THE ORIGIN OF CLASTIC MINERAL ASSOCIATIONS IN THE APTIAN-CENOMANIAN ROCKS OF THE SOUTHWESTERN URAL AND MUGODZHARY REGION

M. E. Demina

Laboratory of Mathematical Geology
Mathematical Institute
Academy of Sciences of the USSR
Leningrad State University
Leningrad, USSR

O. M. Kalinin

A. A. Zhdanov Leningrad State University
Leningrad, USSR

In this paper the origin of clastic mineral associations in the Aptian-Cenomanian rocks of the Ural, Mugodzhary, and northern Ustyurt regions is considered. Methods of discriminant analysis are used in solving the problem.

The Aptian-Cenomanian clastic strata in the region bordered by Ustyurt, the Mugodzhary, and the Southern Urals have repeatedly drawn attention to possible bauxite occurrence in the foothills districts and to oil occurrence in the region adjacent to Ustyurt. In this connection it has become necessary to solve the problem of provenance of the clastic material.

This provenance problem has been solved by many geologists of the present century. It has been shown that the clastic material has come from regions where, at the present time, the deeply eroded uplands of the Urals and Mugodzhary now exist [Koltypin, 1957; Yanshin, 1948, 1953], and from which, as it has been stated, a broad piedmont slope developed, extending as far as Emba and Ustyurt.

This point of view, until recently, has apparently been very widely accepted, but recent investigations of garnet compositions from the Aptian-Cenomanian rocks show that, at least within the northern Ustyurt, the garnets have no relations whatsoever to the Urals [Vistelius and Demina, 1963]. It has thus become necessary to refine the boundaries between the different provenance regions involved in contribution to the investigated clastic strata. In this paper, an attempt is made to shed light on these matters by studying the mineral associations in the heavy fraction by means of discriminant analysis.

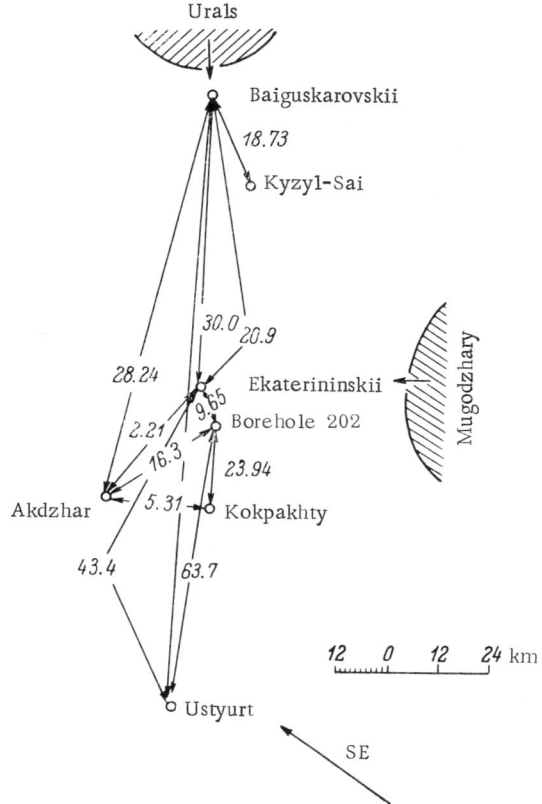

Fig. 1. Sketch of the relative positions of the investigated sections. Figures represent the coefficient Δ, arrows the direction of detrital transport.

Characteristics of the Mineral Associations

In order to solve the problems we have set up, the following sections were studied: Baiguskarovskii, Kyzyk-Sai, Ekaterininskii, borehole 202, Akdzhar, Kokpakhty, and the Ustyurt test hole (Fig. 1). From these sections, samples of Aptian-Cenomanian sandstones and siltstones were selected (Table 1). Regionally the Baiguskarovskii and Kyzyl-Sai sections are in the Ural zone; the Ekaterininskii and Akdzhar sections, and borehole 202 are in the Mugodzhary zone.

In the Ural region the rocks that interest us here are represented by only the Albian stage, which has been studied in connection with possible bauxite and phosphorite occurrences. The stratigraphic assignment was worked out by Yanshin, Bezrukov, and Fokin [1933]. The Albian here consists of cross-bedded quartz sands, of varied grain size, and interbeds of gray clay or of ocherous yellow and red clay.

In the Mugodzhary region, the samples taken for study belong to the Albian and Cenomanian according to the stratigraphic scheme of Koltypin [1957]. The Mugodzhary region has been studied in detail in connection with occurrences of petroleum. The Albian and Cenomanian rocks are chiefly well-sorted coarse-, medium-, and fine-grained friable sands and silts in various tints and shades of gray and yellow. The sands are separated by layers of smoky gray or bluish gray carbonate-free, heavy, sticky clays, which wedge out along the strike.

The cores from the Ustyurt test hole that were studied contain Aptian, Albian, and Cenomanian representatives. The stratigraphic subdivision has been taken from the work of Atanasyan, Grinberg, and Sukhilin [1962], in which the subdivisions are based on faunal evidence. The Aptian-Cenomanian sequence consists of an alternation of clay, sandstone, and siltstone. The sands are gray, of various shades, rarely brownish, fine-grained and varigrained, argillaceous, compact and friable, with carbonized plant remains. Sandstones are gray, fine- and medium-grained, weakly cemented, and argillaceous. Siltstones are gray and greenish gray, sandy, and argillaceous. Clays range from light gray to black; they are silty, sandy, finely elutriated, with carbonized detritus.

It should be noted that the Baiguskarovskii section was adopted by us as a typical Ural section for comparison of its monotypic clastic material with sections of other investigated regions.

The heavy fraction was separated from all sandstones and siltstones and its mineral composition was studied (Table 2). Below we give a brief description of the minerals in the heavy fraction.

Table 1. Sample Localities and Numbers of Investigated Samples

Sample area (section), number of samples	Sample locality	Collector
Baiguskarovskii, 12	Northernmost exposure of Lower Cretaceous (Albian), white quartzose sands; holes 27, 28, 30, 31; depth of 7.5-14.0 m	A. A. Zakharov
Kyzyl-Sai, 10	Yellow argillaceous and yellow coarse-grained Albian sands; holes of the Aktyubinsk prospecting group, sps. 43-52	A. B. Vistelius
Ekaterininskii, 12	Yellow sands, otc. 15, beds 1-3, sps. 196, 196a, 198, 199, 202, 204, 209, 211-214	S. N. Koltypin
Borehole 202, 12	Sands, depth of 70-327 m	S. N. Koltypin
Akdzhar, 15	Sands, otc. 16, beds 1-5, Borehole K-63, depth 66-70 m (Albian)	S. N. Koltypin
Ustyurt, 24	Aptian and Albian deposits, sandstones, Ustyurt test hole, depth 648-1165 m	V. A. Atanasyan
Kokpakhty, 14	Aptian and Albian sandstones and silts, Borehole 2, 9, 10, 15-17, 19, 32, depth 35-303 m	A. B. Vistelius

Hematite forms flat, dull, slightly rounded grains; much of it is strongly limonitized. The streak is reddish brown.

Ilmenite forms lustrous black grains, poorly rounded; the surfaces are pitted and locally altered to leucoxene. Crystal faces are not observed. The streak is black, and grains are opaque.

Rutile grains are bright red, lustrous, elongated, and slightly rounded; elbow twins are rarely encountered. Many grains contain inclusions of an opaque black mineral, probably one of the ores.

Leucoxene grains range from slightly rounded to well rounded, from brown to almost black. In microscope mounts they appear in turbid masses with rare needles of rutile.

Chromite ranges from well-terminated crystals with slightly rounded octahedral faces to slightly rounded and angular fragments of various sizes. The streak is brown, and thin slivers transmit brownish red light. The grain surface may be shiny, dull, or granular.

Magnetite forms black, dull and lustrous, slightly rounded grains. Some grains are hematitized, but they have not lost their magnetic properties.

Kyanite occurs in dense blue to colorless, angular, slightly rounded and well-rounded flat grains. Many grains contain black, dot-like inclusions of ore mineral, randomly scattered or distributed in belts and aureoles. Deformed grains are observed, bent along one of the cleavage faces. Refractive indices are $Ng = 1.728$ and $Np = 1.712$. Pleochroism is present in the colored varieties: from a sharply defined violet along Ng to blue along Np.

Zircon ranges from well-terminated crystals of zircon type and fragments of terminated crystals of zircon and garnet types, of a slightly rose color, to angular fragments, slightly rounded, and markedly green. One may rarely find varieties of a lilac color, generally well rounded. Some grains are filled with needles and brown inclusions; very rarely bubbles are observed. Refractive indices are $Ng = 1.987$ and $Np = 1.928$.

Garnet forms angular and well-rounded fragments, rarely terminated crystals (pentagonal dodecahedrons). Most grains have a faint rosy cast, rarely are a bright rose color or slightly orange; the grains are mostly clean, transparent, with few inclusions (black and dot-like). Some grains exhibit a system of fractures. Very rarely one may find completely colorless transparent grains filled with inclusions of ore mineral. The refractive index ranges from 1.818 to 1.772.

Table 2. Mineral Content

Section	Yield of heavy frac., % of sample	Pyrite	Martite	Hematite	Limonite	Ilmenite	Rutile
Baiguskarovskii							
Sp. 56, h. 27, dep. 14.25 m	1.96	—	—	—	—	—	46.1
Sp. 13, h. 28, dep. 7.40 m	1.48	—	—	0.6	—	—	48.3
Sp. 15, h. 28, dep. 8.00 m	0.68	—	—	2.0	—	—	55.7
Sp. 16, h. 28, dep. 8.50 m	4.86	—	—	2.1	—	—	39.3
Sp. 17, h. 28, dep. 8.80 m	3.18	—	—	1.0	—	0.3	50.2
Sp. 8, h. 30, dep. 11.20 m	5.53	—	—	2.9	—	—	50.0
Sp. 9, h. 31, dep. 12.00 m	1.35	—	—	2.1	—	1.2	81.7
Sp. 11, h. 31, dep. 13.5 m	1.06	0.1	—	0.8	—	—	53.6
Sp. 12, h. 31, dep. 14.00 m	1.25	0.1	—	0.4	—	—	74.9
Sp. 13, h. 31, dep. 14.5 m	1.89	—	—	0.5	—	—	67.9
Sp. 9, h. 53, dep. 10.30 m	0.06	—	—	15.0	—	2.7	16.9
Sp. 55, h. 27, dep. 11.00 m	1.32	—	—	0.6	—	—	40.7
Kyzyl-Sai							
Sp. 43	0.1	—	—	30.7	—	—	3.7
Sp. 44	0.06	—	—	10.0	—	—	6.8
Sp. 45	0.07	—	—	10.9	—	—	5.7
Sp. 46	0.01	0.7	—	3.2	—	—	9.2
Sp. 47	0.03	0.4	—	4.7	—	—	5.9
Sp. 48	0.002	—	—	12.7	—	—	10.6
Sp. 49	0.386	—	—	25.0	—	3.6	3.1
Sp. 50	0.28	—	—	19.3	—	—	3.1
Sp. 51	0.38	—	—	21.6	—	—	5.0
Sp. 52	1.07	—	—	13.7	—	—	7.1
Ekaterininskii							
Sp. 196, otc. 15, bed 1	0.05	R. g.	—	6.8	—	1.2	1.1
Sp. 196a, otc. 15, bed 3	0.58	—	—	6.5	—	1.0	1.3
Sp. 198, otc. 15, bed 1	0.17	0.4	—	6.6	—	1.5	1.1
Sp. 199, otc. 15, bed 1	0.08	—	—	7.8	—	1.3	3.9
Sp. 202, otc. 15, bed 1	0.15	R. g.	—	9.2	—	0.3	3.7
Sp. 204, otc. 15, bed 1	0.38	—	—	7.7	—	3.8	1.4
Sp. 209, otc. 15, bed 1	1.39	—	—	4.6	—	3.0	1.5
Sp. 211, otc. 15, bed 2	0.11	2.0	—	13.4	—	2.8	1.5
Sp. 212, otc. 15, bed 1	0.14	1.1	—	15.0	—	3.7	2.1
Sp. 213, otc. 15, bed 2	0.90	—	—	6.6	—	—	1.0
Sp. 214, otc. 15, bed 3	0.88	—	—	20.2	—	2.7	0.7
Sp. 220, otc. 15, bed 3	0.28	—	—	7.2	—	1.5	2.4
Borehole 202							
Dep. 71–76 m	0.18	—	—	12.3	—	—	1.4
Dep. 70–80 m	0.21	—	—	8.1	—	—	1.0
Dep. 80–84 m	0.26	—	—	11.2	—	—	0.9
Dep. 84–88 m	0.26	—	—	6.0	—	—	0.6
Dep. 84–94 m	0.07	—	—	7.0	—	—	0.6
Dep. 105–110 m	0.23	—	—	3.2	—	1.8	1.2
Dep. 124–144 m	0.18	—	—	6.4	0.5	—	0.5
Dep. 154–174 m	0.16	—	—	5.2	—	—	0.1
Dep. 185–190 m	0.17	—	—	6.4	—	—	1.2
Dep. 279–284 m	0.11	—	—	12.4	—	—	0.2
Dep. 304–327 m	0.18	—	—	5.6	—	1.9	0.6
Dep. 332–342 m	0.08	—	—	8.2	—	—	0.7

of Heavy Concentrates

Anatase	Brookite	Corundum	Leucoxene	Spinel	Chrom-picotite	Chromite	Magnetite	Andalusite	Silliman-ite	Kyanite	Staurolite	Zircon
—	—	—	1.7	—	—	4.6	0.4	—	—	39.2	—	2.0
—	—	R.g.	4.0	—	—	3.8	R.g.	—	—	32.6	—	2.1
—	—	—	3.7	—	—	11.1	R.g.	—	—	23.6	R.g.	1.2
—	—	—	4.1	—	—	33.8	R.g.	—	—	14.8	R.g.	1.8
R.g.	—	—	1.9	—	—	34.8	R.g.	—	—	7.5	0.1	1.1
0.4	—	—	1.3	—	—	2.6	0.2	—	—	33.9	0.5	4.9
—	—	—	1.5	—	0.1	4.8	0.4	—	—	3.5	R.g.	1.6
—	—	—	5.3	—	—	4.8	R.g.	—	—	28.2	—	2.0
—	—	—	2.3	—	—	9.9	R.g.	—	—	7.2	R.g.	2.6
—	—	—	2.4	—	0.2	18.9	R.g.	—	—	2.9	—	2.8
1.0	—	—	36.1	—	—	13.5	1.0	—	—	0.3	1.7	8.0
—	—	—	4.1	—	—	2.0	R.g.	—	—	43.7	—	1.5
0.9	—	—	45.5	—	R.g.	7.8	0.1	—	—	0.5	5.3	1.9
0.7	—	—	39.2	—	—	4.4	13.0	—	R.g.	0.2	5.7	15.7
0.1	—	—	14.2	—	—	8.2	38.7	—	—	—	6.5	12.9
1.5	—	—	30.0	—	—	8.5	6.3	—	—	2.0	2.0	10.0
2.2	—	—	17.7	—	—	5.2	R.g.	—	—	1.4	1.0	16.5
1.4	—	—	42.8	—	—	7.1	R.g.	—	—	—	2.8	13.4
0.6	—	—	28.8	—	—	2.8	16.4	—	—	2.0	4.9	8.1
0.1	—	—	24.3	—	—	19.0	12.6	—	—	0.5	11.7	4.6
0.5	—	—	14.2	—	—	20.4	19.3	—	—	0.1	6.1	11.7
—	—	—	19.0	—	—	9.8	23.9	—	—	0.3	9.8	15.7
0.8	—	—	45.2	—	—	—	25.8	—	—	1.5	0.6	4.6
0.8	—	—	19.7	—	—	1.1	36.4	R.g.	—	0.7	0.6	6.0
0.9	—	—	50.2	—	—	2.7	13.0	—	—	1.5	1.1	6.5
1.3	—	0.1	39.6	—	—	1.5	15.8	—	—	1.9	2.1	9.2
0.6	—	—	45.8	—	—	0.6	19.2	—	—	1.5	0.9	10.2
0.7	—	R.g.	33.6	—	—	0.8	25.9	—	—	2.6	1.2	15.0
0.5	—	R.g.	17.4	—	—	4.0	23.1	R.g.	—	0.8	2.2	5.9
—	—	—	27.4	—	—	1.2	18.6	—	—	17.6	1.6	3.1
—	—	—	29.1	—	—	2.5	18.6	—	—	13.3	0.4	4.8
0.2	—	—	20.8	—	—	0.7	53.3	—	—	4.6	1.7	6.4
0.2	—	—	22.9	—	—	1.2	30.9	—	—	0.7	0.5	15.2
0.8	—	—	18.0	—	—	2.3	9.5	—	—	3.4	0.4	10.3
0.9	—	—	10.9	—	—	15.8	32.1	—	—	0.9	3.1	7.6
0.6	—	R.g.	14.0	—	—	15.2	29.1	0.3	—	1.6	3.0	8.5
0.6	—	—	16.8	—	—	20.6	23.3	R.g.	—	1.8	1.3	7.7
0.4	—	—	8.3	—	—	17.5	34.2	—	—	1.2	4.3	5.8
2.0	—	—	17.3	—	1.2	7.8	27.5	R.g.	—	2.4	2.0	5.1
0.3	—	—	6.6	—	—	11.5	40.3	R.g.	—	1.8	1.2	5.4
1.2	—	—	10.5	—	—	18.9	17.6	0.2	—	3.6	3.5	4.8
0.1	—	—	6.8	—	—	20.2	21.2	—	—	0.1	2.8	1.5
0.5	—	—	15.3	—	—	18.6	17.0	—	—	1.1	4.8	4.6
0.3	—	—	13.0	0.3	—	7.4	15.4	—	—	0.3	4.6	11.6
1.0	—	—	7.4	—	—	17.4	19.6	—	—	0.6	4.2	8.8
0.6	—	—	8.9	—	—	18.2	19.4	—	—	0.4	3.1	6.4

Table 2 (continued)

Section	Actinolite	Amphibole	Rhombic pyroxene	Monoclinic pyroxene	Garnet	Epidote	Tourmaline
Baiguskarovskii							
Sp. 56, h. 27, dep. 14.25 m	—	—	—	—	1.1	—	4.6
Sp. 13, h. 28, dep. 7.40 m	—	—	—	—	0.5	—	5.7
Sp. 15, h. 28, dep. 8.00 m	—	—	—	—	0.5	—	2.0
Sp. 16, h. 28, dep. 8.50 m	—	—	—	—	—	—	2.6
Sp. 17, h. 28, dep. 8.80 m	—	—	—	—	0.5	0.1	1.3
Sp. 8, h 30, dep. 11.20 m	R.g.	—	0.2	—	R.g.	—	2.0
Sp. 9, h. 31, dep. 12.00 m	—	—	—	—	1.1	R.g.	0.2
Sp. 11, h. 31, dep. 13.5 m	—	—	R.g.	—	0.8	—	2.0
Sp. 12, h. 31, dep. 14.00 m	—	—	—	—	1.3	R.g.	0.6
Sp. 13, h. 31, dep. 14.5 m	—	—	—	—	2.7	0.2	2.0
Sp. 9, h. 53, dep. 10.30 m	—	—	—	—	R.g	—	2.0
Sp. 55, h. 27, dep. 11.00 m	—	—	—	—	0.6	—	4.0
Kyzyl-Sai							
Sp. 43	R.g.	0.3	R.g.	—	0.3	R.g.	2.2
Sp. 44	—	—	—	—	0.4	0.9	1.2
Sp. 45	—	—	—	—	0.2	—	1.6
Sp. 46	—	—	—	—	R.g.	25.2	0.9
Sp. 47	—	—	—	—	0.9	35.6	3.1
Sp. 48	—	—	—	—	R.g.	2.3	3.5
Sp. 49	—	—	—	—	0.3	0.3	3.6
Sp. 50	—	—	R.g.	—	0.6	—	3.8
Sp. 51	—	—	—	—	0.3	—	0.5
Sp. 52	—	—	—	—	—	—	0.4
Ekaterininskii							
Sp. 196, otc. 15, bed 1	—	—	—	—	3.3	—	5.3
Sp. 196a, otc. 15, bed 3	—	—	—	—	7.9	16.1	1.7
Sp. 198, otc. 15, bed 1	—	R.g.	—	—	3.5	—	6.6
Sp. 199, otc. 15, bed 1	—	—	—	—	7.2	R.g.	5.0
Sp. 202, otc. 15, bed 1	—	—	—	—	2.5	R.g.	5.5
Sp. 204, otc. 15, bed 1	—	—	—	—	0.4	0.2	5.4
Sp. 209, otc. 15, bed 1	—	—	—	—	6.6	23.8	4.8
Sp. 211, otc. 15, bed 2	—	—	—	—	1.6	2.1	5.4
Sp. 212, otc. 15, bed 1	—	—	—	—	0.8	0.7	3.7
Sp. 213, otc. 15, bed 2	—	—	—	—	0.5	0.5	2.7
Sp. 214, otc. 15, bed 3	—	—	R.g.	—	0.6	0.5	2.1
Sp. 220, otc. 15, bed 3	—	—	—	—	0.9	1.5	5.3
Borehole 202							
Dep. 71– 76 m	—	—	—	—	3.8	6.5	2.7
Dep. 70– 80 m	—	—	—	—	6.7	7.3	4.0
Dep. 80– 84 m	—	R.g.	0.1	—	3.8	7.3	2.9
Dep. 84– 88 m	R.g.	—	R.g.	—	6.9	6.3	4.6
Dep. 84– 94 m	—	—	—	—	5.5	8.1	6.1
Dep. 105–110 m	—	—	—	—	10.6	10.3	2.2
Dep. 124–144 m	—	—	0.3	—	8.1	16.5	3.2
Dep. 154–174 m	—	—	—	—	10.6	26.4	2.3
Dep. 185–190 m	—	—	—	—	7.9	11.4	6.8
Dep. 279–284 m	—	—	R.g.	—	10.0	26.6	5.3
Dep. 304–327 m	—	—	—	—	9.6	18.7	2.8
Dep. 332–342 m	—	—	—	—	9.0	22.0	1.0

ORIGIN OF CLASTIC MINERAL ASSOCIATIONS

Chloritoid	Chlorite	Sphene	Monazite	Xenotime	Apatite	Phosphate	Barite	Celestite	Carbonate	Biotite	Muscovite	Rock fragments	Nigrine
—	—	—	—	—	—	—	—	—	—	—	0.3	—	—
—	—	—	—	—	—	—	—	—	—	—	1.4	0.7	0.3
—	—	—	—	—	—	—	—	—	—	—	0.2	—	—
—	—	R.g.	—	—	—	—	—	—	—	—	0.3	—	1.2
—	—	R.g.	R.g.	—	—	—	—	—	—	—	0.5	—	0.7
—	—	—	—	—	—	—	—	—	—	—	1.0	—	0.1
—	—	R.g.	—	R.g.	—	—	—	—	—	—	—	0.1	1.7
—	—	—	—	—	—	—	—	—	—	—	2.1	—	0.3
—	—	R.g.	0.2	—	—	—	—	—	—	—	—	—	0.5
—	—	—	—	—	—	—	—	—	—	—	R.g.	0.5	1.0
—	—	—	—	—	—	—	—	—	—	—	0.4	0.7	0.7
—	—	—	—	—	—	—	—	—	—	—	2.8	—	—
—	—	—	—	—	0.1	—	—	—	—	—	0.7	0.1	—
—	—	—	—	—	—	—	0.2	—	—	0.6	R.g.	1.5	R.g.
—	—	—	—	—	—	—	—	—	—	R.g.	0.2	0.2	—
—	—	0.2	—	—	—	—	5.2	—	—	—	0.5	—	—
—	—	—	—	—	—	—	2.2	—	—	—	R.g.	0.7	—
—	—	—	—	—	—	—	—	—	—	R.g.	0.3	0.2	R.g.
—	—	R.g.	—	—	—	—	0.3	—	—	—	R.g.	0.1	R.g.
—	—	—	—	—	—	—	—	—	—	0.2	0.1	—	—
—	—	—	—	—	—	—	—	—	—	—	0.1	0.1	0.2
—	0.3	—	—	—	—	—	—	—	—	—	3.3	0.2	—
—	—	—	—	—	—	—	—	—	—	—	—	0.2	—
—	0.3	—	—	—	—	—	—	—	—	0.6	1.7	1.8	—
—	0.1	0.2	—	—	—	—	—	—	—	—	0.7	2.3	—
—	—	—	—	—	—	—	—	—	—	—	—	—	—
—	—	R.g.	—	—	—	—	—	—	—	—	0.8	0.5	—
—	—	—	—	—	—	—	—	—	—	—	1.0	0.8	—
—	0.1	R.g.	—	—	—	—	—	—	—	—	—	1.7	—
—	—	R.g.	—	—	—	—	—	—	—	—	1.0	4.1	—
—	—	—	—	—	—	—	—	—	—	—	—	2.1	—
—	0.2	—	—	—	—	34.6	—	—	—	—	—	1.7	—
—	—	—	—	—	—	—	0.6	—	—	R.g.	—	1.4	—
—	—	R.g.	—	—	R.g.	—	—	—	—	R.g.	0.6	—	—
—	—	—	—	—	—	—	0.8	—	—	R.g.	0.2	0.4	0.3
—	—	—	—	—	—	—	0.3	—	—	R.g.	—	3.6	—
—	0.3	—	—	—	—	—	1.0	—	—	—	R. g.	6.0	0.1
—	—	—	—	—	—	—	—	—	—	—	—	3.6	—
—	0.5	—	—	—	—	—	—	—	—	—	0.6	3.1	—
—	—	—	—	—	—	—	—	—	—	R.g.	R.g.	2.7	—
—	—	—	—	—	—	—	—	—	—	—	R.g.	4.4	—
—	—	—	—	—	—	—	—	—	—	—	0.1	2.5	—
—	—	R.g.	—	—	—	—	—	—	—	—	R.g.	1.8	—
—	—	—	—	—	—	—	—	—	—	—	0.7	1.4	—

Table 2 (continued)

Section	Yield of heavy frac. in % of sample	Pyrite	Martite	Hematite	Limonite	Ilmenite	Rutile
Kokpakhty							
Sp. 15, b.h. K-2, dep. 100—115 m	0.065	—	—	3.9	—	2.9	1.2
Sp. 11, b.h. K-9, dep. 147—160 m	0.003	—	—	2.8	—	2.2	1.8
Sp. 4, b.h. K-10, dep. 35—40 m	0.12	—	—	5.2	—	2.8	3.5
Sp. 7, b.h. K-10, dep. 58—65 m	0.001	—	—	13.4	—	2.0	2.3
Sp. 14, b.h. K-10, dep. 148.8—155.8 m	0.002	—	—	10.6	—	3.2	2.2
Sp. 3, b.h. K-15, dep. 70—73 m	0.002	—	—	12.3	17.3	3.4	0.4
Sp. 6, b.h. K-15, dep. 158—163.2 m	0.008	—	—	10.3	—	3.8	1.2
Sp. 7, b.h. K-15, dep. 188.7—192.1 m	0.001	—	—	8.4	—	2.1	1.3
Sp. 8, b.h. K-15, dep. 200—203 m	0.002	—	—	4.8	—	2.8	2.5
Sp. 2, b.h. K-16, dep. 80—92 m	0.014	—	—	6.5	—	4.7	2.9
Sp. 3, b.h. K-17, dep. 140—144 m	0.003	—	—	8.0	—	1.8	1.7
Sp. 1, b.h. K-19, dep. 45—50 m	0.02	—	—	9.3	—	2.3	4.8
Sp. 2, b.h. K-19, dep. 50—57 m	0.01	—	—	11.0	—	0.7	2.4
Sp. 2, b.h. K-32, dep. 50—54 m	0.004	11.3	—	8.0	—	4.4	2.2
Akdzhar							
Sp. 221, otc. 16, bed 4	0.09	—	—	5.1	—	1.0	3.2
Sp. 222, otc. 16, bed 1	0.07	1.7	—	3.7	—	0.6	4.1
Sp. 223, otc. 16, bed 1	0.02	1.0	—	9.8	—	1.6	3.1
Sp. 224, otc. 16, bed 2	0.06	—	—	5.9	—	2.4	6.0
Sp. 227, otc. 16, bed 1	0.04	—	—	6.6	—	0.4	3.7
Sp. 228, otc. 16, bed 3	0.02	1.5	—	11.6	—	0.5	4.1
Sp. 229, otc. 16, bed 3	0.06	—	—	15.6	—	1.9	4.0
Sp. 230, otc. 16, bed 4	0.11	0.4	—	4.6	—	—	3.1
Sp. 231, otc. 16, bed 4	0.05	—	—	16.8	—	2.5	8.7
Sp. 232, otc. 16, bed 4	0.12	—	—	25.8	—	6.7	4.1
Sp. 233, otc. 16, bed 4	0.09	—	—	11.2	—	3.4	2.3
Sp. 234, otc. 16, bed 5	0.08	—	—	30.7	—	—	3.0
Sp. 235, otc. 16, bed 5	0.06	0.8	—	14.8	—	2.0	4.0
Sp. 1522, b.h. K-63, bed 66—67 m	0.10	—	—	2.2	—	2.4	3.2
Sp. 1523, b.h. K-63, bed 67—70 m	0.13	—	—	18.4	4.6	7.4	0.2
Ustyurt							
Sp. 112, dep. 663—675 m	0.05	2.0	—	7.7	0.4	36.4	0.5
Sp. 113 (2), dep. 675—683 m	0.05	—	—	7.0	—	10.8	0.1
Sp. 2, dep. 684—690 m	0.3	—	—	14.6	—	13.0	0.2
Sp. 4(2) dep. 704.7—708.2 m	0.1	3.7	—	8.1	—	13.0	0.2
Sp. 8 (1), dep. 718.7—722.2 m	0.2	27.1	—	1.0	—	1.0	R.g.
Sp. 6, dep. 731.9—735.4 m	0.5	1.2	1.5	12.7	—	27.8	0.1
Sp. 14 (1), dep. 738.9—742 m	0.2	7.1	—	3.4	—	1.1	0.8
Sp. 58, bed 749.4—752.9 m	0.01	0.6	—	5.6	R.g.	35.7	—
Sp. 19 (1), dep. 755—780 m	0.1	0.6	—	5.3	—	10.6	0.2
Sp. 30 (3), dep. 828—831.5 m	1.2	6.3	—	24.7	—	8.4	0.3
Sp. 44 (3), dep. 877—880.5 m	0.01	24.4	—	3.0	0.1	7.1	—
Sp. 50, dep. 936—940 m	0.02	0.5	—	12.9	1.0	6.2	0.2
Sp. 68 (2a), dep. 961—964.5 m	0.4	0.5	—	11.6	—	23.8	0.3
Sp. 46, dep. 963—971.5 m	0.2	5.6	—	3.2	—	2.8	0.2
Sp. 72 (2), dep. 975—978 m	0.3	2.7	—	6.7	—	10.4	R.g.
Sp. 76 (1), dep. 989—992.5 m	0.2	—	—	6.6	—	15.0	0.6
Sp. 82 (1), dep. 1017—1020.5 m	0.2	—	—	4.5	—	6.0	0.6
Sp. 85, dep. 999.5—100.5 m	0.1	2.2	—	6.0	—	8.4	0.7
Sp. 86 (2), dep. 1031—1034.5 m	0.1	0.3	—	6.1	—	12.6	0.8
Sp. 92 (2), dep. 1061—1063 m	0.02	4.9	—	9.6	0.5	17.3	—
Sp. 99 (1), dep. 1094.5—1098 m	0.06	4.8	—	12.0	R.g.	21.2	0.2
Sp. 107 (1), dep. 1128.7—1132.2 m	0.1	1.1	—	11.3	—	22.3	0.2
Sp. 112 (1), dep. 1146.2—1149.7 m	0.03	1.7	—	11.3	—	27.5	—
Sp. 115 (2), dep. 1161.5—1165 m	0.1	2.3	—	6.7	—	22.0	1.2

Anatase	Brookite	Corundum	Leucoxene	Spinel	Chrom-picotite	Chromite	Magnetite	Andalusite	Sillimanite	Kyanite	Staurolite	Zircon
0.3	—	—	7.5	—	0.8	3.3	30.6	—	—	R.g.	1.8	10.2
0.3	—	—	8.0	—	—	2.8	9.8	—	—	0.2	1.0	8.7
0.6	—	—	19.3	—	—	3.7	31.0	—	—	—	—	12.2
—	—	—	17.3	—	—	8.4	35.2	—	—	0.8	0.3	5.8
1.2	—	—	23.5	—	—	3.2	7.4	—	—	0.8	1.5	18.2
2.1	—	—	15.6	—	—	2.1	16.1	—	—	—	0.6	5.9
1.2	—	—	10.2	—	—	9.3	20.2	—	—	0.4	—	28.7
0.6	—	—	13.0	—	—	6.2	36.2	—	—	—	0.6	9.9
0.4	—	—	19.4	—	—	4.1	24.5	—	—	0.9	0.7	18.0
0.8	—	—	26.1	—	—	3.2	22.0	—	—	2.6	2.5	3.4
2.0	—	—	17.3	—	—	9.7	18.5	—	—	2.2	1.0	10.1
3.3	—	—	33.0	—	—	5.6	10.4	—	—	1.0	1.4	18.2
2.0	—	R.g.	48.6	—	—	4.3	2.5	—	—	1.5	0.7	12.9
4.9	—	—	7.7	—	0.5	3.5	14.7	—	—	1.1	0.3	35.9
0.3	—	R.g.	61.4	—	—	—	3.5	—	—	3.0	1.0	9.7
0.4	—	—	36.5	—	—	1.6	10.2	—	—	6.9	7.7	8.5
1.0	—	—	46.4	—	—	2.9	15.4	—	—	4.2	2.0	1.2
0.7	—	—	35.0	—	—	3.2	20.4	—	—	6.9	3.5	4.5
0.5	—	R.g.	55.1	—	—	0.8	4.5	—	—	4.9	3.5	5.0
0.4	—	—	35.4	—	—	3.8	18.6	—	—	2.8	3.4	2.0
0.7	—	—	28.0	—	—	1.6	40.6	—	—	1.5	0.5	1.6
0.3	—	R.g.	19.5	—	—	3.4	42.4	—	—	2.5	3.5	7.7
0.5	—	—	17.5	—	—	7.0	18.0	0.1	—	2.2	7.8	4.1
0.1	—	—	14.8	—	—	3.1	15.9	—	—	4.6	6.7	7.5
0.4	—	—	21.4	—	—	2.7	20.4	0.2	—	5.1	8.5	13.8
0.2	—	—	19.8	—	—	5.4	19.4	R.g.	—	4.1	9.8	3.1
1.0	—	—	19.9	—	—	3.4	18.7	—	—	8.9	4.7	11.8
R.g.	—	—	21.1	—	—	7.0	13.9	—	—	3.9	1.7	1.9
0.2	—	—	21.2	—	—	2.8	9.2	—	—	1.3	0.9	2.6
—	—	—	13.3	—	—	0.9	—	—	—	0.7	7.7	8.0
—	—	—	53.8	—	—	—	1.7	—	—	0.4	1.6	4.5
—	—	—	36.4	—	—	0.6	—	—	—	0.4	3.7	7.4
—	—	—	30.5	—	—	0.1	3.8	R.g.	—	0.5	4.5	7.2
—	—	—	13.9	—	—	—	5.7	0.1	—	0.4	1.9	9.3
—	—	—	2.0	—	—	—	1.2	—	—	0.2	1.7	5.6
—	—	—	30.8	—	—	0.5	4.9	—	—	0.8	1.6	7.6
—	—	—	22.8	—	—	—	1.5	—	—	0.2	6.6	4.8
0.1	—	—	26.8	—	—	0.7	4.1	0.1	—	0.3	3.2	4.6
—	—	—	21.0	—	—	0.7	3.4	0.1	—	0.2	2.4	7.6
—	—	—	27.0	—	0.1	0.3	2.3	—	—	—	1.2	8.9
—	—	—	15.9	—	—	—	6.0	—	—	0.2	—	6.2
—	—	—	25.6	—	—	—	2.2	—	—	0.8	4.0	12.6
0.4	—	—	38.6	—	—	0.8	13.5	—	—	0.8	2.3	8.1
—	—	—	43.7	—	—	0.5	7.2	0.1	—	0.2	2.6	3.7
0.1	—	—	23.0	—	—	0.5	1.6	—	—	0.5	2.0	5.5
R.g.	—	—	36.3	—	—	—	10.0	—	—	0.6	1.7	19.3
—	—	—	40.7	—	—	—	5.5	—	—	1.5	1.0	11.0
—	—	0.1	23.1	—	—	0.7	2.5	0.2	—	0.6	6.5	12.2
—	—	—	22.0	—	—	—	1.5	—	—	0.6	2.0	8.6
0.1	—	—	22.2	—	—	—	1.5	—	—	—	2.7	9.5
—	—	—	23.7	—	—	0.3	6.2	—	—	R.g.	2.8	6.8
—	—	—	22.5	—	—	0.4	3.4	—	—	0.7	2.3	10.6
0.1	—	—	22.0	—	—	—	6.2	—	—	0.5	—	13.0

Table 2 (continued)

Section	Actinolite	Amphibole	Rhombic pyroxene	Monoclinic pyroxene	Garnet	Epidote	Tourmaline
Kokpakhty							
Sp. 15, b.h. K-2, dep. 100—115 m	—	—	—	—	10.0	25.7	0.8
Sp. 11 b.h. K-9, dep. 147—160 m	—	—	—	—	8.2	13.8	0.4
Sp. 4, b.h. K-10, dep. 35—40 m	—	—	—	—	3.8	14.8	2.3
Sp. 7, b.h. K-10, dep. 58—65 m	—	—	—	—	10.3	1.3	—
Sp. 14 b.h. K-10, dep. 148.8—155.8 m	—	—	—	—	19.5	5.5	1.3
Sp. 3, b.h. K-15, dep. 70—73 m	—	—	—	—	17.8	4.8	—
Sp. 6, b.h. K-15, dep. 158—163.2 m	—	0.1	—	—	10.0	0.6	1.5
Sp. 7, b.h. K-15, dep. 188.7—192.1 m	—	—	—	—	17.3	4.2	—
Sp. 8, b.h. K-15, dep. 200—203 m	—	—	—	—	18.8	1.0	0.6
Sp. 2, b.h. K-16, dep. 80—92 m	—	—	—	—	5.8	7.2	5.1
Sp. 3, b.h. K-17, dep. 140—144 m	—	0.1	—	—	10.9	13.8	1.0
Sp. 1, b.h. K-19, dep. 45—50 m	—	—	—	—	2.3	2.3	5.0
Sp. 2, b.h. K-19, dep. 50—57 m	—	—	—	—	2.5	5.4	3.9
Sp. 2, b.h. K-32, dep. 50—54 m	—	—	—	—	3.5	0.5	0.4
Akdzhar							
Sp. 221, otc. 16, bed 4	—	—	R.g.	—	2.5	R.g.	5.6
Sp. 222, otc. 16, bed 1	—	—	R.g.	—	1.8	0.2	7.6
Sp. 223, otc. 16, bed 1	—	0.2	—	—	2.9	2.0	4.5
Sp. 224, otc. 16, bed 2	—	—	—	—	3.1	0.4	3.1
Sp. 227, otc. 16, bed 1	—	—	0.4	—	3.9	—	7.4
Sp. 228, otc. 16, bed 3	—	—	—	—	4.3	0.4	5.6
Sp. 229, otc. 16, bed 3	—	0.2	—	—	1.1	0.9	1.3
Sp. 230, otc. 16, bed 4	R.g.	—	R.g.	—	3.5	1.4	6.9
Sp. 231, otc. 16, bed 4	—	0.8	—	—	3.3	0.4	10.3
Sp. 232, otc. 16, bed 4	—	—	—	—	3.1	—	6.7
Sp. 233, otc. 16, bed 4	—	R.g.	—	—	1.0	1.4	7.5
Sp. 234, otc. 16, bed 5	—	R.g.	—	—	1.2	R.g.	3.0
Sp. 235, otc. 16, bed 5	—	R.g.	—	—	1.3	0.8	3.5
Sp. 1522, otc. K-63, bed 66—67 m	—	0.2	—	—	4.1	32.0	2.9
Sp. 1523, otc. K-63, bed 67—70 m	—	—	0.1	—	10.1	18.1	1.8
Ustyurt							
Sp. 112, dep. 663—675 m	—	—	—	—	8.2	2.8	7.2
Sp. 113 (2), dep. 675—683 m	—	0.7	0.2	—	9.0	1.1	3.4
Sp. 2, dep. 684—690 m	—	—	—	—	9.0	6.5	3.7
Sp. 4 (2), dep. 704.7—708.2 m	—	—	0.2	—	5.0	10.3	4.4
Sp. 8 (1), dep. 718.7—722.2 m	—	—	—	—	16.0	10.2	2.8
Sp. 6, dep. 731.9—735.4 m	—	—	0.2	—	15.6	7.7	2.9
Sp. 14 (1), dep. 738.9—742 m	—	—	0.3	1.1	10.9	8.2	2.1
Sp. 58, dep. 749.4—752.9 m	—	—	—	—	3.8	2.7	6.3
Sp. 19 (1), dep. 755—780 m	—	0.3	—	—	12.5	8.1	6.0
Sp. 30 (3), dep. 828—831.5 m	—	0.4	0.2	1.1	7.7	—	1.6
Sp. 44 (3), dep. 877—880.5 m	—	—	—	—	6.8	—	3.1
Sp. 50, dep. 936—940 m	—	—	—	—	3.6	—	1.5
Sp. 68 (2a), dep. 961—964.5 m	—	—	0.2	—	9.4	—	1.8
Sp. 46, dep. 963—971.5 m	—	—	—	0.2	9.0	0.7	2.2
Sp. 72 (2), dep. 975—978 m	—	0.3	0.1	—	7.5	1.3	3.3
Sp. 76 (1), dep. 989—992.5 m	—	0.7	0.2	—	7.5	—	2.0
Sp. 82 (1), dep. 1017—1020.5 m	—	—	0.1	0.4	4.9	—	2.8
Sp. 85, dep. 999.5—100.5 m	—	0.3	0.7	—	8.5	—	1.3
Sp. 86 (2), dep. 1031—1034.5 m	—	—	0.4	0.3	6.8	—	1.7
Sp. 92 (2), dep. 1061—1063 m	—	—	0.3	—	11.6	—	5.6
Sp. 99 (1), dep. 1094.5—1098 m	—	0.4	0.1	—	9.1	0.4	1.8
Sp. 107 (1), dep. 1128.7—1132.2 m	—	—	—	—	4.9	—	3.2
Sp. 112 (1), dep. 1146.2—1149.7 m	—	—	0.2	0.2	6.1	—	1.8
Sp. 115 (2), dep. 1161.5—1165 m	—	0.5	0.3	—	13.0	—	2.4

Chloritoid	Chlorite	Sphene	Monazite	Xenotime	Apatite	Phosphate	Barite	Celestite	Carbonate	Biotite	Muscovite	Rock fragments	Nigrine
—	—	R.g.	—	—	—	—	—	—	—	0.3	0.1	0.6	—
—	—	—	—	—	—	—	36.6	—	—	0.5	2.9	—	—
—	—	R.g.	—	—	—	—	—	—	—	—	0.8	—	—
—	—	—	—	—	—	—	—	—	—	—	—	2.9	—
—	—	—	—	—	0.3	—	—	—	—	0.5	1.1	—	—
—	—	—	—	—	—	—	—	—	—	—	—	1.6	—
—	0.2	—	—	—	—	—	—	—	—	0.1	0.1	2.1	0.2
—	0.1	0.2	—	—	—	—	—	—	—	—	1.2	—	—
—	—	—	—	—	—	—	—	—	—	—	1.0	6.2	—
—	—	—	—	—	0.2	—	—	—	—	—	1.3	0.4	—
—	—	—	—	—	—	—	—	—	—	0.3	0.8	—	—
—	—	—	—	—	—	—	—	—	—	—	0.9	0.7	—
—	—	—	—	—	—	—	—	—	—	—	—	1.1	—
—	0.5	—	—	—	—	—	—	—	—	—	R.g.	3.2	—
—	0.2	—	—	—	—	—	—	—	—	—	0.1	8.2	—
—	R.g.	—	—	—	—	—	—	—	—	—	0.4	1.4	—
—	—	—	—	—	—	—	—	—	—	—	—	4.9	—
—	1.0	—	—	—	—	—	—	—	—	—	—	2.3	—
—	0.5	—	—	—	—	—	—	—	—	—	—	4.8	0.3
—	R.g.	—	—	—	—	—	—	—	—	—	R.g.	0.5	—
—	—	—	—	—	R.g.	—	—	—	—	—	—	0.8	—
—	—	—	—	—	0.4	—	—	—	—	—	—	0.5	—
—	—	—	—	—	—	—	—	—	—	—	—	0.7	—
—	—	—	—	—	R.g.	—	—	—	—	—	—	0.3	—
—	—	—	—	—	—	—	—	—	—	—	—	4.4	—
—	—	—	—	—	—	—	—	—	—	—	—	3.5	—
—	—	—	—	—	—	—	—	—	—	—	0.1	1.0	—
—	0.1	—	—	—	0.4	—	—	—	1.4	—	2.2	—	—
—	—	—	—	R.g.	—	P. a.	—	2.4	3.2	—	1.1	—	—
—	0.1	—	—	0.2	—	0.6	—	0.3	1.3	—	—	—	—
—	—	—	—	0.1	—	2.4	—	0.4	3.9	—	1.7	—	—
—	—	—	—	0.2	—	7.2	—	—	2.1	—	0.6	—	—
—	—	—	—	0.4	—	0.2	—	—	9.1	—	9.8	—	—
—	0.1	—	—	0.3	0.2	2.7	—	4.6	5.1	—	5.8	—	—
—	—	—	—	—	—	8.8	—	—	—	—	0.6	—	—
—	—	—	—	0.2	—	0.2	—	1.3	9.7	—	5.1	—	—
—	0.2	—	—	0.3	—	5.4	0.1	—	0.2	—	7.3	—	—
—	0.3	—	—	—	—	1.9	—	0.3	9.9	—	3.3	—	—
—	—	—	—	—	0.5	24.2	—	—	18.5	—	2.6	—	—
—	0.3	—	—	0.2	0.2	2.2	—	1.0	3.3	—	—	—	—
—	0.2	—	—	0.2	—	1.7	—	0.8	3.5	—	5.2	—	—
—	—	—	—	0.1	—	0.4	—	0.2	3.0	—	5.7	—	—
1.5	0.1	—	—	0.6	1.5	1.3	—	1.0	24.3	—	3.9	—	—
—	—	—	—	0.2	0.1	4.2	0.2	0.3	2.3	—	4.8	—	—
—	—	—	—	0.9	0.2	5.5	0.2	3.7	0.9	—	0.8	—	—
4.1	0.2	—	—	0.8	0.5	1.6	—	—	14.6	—	3.3	—	—
1.0	0.1	—	—	0.2	R.g.	R.g.	—	—	9.0	—	5.0	—	—
—	0.2	—	—	—	—	R.g.	—	1.1	6.7	—	3.3	—	—
1.4	—	—	—	0.1	0.1	2.7	—	2.5	4.0	—	7.7	—	—
2.3	0.2	—	—	0.4	—	1.3	—	—	6.6	—	0.3	—	—
—	—	—	—	0.5	—	3.5	0.5	1.8	2.4	—	0.8	—	—

Epidote forms pistachio, angular, and slightly rounded aggregates of grains, rarely clean transparent fragments with refractive indices of Ng = 1.751 and Np = 1.726.

Tourmaline ranges from well-terminated crystals to slightly rounded fragments, brown and green in color. The prismatic faces are well preserved in the crystals. Point-like black inclusions and bubbles may be occasionally observed. Very rarely grains may be found with zonal coloration, the center of the grain being green, the margin brown. Pleochroism is pronounced, brown along Ng, and light yellow along Np. Refractive indices are Ng = 1.636 and Np = 1.612.

Mica is represented by fresh, transparent, colorless plates of muscovite. Brownish coloration is rarely observed, occurring in zones of varying intensity. Some grains are entirely free of inclusions, but some contain large numbers of ore inclusions, distributed in bands and aureoles. The refractive index is Ng = 1.587-1.602. In addition, opaque, dirty green plates of chloritized biotite are present.

Staurolite occurs in orange-brown, variably formed clastic grains, angular and subrounded forms more or less equally abundant. Black, point-like inclusions are numerous. The indices of refraction are Ng = 1.741 and Np = 1.736. The mineral is pleochroic from brown along Ng to slightly yellow along Np.

From Table 2 it may be seen that the mineral association of kyanite, rutile, and chromite is present chiefly in the region of the Baiguskarovskii section. To the south (starting in the Kyzyl-Sai region), the amount of chromite and kyanite declines sharply and the rutile content diminishes gradually. Ilmenite reaches its highest concentration in the vicinity of the Ustyurt test hole. Garnet, and also zircon, increase in content somewhat to the south.

Deposits in the Mugodzhary region are rich in epidote. The muscovite content in these rocks gradually declines toward the south, and in the Ustyurt region only chloritized biotite is found.

Analysis of Observations

Discriminant analysis has been used in order to compare the investigated mineral associations, specifically the linear discriminant function [Anderson, 1963] and the "generalized distance" of Mahalanobis [Roy and Mitra, 1956]. The basic idea of linear discriminant analysis involves the search for some new test character

$$z = \lambda_1 x_1 + \lambda_2 x_2 + \ldots + \lambda_m x_m,$$

representing a linear combination of the original characters $x_1, x_2, x_3, \ldots, x_m$ which, so far as possible, would include all the information of the original characters.

The linear discriminant function in geology was first applied by Vistelius [1950]. Methods of discriminatory analysis are now extensively used in the literature of cybernetics, in the so-called theory of identification systems.

Before coefficients of the linear discriminant function were calculated, the original material was analyzed. Point diagrams (Figs. 2 and 3) and histograms (Fig. 4) show the distribution of characters for the existing samples of different minerals in the heavy fraction. The points (or solid lines) indicate the value of the character for one sample; the crosses (or dashed lines) give the values for a second sample. For illustration, we have examined two pairs of regions: Baiguskarovskii and Kyzyl-Sai, which are similar in setting and mineral association, and Baiguskarovskii and borehole 202, which differ markedly from each other, more so than the remaining pairs, which occupy intermediate positions. As we may see from the figures, the

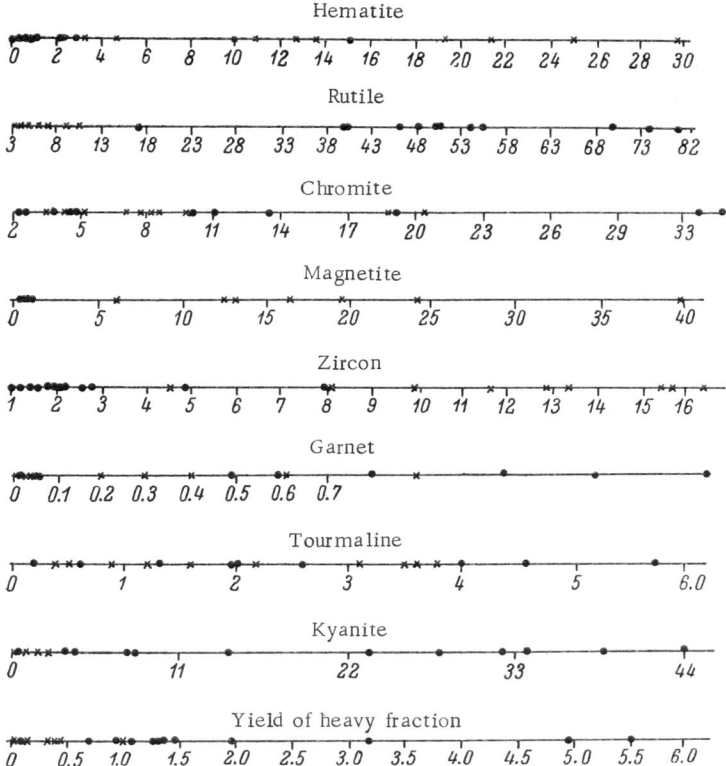

Fig. 2. Point diagrams of distribution of minerals in the heavy fraction for the Baiguskarovskii—Kyzyl-Sai regions. Figures along the axes represent percentage content of minerals in the heavy fraction of the concentrate.

regions may be distinguished by rutile content, more rarely by magnetite and garnet. Most pairs of regions cannot be reliably separated by initial characters.

From the original data, we have computed the coefficients of the discriminant function z on a BESM-2 computer by a program prepared by M. D. Belonin.

From the results of the computation it follows (Fig. 5) that the Ekaterininskii, Akdzhar, and Kokpakhty regions are indistinguishable, whereas all remaining paired combinations of investigated regions may be distinguished from each other. We know that the Baiguskarovskii region is near the Southern Urals and that the Ustyurt 1 test hole occupies the southernmost position. Between these extreme points lies an intermediate zone close to the Mugodzhary region. Districts within this intermediate zone are indistinguishable from each other.

For a quantitative expression of the difference between regions, we used the generalized distance of Mahalanobis, and the results of this computation are summarized in Table 3. As seen from the table, the Mahalonobis distance in similar regions gives figures from 4.0 to 16.0, and in regions differing strongly gives values from 89.0 to 106.0.

As everyone knows, it is a complex problem to obtain the coefficients of the discriminant function without a programmed procedure and a computer, since a tremendous expenditure of time is required. Manual computations by geologists, who are commonly unacquainted with approximation calculations, are sometimes accompanied by arithmetic errors.

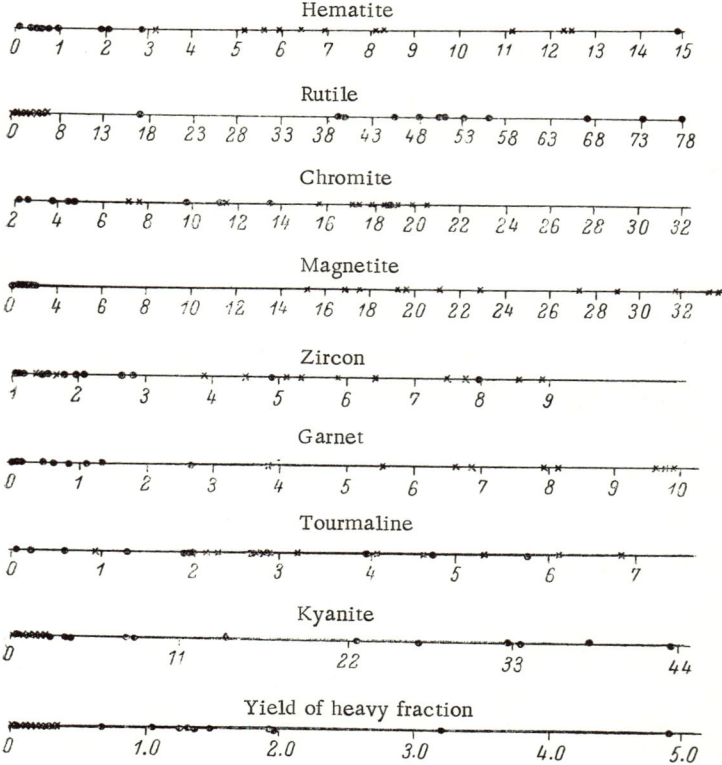

Fig. 3. Point diagrams of distribution of minerals in the heavy fraction for the Baiguskarovskii—borehole 202 regions. Meaning of figures as in Fig. 2.

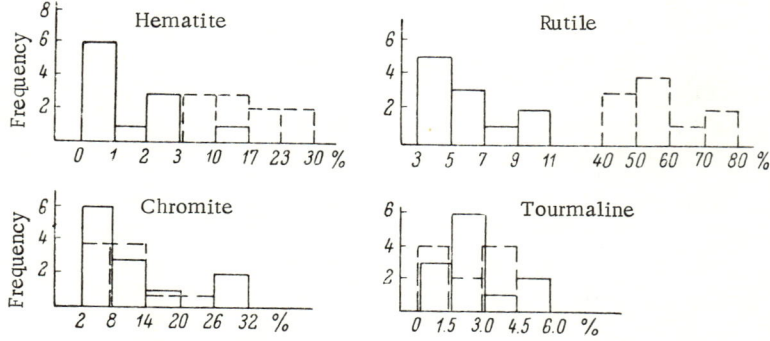

Fig. 4. Histograms for mineral distributions in the Baiguskarovskii—Kyzyl-Sai regions.

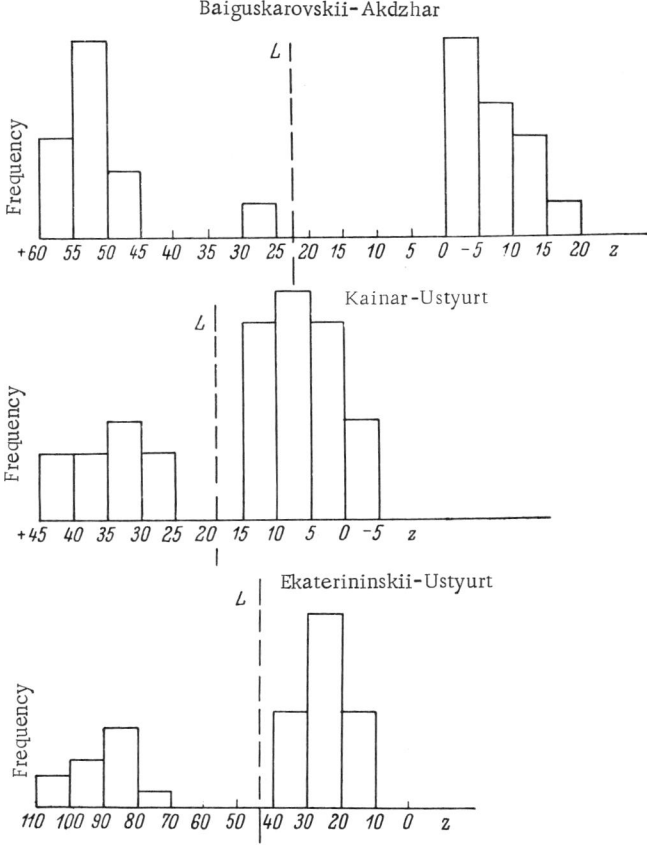

Fig. 5. Distribution histograms of the coefficient z of the linear discriminant function.

Table 3. Generalized Distance of Mahalonobis for the Investigated Regions

Region	Baiguska-rovskii	Kysyl-Sai	Ekaterinin-skii	Borehole 202	Akdzhar	Kokpakhty	Ustyurt
Baiguskarovskkii		32.1	56.9	88.2	57.4	73.8	106.1
Kyzyl-Sai			25.5	39.8	31.2	18.3	78.8
Ekaterininskii . .				60.0	4.0	16.4	61.6
Borehole 202 . .					32.1	34.8	102.8
Akdzhar						15.8	28.3
Kokpakhty							27.3

If an electronic computer is unavailable, the discriminator z_0 [Kalinin and Doktorov, 1966] may be obtained from the linear discriminant function on the assumption that all correlation coefficients are zero. This condition is not met, of course, but experience has shown that the discrimination procedure is not greatly worsened. The formula for the discriminator is the following:

$$z_0 = \sum_{i=1}^{m} \frac{\bar{x}_{1i} - \bar{x}_{2i}}{s_i^2} x_i$$

Table 4. Computational Scheme for Standard Error of Difference

$\bar{x}_{1i}$	$\bar{x}_{2i}$	$\bar{x}_{1i} - \bar{x}_{2i}$	$(\bar{x}_{1i} - \bar{x}_{2i})^2$	x_{1i}	$(x_{1i} - \bar{x}_{1i})^2$	x_{2i}	$(x_{2i} - \bar{x}_{2i})^2$	s_i^2
12.05	15.76	3.71	13.75	−7.45	55.50	0.04	0	—
				−8.25	68.06	−0.56	0.31	—
				−0.95	0.90	4.84	23.43	—
				21.75	473.06	1.74	3.03	—
				22.75	517.56	−7.96	63.36	76.76
				−9.45	89.30	−4.26	18.15	—
				−7.25	52.56	3.14	9.86	—
				−7.25	52.56	4.44	19.71	—
				−2.15	4.62	2.84	8.07	—
				6.85	46.92	−8.36	69.89	—
				1.45	2.10	1.64	2.69	—
				−10.05	101.10	2.44	5.95	—
					1464.24		244.45	

with the threshold

$$\frac{1}{2} \sum_{i=1}^{m} \frac{(\bar{x}_{1i})^2 - (\bar{x}_{2i})^2}{s_i^2},$$

where z_0 is the value of the discriminator; m is the number of characters; $\bar{x}_{1i}$ is the average value of one character for the first sample; $\bar{x}_{2i}$ is the average value of the same character for the second sample; s^2 is the standard error of difference for one character for the two samples; and x_i are the observed values of the character.

The threshold is a point on the straight line having plotted values of the discriminator, taken as the boundary between the investigated samples.

The standard error of difference is computed by the formula:

$$s^2 = \frac{1}{n_1 + n_2 - 2} \sum_{j=1}^{n_1} (x_{1ij} - \bar{x}_{1i})^2 + \sum_{j=1}^{n_2} (x_{2ij} - \bar{x}_{2i})^2,$$

where x_{1ij}, x_{2ij} are the values of the character for the first and second samples; $\bar{x}_{1i}$, $\bar{x}_{2i}$ are the average values of the character for the first and second samples; n_1 is the size of the first sample, n_2 the size of the second sample. In Table 4 we have given the calculation scheme for the standard error of difference, and in Table 5 for the discriminator. The standard error of difference was computed on the basis of the data in Table 2 for chromite content in the regions of Baiguskarovskii and borehole 202.

When we combine $(x_{1i} - \bar{x}_{1i})^2$ and $(x_{2i} - \bar{x}_{2i})^2$, we obtain: 1464.17 + 224.45 = 1688.69. Let us then divide this sum by $n_1 + n_2 - 2$ (in this case, 12 + 12 − 2 = 22), and we obtain the standard error of difference: 1688.69/22 = 76.76.

Having calculated the standard error of difference, let us go on to computation of the discriminator. In the example below, computations of the discriminator are made for seven characters in the Baiguskarovskii—borehole 202 regions.

Table 5. Computational Scheme for the Discriminator z_0

Baiguskarovskii section

Character	Value of $x_i \cdot \left(\frac{\bar{x}_{1i} - \bar{x}_{2i}}{s^2}\right)$											
	Sp. 56, h. 27, dep.14.25 m	Sp. 13, h. 28, dep.7.40 m	Sp. 15, h. 28, dep. 8.0 m	Sp. 16, h. 28, dep.8.50 m	Sp. 17, h. 28, dep.8.80 m	Sp. 8, h. 30, dep.11.2 m	Sp. 9, h. 31	Sp. 11, h. 31, dep.13.5 m	Sp. 12, h. 31, dep.14.0 m	Sp. 13, h. 31, dep.14.5 m	Sp. 9, h. 53	Sp. 55, h. 27
Hematite	0	−0.26	−0.86	−0.9	−0.43	−1.25	−0.9	−0.34	−0.17	−0.21	−23.65	−0.26
Rutile	+16.13	+16.90	+19.49	+13.75	+17.57	+17.50	+28.59	+18.76	+26.21	+23.76	+5.91	+14.24
Chromite	−0.23	−0.19	−0.55	−1.69	−1.74	−0.13	−0.24	−0.24	−0.49	−0.94	−0.67	−0.10
Magnetite	−0.32	0	0	0	0	−0.16	−0.32	0	0	0	−0.79	0
Zircon	−1.26	−1.32	−0.76	−1.13	−0.69	−3.09	−1.01	−1.26	−1.64	−1.76	−5.04	−0.94
Garnet	−2.18	−0.99	−0.99	0	−0.99	0	−2.18	−1.58	−2.57	−5.35	0	−1.19
Tourmaline	−1.89	−2.34	−0.82	−1.07	−0.53	−0.82	−0.08	−0.82	−0.25	−0.82	−0.82	−1.64
	+10.25	+11.80	+15.51	+8.96	+13.19	+12.05	+23.86	+14.52	+21.09	+14.68	−25.06	+10.11

Borehole 202

Character	Value of $x_i \cdot \left(\frac{\bar{x}_{1i} - \bar{x}_{2i}}{s^2}\right)$											
	71–76 m	70–80 m	80–84 m	84–88 m	88–94 m	105–110 m	124–144 m	154–174 m	185–190 m	279–284 m	304–327 m	332–342 m
Hematite	−5.29	−3.48	−4.82	−2.58	−3.01	−1.38	−2.75	−2.24	−2.75	−5.33	−2.41	−3.53
Rutile	+0.49	+0.35	+0.32	+0.21	+0.21	+0.42	+0.18	+0.03	+0.42	+0.07	+0.21	+0.24
Chromite	+0.79	−0.76	−1.03	−0.87	−0.39	−0.57	−0.94	−1.01	−0.93	−0.37	−0.87	+0.91
Magnetite	−25.36	−22.99	−18.41	−27.02	−21.72	−31.84	−13.90	−16.75	−13.43	−12.17	−15.48	−15.33
Zircon	−4.79	−5.35	−4.85	−3.65	−3.21	−3.40	−3.02	−0.94	−2.90	−1.01	−5.54	−4.03
Garnet	−7.52	−13.27	−7.52	−13.66	−10.89	−20.99	−16.04	−20.99	−15.64	−19.8	−19.01	−17.82
Tourmaline	−1.11	−1.64	−1.19	−1.89	−2.50	−0.90	−1.31	−0.94	−2.79	−2.17	−1.15	−0.41
	−44.37	−47.14	−37.50	−49.46	−41.51	−58.66	−37.78	−42.84	−38.02	−40.78	−44.25	−41.79

Table 6. The Separation Coefficient Δ for Each Character

Section	Kyzyl-Sai									Ekaterininskii									Borehole 202								
	Hematite	Rutile	Chromite	Magnetite	Zircon	Garnet	Tourmaline	Yield of heavy frac.	Kyanite	Hematite	Rutile	Chromite	Magnetite	Zircon	Garnet	Tourmaline	Yield of heavy frac.	Kyanite	Hematite	Rutile	Chromite	Magnetite	Zircon	Garnet	Tourmaline	Yield of heavy frac.	Kyanite
Baiguskarovskii	3.71	12.8	0.1	2.4	5.5	0.6	0.1	2.1	2.7	2.6	17.1	1.6	8.2	3.1	1.2	1.7	1.8	1.8	2.3	17.8	0.2	19.5	1.9	15.0	0.6	2.6	2.8
Kyzyl-Sai										0.7	5.2	3.7	0.8	0.5	1.7	2.7	0.2	0.7	1.4	9.5	1.6	1.3	2.1	16.8	1.0	0.1	0.5
Ekaterininskii																			0.8	1.8	18.6	0.0	0.6	3.3	0.2	0.7	0.5

Section	Akdzhar									Kokpakhty									Ustyurt								
	Hematite	Rutile	Chromite	Magnetite	Zircon	Garnet	Tourmaline	Yield of heavy frac.	Kyanite	Hematite	Rutile	Chromite	Magnetite	Zircon	Garnet	Tourmaline	Yield of heavy frac.	Kyanite	Hematite	Rutile	Chromite	Magnetite	Zircon	Garnet	Tourmaline	Yield of heavy frac.	Kyanite
Baiguskarovskii	2.1	17.7	1.3	4.8	0.9	1.9	1.6	3.3	2.2	2.6	18.3	0.8	6.4	2.9	4.1	0.2	3.3	3.2	1.7	28.0	3.2	2.0	3.6	7.9	0.2	3.7	4.8
Kyzyl-Sai	0.1	1.1	2.3	0.2	1.5	2.6	2.0	0.6	4.0	1.3	4.5	1.1	0.4	0.2	4.2	0.1	1.1	0.0	1.2	17.7	5.9	1.6	0.4	8.5	0.4	0.0	0.2
Ekaterininskii	0.2	1.7	1.0	0.3	0.4	0.0	0.1	1.6	0	0.1	0.1	2.9	0.7	0.7	2.1	2.9	2.0	0.8	0.0	5.4	3.3	7.8	0.0	3.1	0.7	0.6	1.4
Borehole 202	0.5	4.8	14.0	0.5	0	3.9	0.5	4.3	2.7	0	2.8	9.1	0.3	1.5	0.2	1.4	10.1	0.3	0.0	1.6	35.9	15.9	0.8	0.1	0.1	0.0	1.8
Akdzhar										0.4	1.1	0.6	0.0	1.5	2.3	2.6	2.7	4.2	0.3	9.1	5.6	3.8	0.6	3.4	1.0	0.3	7.7
Kokpakhty																			0	7.0	9.3	5.4	0.9	0.1	0.8	0.7	0.4

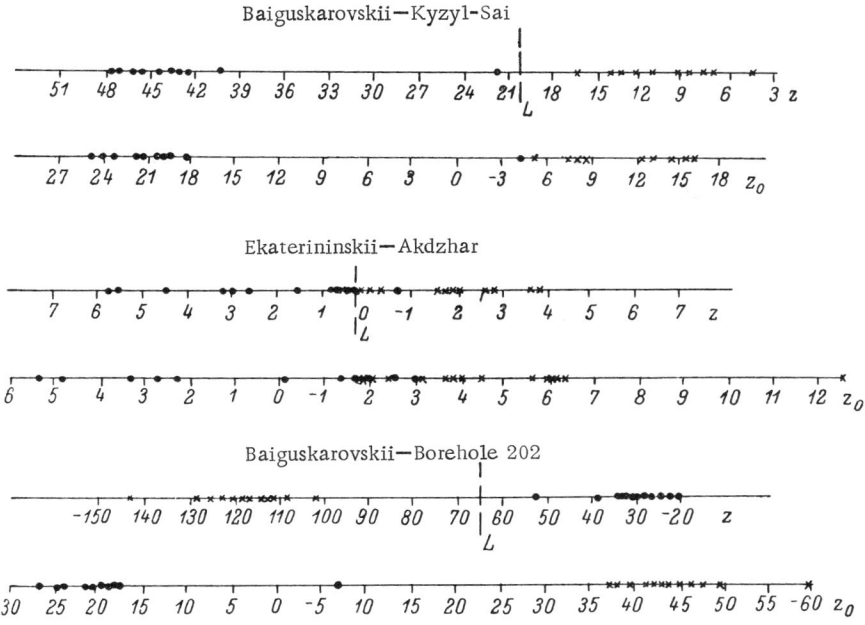

Fig. 6. Point distribution diagrams of the discriminators z and z_0.

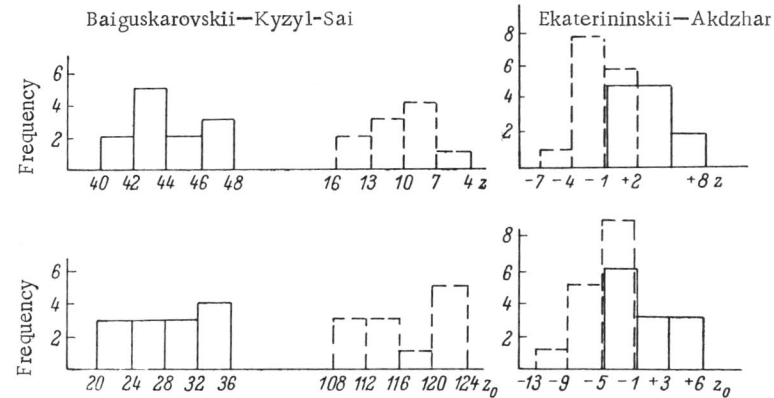

Fig. 7. Distribution histograms of the discriminators z and z_0.

1. We have computed the value of $(\bar{x}_{1i} - \bar{x}_{2i})/s^2$ for each character of the given pair of regions. Thus, the average value of hematite from the Baiguskarovskii region is 2.3, and for the region of borehole 202 it is 7.67. We obtain the difference: $2.3 - 7.67 = -5.37$. This number is divided by the standard error of difference for the character. Since the standard error is 12.6, for hematite the expression $(\bar{x}_{1i} - \bar{x}_{2i})/s^2$ will be equal to $-5.37/12.6 = -0.43$. For the other characters it will be: 0.35 for rutile, 0.05 for chromite, -0.79 for magnetite, -0.63 for zircon, -1.98 for garnet, and -0.41 for tourmaline.

2. The coefficients thus obtained have been multiplied consecutively by the original values of each character.

Table 7. The Separation Coefficient Δ Computed for
Data of the Discriminator z_0

Region	Kyzyl-Sai	Ekaterininskii	Borehole 202	Akdzhar	Kokpakhty	Ustyurt
Baiguskarovskii....	23.6	18.7	12.6	25.5	28.2	58.13
Kyzyl-Sai		30.4	31.0	20.62	12.6	41.5
Ekaterininskii.....			9.6	2.2	13.0	43.0
Borehole 202				16 3	24.0	64.0
Akdzhar					5.3	30.4
Ustyurt						77.0

Thus, the coefficient obtained for hematite has been multiplied by the percentage content of hematite in the Baiguskarovskii region and then in the region of borehole 202. The other characters have been multiplied by their corresponding coefficients. The computations have been made by the convenient scheme shown in Table 5.

From the data we have obtained in this way, it may be seen that the investigated samples must be considered different.

The discriminator z_0 may be satisfactorily used in practical work. In comparison with the linear discriminant function, it gave poorer results in several studies by a factor of but 2-3%.

Results of using the discriminator z_0 and the linear discriminant function have proved rather comparable. All pairs of regions, except Ekaterininskii, Akdzhar, and Kokpakhty, are distinguishable by means of either method. A comparison of the operation of the discriminators z and z_0 is shown by point diagrams and histograms (Figs. 6 and 7). The symbols used are the same as for Figs. 2, 3, and 4. It may be seen from the figures that the Baiguskarovskii—Kyzyl-Sai and Baiguskarovskii—borehole 202 regions may be distinguished by means of the discriminators, but in the Ekaterininskii—Akdzhar region, discrimination is completely lacking.

For a quantitative expression of the difference between regions along with the generalized distance of Mahalonobis, the separation coefficient Δ was used. This coefficient was obtained from the generalized distance at a number of characters equal to 1. The formula for the coefficient has the following form:

$$\Delta = \frac{(\bar{x}_{1i} - \bar{x}_{2i})^2}{s^2}.$$

This separation coefficient Δ is intimately related to Kullback information measures [Kullback 1959]. By means of the coefficient, the difference between original characters and values of the discriminator z_0 have been computed (the formula given is for computing difference for original characters; for computing difference for the discriminator z_0, $\bar{x}_1$ and $\bar{x}_2$ in the formula are replaced by $\bar{z}_{01}$ and $\bar{z}_{02}$, which correspond to the average values of the discriminator z_0 for the first and second samples).

In Table 6 are shown the separation coefficients Δ computed for the original characters. They show the degree of variation between the compared pairs of characters.

Table 7 gives the separation coefficient Δ computed for data of the discriminators z_0 for each pair of regions. From this table it follows that the separation coefficient is 2.2-5.3 for regions that are indistinguishable by the discriminator z_0, 10-16 for regions distinguishable for z_0 but exhibiting little difference on the graphs for average values of the samples, and as much as 43.0-64.0 for regions that are markedly different.

The separation coefficient computed for the discriminator z_0 in the present example practically coincides with the generalized distance of Mahalonobis.

Conclusion

The investigation of similarity between mineral associations in the regions of the Southern Urals, the Mugodzhary region, and Ustyurt has been made by means of the linear discriminant function and the generalized distance of Mahalonobis. As a result, it has been shown that a number of the regions differ among themselves in mineral associations. The linear discriminant function and the generalized distance are simple to use when an electronic computer is available for the calculations. When a computer is not available, the application becomes much more difficult. In view of this, it has been proposed that the same problems may be solved by methods that are cruder but that are simpler and more widely applicable in practice. In this paper we have examined a number of these methods, and have found that they give results that in the present example are practically the same as those obtained by the more complex method.

Geologically, the results of this work permit us to make the following conclusions. Three regions have been distinguished, differing among themselves in their mineral associations. One region is the Baiguskarovskii district, i.e., the Southern Urals. South of it occur a whole series of sections having similar mineral associations among themselves but differing markedly from the Southern Urals association. Lastly, far to the south, the mineral association of the section cut by the northern Ustyurt test hole may be distinguished. All the investigated regions may be divided into three, approximately equal segments in their extent from north to south. The extreme sections differ strongly from the middle, but the middle sections are practically uniform within themselves. If we assume that the detrital material came from the Urals, it is then incomprehensible why the mineral associations at first changed very strongly, then over an equivalent distance changed hardly at all, and, finally, also in a similar distance, changed very markedly again. If we assume two sources for the detritus, a Ural and a Kyzylkum, it then becomes natural to suppose that the stream of material would change rapidly along a line from one to the other, but within zones of mixing might exhibit no differences. This appears even more natural when we consider the possible effect of Mugodzhary in the intermediate zone; the introduction of clastic material from this region played the role of a distinctive buffer.

The above discussion shows that the previously obtained data [Vistelius and Demina, 1963] to the effect that material of northern Ustyurt was unrelated to the Southern Urals finds confirmation in the results of analyzing the mineral association as a whole.

REFERENCES

Anderson, T. W., An Introduction to Multivariate Statistical Analysis, John Wiley and Sons, New York (1958).

Atanasyan, V. A., Grinberg, I. G., and Sukhilin, V. G., The Northern Ustyurt Test Hole No. 1, Izd. Nedra, Leningrad (1965).

Kalinin, O. M., and Doktorov, B. Z., "Discriminant functions," in: The Application of Mathematical Methods in Biology, No. 4, Tr. Vychislitel'nyi Tsentr, Leningr. Gos. Univ. (1967).

Koltypin, S. N., Upper Cretaceous Deposits of the Ural-Emba Salt-Dome Region, Southeastern Part of the Western Ural and Mugodzhary Region, Gostoptekhizdat, Leningrad (1957).

Kullback, S., Information Theory and Statistics, John Wiley and Sons, New York (1959).

Roy, S. N., and Mitra, S. K., "An introduction to some nonparametric generalizations of analysis of variance and multivariate analysis," Biometrica, No. 43, pp. 361-376 (1956).

Vistelius, A. B., "Mineral content of the heavy sand fraction in the lower part of the Productive sequence of the Apsheron Peninsula, the Chokrak strata of southern Dagestan, and alluvium of the Volga," Dokl. Akad. Nauk SSSR, Vol. 71, No. 2, pp. 367-370 (1950).

Vistelius, A. B., and Demina, M. E., "Dispersion of clastic material in the Aptian-Cenomanian basin of southeastern USSR," Dokl. Akad. Nauk SSSR, Vol. 150, No. 6, pp. 1319-1322 (1963).

Yanshin, A. L., "Methods of studying buried folds as exemplified in an investigation of relations in the Urals, Tien-Shan, and Mangyshlak," Izv. Akad. Nauk SSSR, Ser. Geol., No. 5, pp. 135-155 (1948).

Yanshin, A. L., Geology of the Northern Aral Region, Izd. Mosk. Obshchestva Ispytatelei Prirody, Moscow (1953).

Yanshin, A. L., Bezrukov, P. A., and Fokin, A. G., Geology and Mineral Deposits of the Mesozoic and Tertiary Deposits of the Southern Urals, Part I, Stratigraphy, Tr. NIUIF, No. 125 (1934).

ON LOCATING FIELD BOUNDARIES IN SIMPLE PHASE DIAGRAMS BY MEANS OF DISCRIMINANT FUNCTIONS

F. Chayes

Geophysical Laboratory
Carnegie Institution of Washington
Washington, D.C.

The traces of discriminant functions computed from the data are in excellent agreement with field boundaries shown in a number of published phase diagrams. The substitution of a systematic numerical procedure for the conventional inspection technique by which field boundaries are located might reduce the amount of time and effort now sometimes devoted to obtaining points very close to the supposed locus of the field boundary. The principal advantage of the numerical procedure, however, is that it may be applied in multicomponent space, in which the location of field boundaries by graphical inspection is impossible.

Introduction

Phase equilibrium studies were among the first systematic experimental researches in petrology, and in most work of this type the actual data gathering is characterized by strong emphasis on quantitative control. In the course of the past half-century there have been vast improvements in the techniques by which temperatures are controlled and recorded, x-ray diffraction has made possible the study of products too fine for effective identification by microscope, and devices permitting generation and measurement of pressure have developed to such an extent that in many laboratories P—T conditions comparable to those at the base of the crust or well into the upper mantle can be reached as a matter of routine.

Although the whole procedure by which the investigator determines what phase assemblage is stable for a particular experimental combination of temperature, pressure, and composition has indeed improved enormously, there has been no comparable improvement in the graphical technique by which he blocks out, from examination of the assemblage of data, the limits of a stability field or the boundary between two such fields. In this crucial operation, in fact, there seems to have been no change at all.

There are many reasons — good and bad — for this conservatism in an otherwise rapidly developing discipline. Probably the most important is just that the location of field boundaries by graphical inspection is simple and usually adequate. Further, although the sentiment may seem heretical to many practitioners of such an eminently quantitative art (petrology), the delineation of field boundaries is essentially qualitative. This does not mean that it is not susceptible to numerical analysis, but it does mean that the curve-fitting techniques familiar to everyone, techniques based on regression statistics of one kind or other, are essentially inapplicable or irrelevant. Whatever its final interpretation, in the actual experimental situation the field boundary is not a line of central tendency. Points lying on opposite sides of a theoretically correct or "true" field boundary differ from each other in an important qualitative sense which finds no expression in regression analysis. In fact, regression analysis presumes that there is no qualitative distinction between deviations of opposite sign, whereas the existence of just such a distinction is fundamental to the definition of the field boundary.*

Numerical construction — more precisely, reconstruction — and analysis of qualitative classifications is the objective of discriminant function analysis, and although the procedure is usually applied to multivariate arrays so that graphical representation is impossible, the principles of the method are unaffected by the number of variables. It is the object of this note to show that in a number of examples the graphs of discriminant functions computed from the data are in good agreement with boundaries shown graphically in the original publications, and usually located by inspection. The discussion begins with a brief description of the key statistic of discriminant function analysis and the way in which the function is calculated.

Definition and Calculation of the Discriminant Function

It is presumed that each item in a sample may be unequivocally assigned to one of two mutually exclusive groups by means of some initial set of properties, and that for each item observed values of each of a second set of properties, the variables $[X_{aj}]$, $j = 1, 2, \ldots, w$ are available. In our case the "initial" set is the phase assemblage, found at the conclusion of the run, which determines whether an item is a member of class 1 or class 2, while the "second" set consists of the recorded temperature and pressure of the run and the composition of the charge; in addition, we shall also use whatever powers and cross-products of these latter values may seem convenient. (If composition is the same for all runs, it is of course not a variable, and similarly for pressure).

From the assemblage of $[X_{aj}]$, a set of weighting coefficients λ is calculated in such fashion as to maximize the "distance" between the sample statistics $\bar{z}_1 = \bar{x}_1' \lambda$ and $\bar{z}_2 = \bar{x}_2' \lambda$, this distance being defined as the ratio

$$G = \frac{(\bar{z}_1 - \bar{z}_2)^2}{n_1 \sigma_1^2 + n_2 \sigma_2^2}, \tag{1}$$

where n_i is the number of items and σ_i^2 is the variance of z in group i, and barred values denote group averages.

The procedure by which G is maximized is admirably outlined by Hoel [1962], much of whose description is paraphrased by Chayes and Velde [1965]. Briefly, the a-th item is characterized by a row vector

$$x_a' = \{x_1, x_2, \ldots, x_n\} \quad a = 1, 2, \ldots, (n_1 + n_2),$$

---

* For an attempt to escape this dilemma by confining the regression analysis to mean values for pairs of "bracketing points," see Boyd, England, and Davis (1964, pp. 2105-2106).

in which each element is measured from its own group mean, and the matrix **X** is the assemblage of such row vectors; the required weighting coefficients are then the elements of the column vector

$$\lambda = (X'X)^{-1}d, \tag{2}$$

where

$$[d_j] = (\bar{x}_{1,j} - \bar{x}_{2,j}).$$

No other set of weighting coefficients for these particular variables can yield a larger G. Finally, the discriminant $\hat{z}$ is the unweighted average $(\bar{z}_1 + \bar{z}_2)/2$ if the parent variance of z is the same in both groups, and the weighted average $(\sigma_2 \bar{z}_1 + \sigma_1 \bar{z}_2)/(\sigma_1 + \sigma_2)$ if it is not [Fisher, 1936; quoted in Chayes and Velde, 1965]. If $\bar{z}_1 < \bar{z}_2$, the k-th item is assigned to class 1 if $z_k < \hat{z}$ and to class 2 if $z_k > \hat{z}$.

A linear discriminant function based on temperature and pressure, for instance, is simply

$$\lambda_p P + \lambda_t T = \hat{z}, \tag{3}$$

and since λ_p, λ_t, and $\hat{z}$ are computed from the data, the locus of (3) for any assigned values of either P or T can be found directly. The equation may of course be written

$$P = \frac{\hat{z}}{\lambda_p} - \frac{\lambda_t}{\lambda_p} T, \tag{4}$$

in which $(\hat{z}/\lambda_p)$ and $(-\lambda_t/\lambda_p)$ are, respectively, the intercept and slope of a straight line in P–T space. The resemblance to an ordinary regression line is obvious but only superficial. There is only one discriminant function for any set of data; its locus need not include the mean of either variable; there is no underlying assumption that either variable is dependent or independent in the regression sense; and the scatter of data points from the line is not indicative of the goodness or badness of the "fit." The purpose of the line is to recapture the original partition of the data into two qualitatively distinct classes, and its success or failure is to be judged only by the number of correct assignments it makes.

In the following section some practical examples are described in the order of what is probably increasing difficulty (for the discriminant function, not for the reader!). In the work of Boyd and England, and of Boyd, England, and Davis, each published record includes only data points lying very near the field boundary. In that of Akimoto, Fujisawa, and Katsura the points show considerably more dispersion, though location of the field boundary remains the only object of the experiment. In the work of both sets of authors, the original classes in each example contain nearly the same numbers of observations and the curvature of the field boundary is either gentle or negligible.

Some Practical Examples

A. The Quartz-Coesite Field Boundary
[Boyd and England, 1960]

The discriminant function based on P and T may be written

$$P = 20.18 + 0.1069T. \tag{5}$$

It is shown, together with the data, in Fig. 1. Quartz charges which remained unchanged and coesite charges in which quartz appeared in the product are shown by open circles; coesite charges which remained unchanged and quartz charges in which coesite appeared in the product

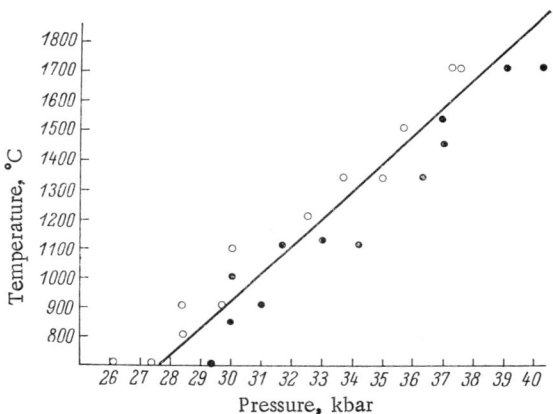

Fig. 1. The quartz—coesite field boundary.

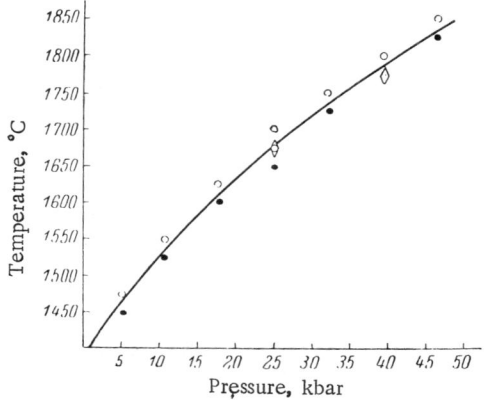

Fig. 2. The melting curve of $CaMgSi_2O_6$.

are shown by solid circles. The discriminant function and the field boundary drawn by Boyd and England partition the data practically identically; at the scale of the illustration, in fact, the two lines cannot be distinguished.

B. The Melting Curve of $CaMgSi_2O_6$ [Boyd and England, 1963]

To permit comparison with the Simon equation shown by Boyd and England, the discriminant function has been calculated from P, T, and T^2; it is

$$P = 123.68 - 0.2337T + 0.0001044T^2, \quad (6)$$

and is shown, together with the data, in Fig. 2. Open circles in the figure represent charges in which the product contained only glass or quench crystals, solid circles those in which neither glass nor quench crystals are recorded in the product.* Two products consisting mostly of stable Di crystals but containing as well appreciable amounts of quench crystals are shown by diamonds in the figure but were not used in the calculation. The curve is again virtually identical with that shown by Boyd and England, and yields an identical data partition.

If one wished the field boundary to be defined by the effect of pressure on the melting point, presumably one would use the variables T, P, and P^2. For these the function is

$$T = 1402.06 + 12.8276P - 0.0662P^2. \quad (6a)$$

which yields the same partition of the data as (6) and cannot be shown separately at the present scale. It extrapolates to a melting point of 1402° at atmospheric pressure, and at this pressure (P = 0.001) one of the solutions of equation (6) is T = 1380°. The melting point of Di is given by Schairer and Bowen [1938] as 1391.5°, almost exactly midway between these values.

C. The Melting Curve of $MgSiO_3$ [Boyd, England, and Davis, 1964]

In this, as in the preceding example, both the data and the original interpretation suggest significant departure from linearity. For the $MgSiO_3$ data a discriminant function based only on P and T misclassifies 5 of the 24 points. If either P^2 or T^2 is included in the calculation, however, the number of misclassifications drops to 2, as in the diagram of Boyd et al.; in fact, these same two points are misclassified also by the discriminant functions based on either (P, T, T^2) or (T, P, P^2). These functions are:

---

* In all illustrations, reference axes are taken as shown in the original publications. Throughout, solid circles denote assemblages stable at higher pressures.

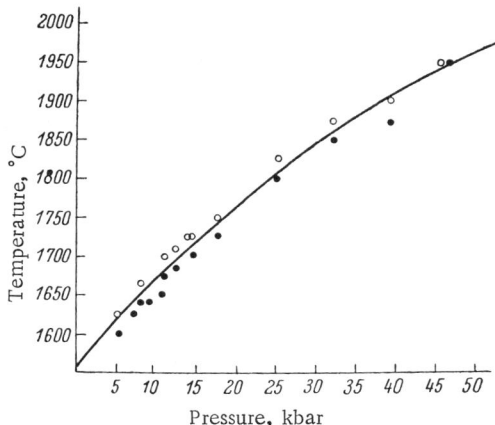

Fig. 3. The melting curve of $MgSiO_3$.

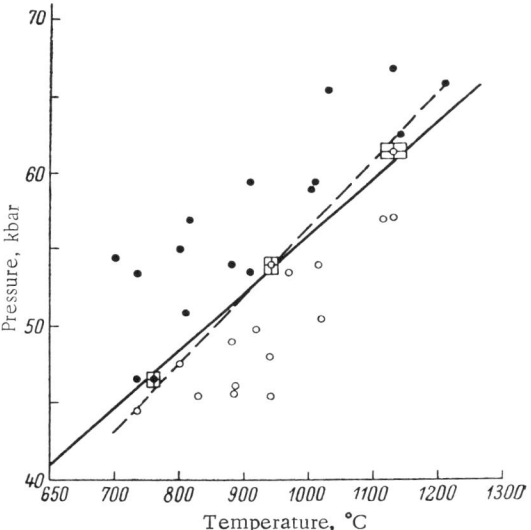

Fig. 4. The Ol—Sp field boundary for Fe_2SiO_4.

$$P = 153.36 - 0.2721T + 0.0001115T^2 \qquad (7)$$

and

$$T = 1560.59 + 11.5361P - 0.07060P^2. \qquad (8)$$

The trace of (7) is shown, together with the data, in Fig. 3, in which open and solid circles have the same significance as in Fig. 2. At the scale of the drawing the traces of (7), (8), and the Simon function of Boyd et al. are indistinguishable. At atmospheric pressure T is 1560.6° in (8) and 1556.9° in (7), as compared to the measured value of 1557° at which $MgSiO_3$ melts incongruently.

In all three examples so far discussed, the investigators record only points lying near the field boundary* but it is one of the most striking properties of the discriminant function, as shown in the next two examples, that results in good agreement with those reached by inspection may still be obtained even when there is considerable scatter of the data.

D. The Ol—Sp Field Boundary for Fe_2SiO_4 [Akimoto, Fujisawa, and Katsura, 1965]

This boundary was recently determined by Akimoto et al., whose data are shown in Fig. 4, in which the dashed line is intended as a copy of the line shown in their graph. The solid line is the trace of the discriminant function

$$P = 18.41 + 0.0375T. \qquad (9)$$

calculated from their data. Solid circles in Fig. 4 denote P—T coordinates at which spinel charges were unaffected or spinel grew in olivine charges; P—T coordinates at which olivine charges were unaffected or olivine grew in spinel charges are shown by open circles. One of the solid circles falls below the trace of (9) and two of the open circles lie above it. The edges of the rectangles drawn about these points are proportional to the stated precision of the observations and it will be noted that the trace of (9) intersects each of the error rectangles. Despite the rather considerable scatter of the data, the discriminant function achieves an almost optimum partition. As the dashed line in Fig. 4 shows, Akimoto et al. have found, presumably by inspection, an even more consistent one, but the line they draw is not precisely the trace of the equation given in their text. For most purposes quite trifling, the difference is critical in the present context; the two solid circles farthest to the right in Fig. 4, properly classified by the line in their graph, lie below the plot of their text function. In any event, it is clear once more that the trace of the discriminant function is in good agreement with the published field boundary.

---

*Excepting the two points by which Boyd, England, and Davis fix the protoenstatite—rhombic-enstatite boundary; these points were not used in the present calculations.

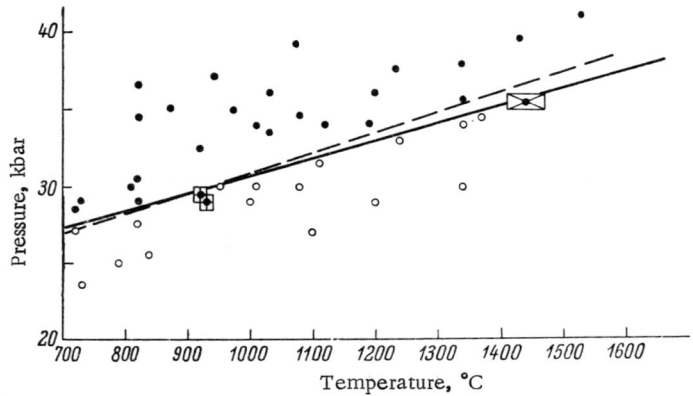

Fig. 5. The Ol—Sp field boundary for Ni_2SiO_4.

E. The Ol—Sp Field Boundary for Ni_2SiO_4
[Akimoto, Fujisawa, and Katsura, 1965]

In Fig. 5 are plotted, with the same symbols as in Fig. 4, the data of Akimoto et al. and the field boundary shown by them for the Ol—Sp transformation of Ni_2SiO_4, together with the trace of the calculated discriminant function

$$P = 19.3073 + 0.01144T. \tag{10}$$

Despite the very wide scatter of points in this example, the discriminant function partitions the data with the same efficiency as the published field boundary; indeed, if we count an intersection with the error rectangle as a "correct" classification, it even does a little better. In this example, as in all the others, the difference between the published field boundary and the trace of the computed discriminant function is trifling. It is evident that the function finds numerically just about the same result the investigator has chosen by inspection, though agreement of this quality is hardly to be expected unless sufficient data are available and the second derivative of the field boundary is everywhere small.

A Limitation on *Ex Post Facto* Comparisons between Field Boundaries and the Traces of Discriminant Functions in More Complicated Systems

From the examples so far described, we may surely conclude that where data are collected for the purpose of delineating a single linear or gently curved field boundary, the discriminant function provides an admirable numerical substitute for the graphical procedures by which such boundaries are usually constructed.

The assignment of unit weight to all points regardless of their distances from the region in which the boundary is though to lie may be inefficient or unrealistic, but does not seem to be wrong in principle if, as in all these examples, the data were in fact collected for the specific purpose of locating one particular field boundary. Even in systems of very modest complexity, however, this condition will not hold. In a system which is strictly binary at liquidus temperatures there will be two field boundaries of interest. Considering only the intersections of the various liquidus surfaces, as is often done in "dry" work at 1 atm, there will be at least two field boundaries to be located in a ternary system in which each component enters only one phase, up to three additional boundaries for each binary compound, and one more for each polymorphic transition that extends to liquidus temperatures. The bearing of data points in any field on the problem of locating its boundaries will then vary widely. Some of the data in a field will have been collected largely for the purpose of finding the liquidus surface itself, others for locating one or another of its interesections with adjacent liquidus surfaces.

The unequal weighting implicit in this situation would offer no particular obstacle to an investigator who intended from the outset to use discriminant function analysis for the location of some of the more important field boundaries in a system and planned his work accordingly. It does preclude unique location of field boundaries by *ex post facto* calculations based upon the data, however, and accordingly renders comparison between published boundaries and those obtained by discriminant function calculations inconclusive. About all one can do is carry through the calculations for as many groupings of the data as seem reasonable. If the results of all calculations are in good agreement with the published diagram the interpretation is clear and unavoidable; otherwise it may be neither. Work of this sort is now in progress.

Advantages of a Systematic Numerical Procedure for Locating Field Boundaries

It has now been shown that a numerical procedure, one which may seem strange, difficult, and perhaps even repulsive to many petrologists, yields results closely comparable to those readily obtained by simple inspection. What is the purpose of this exercise?

There is, first of all, the intrinsic theoretical interest of the problem, an interest which will be apparent to anyone concerned with the progressive "quantification" of the more traditional earth sciences. Quite aside from any practical consequences, one would simply like to know if the thing can be done. It can.

Whether the procedure will be of practical use in studies of systems whose thermodynamic variance is such as to permit graphical construction of field boundaries is an open question. In most phase equilibrium work data accumulate rather slowly and it is natural for the investigator to plan his work sequentially; the original design is neither detailed nor specific, and may be drastically modified as the results of each run become available. If field boundaries are to be established by inspection, a considerable concentration of data in the immediate vicinity of each boundary will ultimately be required, for otherwise there will be many possible solutions and no particular reason for preferring one to another. If, on the other hand, the boundaries are to be constructed by discriminant function analysis, it would often be desirable to have somewhat more data than are usually reported, but the requirement that these data be concentrated in the immediate vicinity of the boundary — the definition, that is to say, of "immediateness" — would be far less stringent. The overall effect might well be a considerable economy of time and effort, for it is an open secret that students of phase equilibrium often accumulate far more information than they publish, preferring to record only results bearing directly and strongly on the final interpretation.

If the number of variables is such as to preclude graphical representation, however, the location of field boundaries by inspection is obviously impossible and a numerical procedure which will perform this operation becomes indispensable. It is in this context that discriminant function analysis is likely to prove of most practical importance. Given the knowledge that results obtained by it are in good agreement with those reached by inspection where inspection is possible, it seems reasonable to suggest that field boundaries could be "located" by means of discriminant functions in multicomponent space, where inspection is impossible.

Acknowledgments. It is gratifying to be included among the contributors to this volume, dedicated to one who has contributed so much and so well to the long-overdue modernization of the naturalistic aspects of geology. All calculations were performed at the computer facility of the US National Bureau of Standards, Washington, D. C.

REFERENCES

Akimoto, S., Fujisawa, H., and Katsura, T., "The olivine—spinel transition in Fe_2SiO_4 and Ni_2SiO_4," J. Geophys. Res., Vol. 70, pp. 1969-1978 (1965).

Boyd, F. R., and England, J. L., "The quartz—coesite transition," J. Geophys. Res., Vol. 65, pp. 749-756 (1960).

Boyd, F. R., and England, J. L., "Effect of pressure on the melting of diopside, $CaMgSi_2O_6$, and albite, $NaAlSi_3O_8$, in the range up to 50 kilobars," J. Geophys. Res., Vol. 68, pp. 311-323 (1963).

Boyd, F. R., England, J. L., and Davis, B. T. C., "Effects of pressure on the melting and polymorphism of enstatite, $MgSiO_3$," J. Geophys. Res., Vol. 69, pp. 2101-2109 (1964).

Chayes, F., and Velde, D., "On distinguishing basaltic lavas of circumoceanic and oceanic-island type by means of discriminant functions," Am. J. Sci., Vol. 263, pp. 206-222 (1965).

Fisher, R. A., "The use of multiple measurements in taxonomic problems," Annals of Eugenics, Vol. 8, Pt. 2, pp. 179-188 (1936).

Hoel, P. G., Introduction to Mathematical Statistics, 3rd edition, John Wiley and Sons, New York, 428 pp. (1962).

Schairer, J. F., and Bowen, N. L., "The system leucite—diopside—silica," Am. J. Sci., Ser. 5, Vol. 35-A, pp. 289-309 (1938).

TWO-CLUSTER DISCRIMINATION IN ANALYTICAL GEOCHEMISTRY USING THE DISTANCE COEFFICIENT

D. M. Shaw

Department of Geology
McMaster University
Hamilton, Ontario, Canada

The distance coefficient d is defined as the Pythagorean distance between two sample item points in k-space in the R-mode, where k is the number of variables. This quantity resembles the taxonomic distance measure in zoology. Values of d may be profitably used in analytical geochemistry to help distinguish samples drawn from different multivariant populations. Qualitative discrimination between groups of amphibolites and of quartzofeldspathic gneisses is aided by using d.

Introduction

The classification of sample objects, according to a preestablished system of categories, and using instrumental measurements of variables, is a common procedure in analytical geochemistry. For example, it might be desirable to know whether a given metamorphic rock had its origin in an igneous or in a sedimentary environment, or whether a particular shale was of marine or fresh-water origin, the basis of decision in each case to be determinations of the major and minor constituents of the rock. The essential problem here is to appraise the degree of similarity between two or more samples drawn from multivariate populations. This situation is common to a number of disciplines (e.g., archaeology, psychology) as well as the earth sciences, and is commonly approached by the method of multivariate discrimination analysis. The advantages of this procedure are its flexibility and the fact that it permits an estimation of the value of the classification decision in probabilistic terms. The mathematics is, however, somewhat complex, and the computational methods, which involve matrix inversion, are prohibitive unless machine methods are available.

For an elementary appraisal of the similarity between two multivariate samples, it has proved convenient to use the taxonomic distance measure (TDM). This quantity has been extensively used in modern zoological and paleontological problems [e.g., Sokal and Sneath, 1963] and may be defined as the Pythagorean distance between two points in k-space in the R-mode (k is the number of variables). The method has not apparently been used hitherto in geochemical situations, which is rather surprising, since these commonly involve numerous analytical data which have been obtained for the purpose of making a classification of the kind referred to above.

It should be stressed at the outset that the TDM is not quantitative, in the sense that it does not lead to probabilistic statements: perhaps it should rather be stated that its modification for probabilistic confidence limits has resulted in discrimination functions. The use proposed here of the TDM is to help visualize the geometrical or topological relationships of point clusters.

Scatter diagrams which show the relationship of sample-item points for one, two, or even three variables have long been used in mineralogy, petrology, and sedimentology as a basis for plotting "composition fields" of different substances. Such diagrams permit the assignment of unknowns to different classification categories, provided that those categories are clearly separated one from another on the diagram, and provided that one has confidence that they have some real existence.* In order to apply this method, it is necessary to have a basis for recognizing clusters of item points. This may be managed visually in simple cases, but where the number of variables exceeds three, it becomes necessary to use ratios, or some more or less arbitrary mineral convention in order to project the diagrams on two dimensions. Often the number of variables (e.g., trace element analyses) is so large that such methods break down completely: the scientist must make the choice of either disregarding many of his hard-won and expensive data or of considering them in groups of two or three at a time. Usually he prefers the latter alternative and ends by trying to reconcile a group of diagrammatic trends or clusters whose interrelationships are very complex. Where the problem is one of discrimination between clusters the TDM permits, at least in favorable circumstances, an appraisal of the diagram which would be obtained by plotting all the variables simultaneously, i.e., the situation which is the most relevant to nature. The TDM, although tedious to compute, may normally be managed comfortably on a desk calculator.

Theory and Procedure

The principle involved in separating point-clusters on a scatter diagram is that points within one cluster lie close together, but are distant from points in a separate cluster. Thus, in Fig. 1a, the distance Δ_{AA} between two points in cluster A will tend to be less than the distance Δ_{AB} between two points, each belonging to a separate cluster.

In a one-dimensional plot, where only one variable i is concerned, the distance between two points with values x_{i1} and x_{i2} is simply Δ_{12}, where

$$\Delta_{12} = x_{i1} - x_{i2}.$$

In a two-variable case, the points 1 and 2 are determined by values x_{11}, x_{21} and x_{12}, x_{22}, and the distance (Fig. 1b) is

$$\Delta_{12} = \sqrt{(x_{11} - x_{12})^2 + (x_{21} - x_{22})^2} = \sqrt{\sum_{i=1}^{2}(x_{i1} - x_{i2})^2}.$$

This relationship remains the same if there are more than 2 variables, say k:

$$\Delta_{12} = \sqrt{\sum_{i=1}^{k}(x_{i1} - x_{i2})^2}.$$

---

* In the sense that these fields would not change markedly if the number of data points were to be doubled.

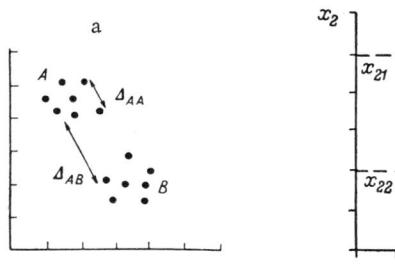 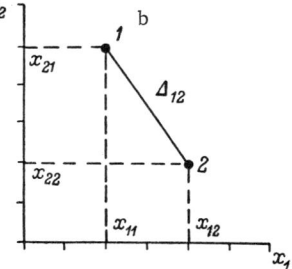

Fig. 1. Distance between two points on one-dimension (a) and two-dimension (b) plots. a) The distance (Δ_{AA}) between two points within a cluster tends to be less than the distance (Δ_{AB}) between points belonging to different clusters; b) the distance between points 1 and 2.

We thus obtain the distance between two points in k-dimensional space, dimensionally identical with the variable, which permits extension of the scatter diagram principle beyond the limits of visual geometry.

The distance Δ_{12} is of course a vector and the present treatment is only concerned with its scalar magnitude. To be completely equivalent to the two-dimensional scatter diagram it would be necessary to specify the angles made by Δ with each coordinate axis, but this is unnecessary here.

Sokal and Sneath [1963, Chapter 6] and Harbaugh [1964] recommend that the variables be coded before calculating Δ. One procedure is to reduce each variable to values between 0 and 1, by dividing by the maximum values. In this case, the maximum possible value of the sum of squares will be

$$1^2 + 1^2 = 2 \text{ (for the 2-dimensional case)},$$
$$1^2 + 1^2 + 1^2 = 3 \text{ (for the 3-dimensional case) and}$$
$$1^2 + 1^2 \ldots + \ldots = k \text{ (for the k-dimensional case)}.$$

If the average sum of squares d^2 is computed, where

$$d^2 = \frac{\Delta^2}{k},$$

then d will lie within the range −1 to +1. Harbaugh [1964] further recommends that the measure of average distance be

$$d'' = 1 - d'$$

in order that the significance of distance measures will resemble the behavior of the correlation coefficient (i.e., maximum similarity between points gives d" = ±1, whereas minimum similarity or maximum distance corresponds to d" = 0).

A disadvantage of coding to values between 0 and 1 is the necessity for recoding if an additional point with a higher absolute value of one of the variables is added to the set. Since some proportional coding is, however, necessary if variables are to be commensurate, it is recommended here that each variable be normalized in the usual manner, i.e.,

$$z_{ij} = \frac{x_{ij} - \bar{x}_i}{s_i},$$

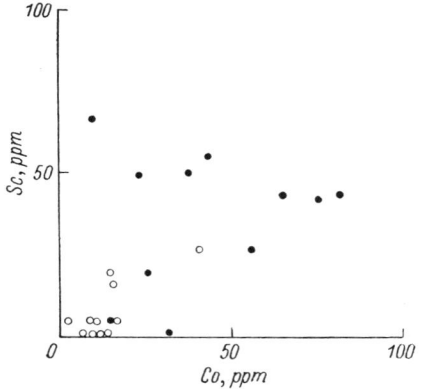

Fig. 2. Sc—Co scatter diagram for 22 amphibolites.

where $\bar{x}_i$ and s_i are computed from all the datapoints in both clusters (or all clusters, if there appear to be three or more). The TDM or distance coefficient between points 1 and 2 is thus

$$d = d_{12} = \sqrt{\frac{\sum_{i=1}^{k}(z_{i1}-z_{i2})^2}{k}}.$$

For computation this is best expanded as follows:

$$d = \sqrt{\frac{\sum_i z_{i1}^2 - 2\sum_i z_{i1}z_{i2} + \sum_i z_{i2}^2}{k}},$$

which reduces to

$$d = \sqrt{\frac{\sum_i \frac{x_{i1}^2}{s_i^2} - 2\sum_i \frac{x_{i1}x_{i2}}{s_i^2} + \sum_i \frac{x_{i2}^2}{s_i^2}}{k}}. \quad (1)$$

It is convenient to rewrite the numerator in this expression, using the following symbols:

$$a_i(1,1) = \frac{x_{i1} \cdot x_{i1}}{s_i^2}, \quad a_i(1,2) = \frac{x_{i1} \cdot x_{i2}}{s_i^2},$$

$$a_i(p,q) = \frac{x_{ip} \cdot x_{iq}}{s_i^2}.$$

This gives

$$d_{pq} = \sqrt{\frac{\sum_i a_i(p,p) - 2\sum_i a_i(p,q) + \sum_i a_i(q,q)}{k}}, \quad (2)$$

and the terms $a_i(p,q)$ denote the squares and cross-product terms for variable i.

An Application to Amphibolites

The data presented in Table 1 have already been published [Shaw and Kudo, 1965] and refers to two groups of amphibolites, whose derivation from igneous rocks and from sediments, respectively, had previously been established on geological grounds. Since the article referred to established the difference between the two multivariate populations by use of the discriminant function, the example is somewhat artificial: it will, nevertheless, serve as an example. Two-dimensional scatter diagrams (Figs. 2 and 3) show that the two groups of amphibolites form nearly separate clusters both on the Co—Sc and TiO_2—P_2O_5 diagrams. This suggests, but does not necessitate, that the two clusters might be distinct in 4-space (they could very well be interpenetrating): differences among other variables too indicate the desirability of a multivariate approach. Initially, therefore, the 22 rocks were treated as one sample and distance coefficients (d) computed (Table 2).

Table 1. Composition of 11 Ortho-Amphibolites (69-28-2 to T1-61-7S) and 11 Para-Amphibolites (69-171-1 to 178-8) Including the Average for Each Group

Component	69-28-2	69-28-5	69-29-5	69-30-2	69-30-6	69-34-7	69-35-7	T1-61-3S	T1-61-4S	T1-61-6S	T1-61-7S	Orthomean
SiO$_2$	50.50	48.50	42.00	53.30	48.70	50.00	48.50	47.10	46.00	47.50	46.10	48.0180
TiO$_2$	3.03	0.940	4.880	1.92	1.47	2.15	1.86	2.75	4.36	5.04	5.21	3.0550
Al$_2$O$_3$	14.50	13.20	13.00	13.80	15.70	15.10	15.10	13.00	13.00	13.60	14.20	14.0730
Fe$_2$O$_3$	3.00	0.72	3.50	3.00	1.80	5.50	2.20	3.80	3.90	3.50	3.80	3.1560
FeO	13.85	4.83	16.54	12.10	9.53	5.23	10.46	16.18	14.15	13.68	13.88	11.8570
MnO	0.32	0.07	0.38	0.29	0.18	0.11	0.19	0.37	0.32	0.29	0.32	0.2580
MgO	2.60	3.80	4.80	3.60	6.80	3.60	7.10	3.30	4.30	5.20	4.80	4.5360
CaO	6.30	16.80	8.60	6.60	10.40	7.00	9.50	6.40	8.60	7.80	7.90	8.7180
Na$_2$O	4.58	2.75	2.71	3.44	2.69	3.67	2.84	2.92	3.10	2.82	2.96	3.1350
K$_2$O	0.44	1.33	0.52	0.40	0.32	3.25	0.20	1.43	0.50	0.71	0.51	0.8740
P$_2$O$_5$	1.31	0.34	2.43	0.43	0.09	1.48	0.16	1.52	2.33	0.54	0.39	1.0020
CO$_2$	0.88	0.73	1.53	0.97	1.19	0.82	1.19	1.36	1.20	1.19	0.70	1.0690
H$_2$O	0.06	6.95	0.96	0.74	0.49	0.33	0.67	0.00	0.04	0.75	0.00	0.9990
Cr	1.50	75.00	23.00	23.00	446.00	1.50	242.00	1.50	1.50	1.50	1.50	74.3640
V	20.00	150.00	376.00	143.00	263.00	207.00	362.00	52.00	287.00	346.00	472.00	245.4550
Ni	10.00	17.00	1.50	54.00	39.00	53.00	159.00	90.00	53.00	56.00	81.00	55.7730
Co	26.00	16.00	10.00	24.00	38.00	32.00	44.00	56.00	66.00	81.00	75.00	42.5450
Sc	19.00	5.00	66.00	49.00	50.00	1.00	55.00	26.00	43.00	43.00	42.00	36.2730
Sr	319.00	1420.00	426.00	633.00	189.00	3500.00	264.00	82.00	182.00	162.00	115.00	662.9090
Ba	146.00	754.00	276.00	197.00	91.00	6500.00	73.00	220.00	127.00	270.00	174.00	802.2730
Zr	420.00	174.00	142.00	575.00	151.00	193.00	190.00	264.00	340.00	255.00	416.00	283.6360

Component	67-171-2	68-74-16	68-77-1	68-78-1	68-81-10	68-81-118	69-27-2	69-32-1	70-139-1	70-146-6	178-8	Paramean
SiO$_2$	61.50	48.70	42.70	60.20	62.70	39.80	58.80	56.50	57.30	55.50	61.40	55.0090
TiO$_2$	0.27	1.60	0.53	0.80	0.91	0.32	0.76	0.93	0.78	0.76	0.78	0.7670
Al$_2$O$_3$	4.20	15.60	10.90	14.90	15.40	7.90	15.60	16.40	13.50	15.80	15.80	13.2730
Fe$_2$O$_3$	0.17	3.30	0.57	0.76	2.10	0.66	0.88	2.90	0.75	2.10	1.60	1.4350
FeO	1.67	8.74	3.48	5.72	4.58	2.64	5.69	4.30	3.43	4.55	4.27	4.4610
MnO	0.25	0.20	0.06	0.02	0.10	0.08	0.04	0.15	0.05	0.07	0.13	0.1050
MgO	17.40	5.40	2.80	4.60	2.60	1.90	4.50	2.90	3.40	4.30	2.40	4.7450
CaO	9.80	10.30	22.80	5.40	4.40	24.90	5.80	7.50	13.10	10.10	8.20	11.1180
Na$_2$O	1.11	3.76	2.64	4.29	4.19	1.64	5.31	4.88	2.64	2.04	3.58	3.2800
K$_2$O	1.61	0.26	0.68	0.91	1.98	0.41	0.45	0.89	0.33	2.24	0.53	0.9350
P$_2$O$_5$	0.06	0.15	0.13	0.13	0.26	0.05	0.19	0.27	0.21	0.22	0.16	0.1660
CO$_2$	1.41	1.19	0.57	1.02	0.78	0.45	0.94	0.69	0.69	0.63	0.68	0.8230
H$_2$O	0.02	0.67	12.98	0.34	0.05	18.70	1.22	1.10	3.84	1.09	0.38	3.6720
Cr	21.00	150.00	44.00	124.00	37.00	1.50	115.00	37.00	0.30	1.50	30.00	51.0270
V	41.00	433.00	72.00	267.00	101.00	48.00	128.00	101.00	136.00	133.00	94.00	141.2730
Ni	8.00	46.00	26.00	17.00	21.00	52.00	60.00	14.00	11.00	23.00	9.00	26.0910
Co	8.00	41.00	11.00	16.00	13.00	10.00	15.00	16.00	10.00	3.00	10.00	13.9090
Sc	1.00	26.00	5.00	5.00	1.00	1.00	19.00	16.00	1.00	5.00	5.00	7.7270
Sr	1240.00	240.00	584.00	468.00	569.00	201.00	373.00	440.00	192.00	1250.00	676.00	566.6360
Ba	300.00	46.00	165.00	2160.00	431.00	97.00	260.00	446.00	161.00	345.00	258.00	424.4550
Zr	103.00	222.00	128.00	140.00	166.00	84.00	183.00	256.00	23.00	186.00	10.00	136.4550

Note. First 13 components in %, last 8 in ppm.

Table 2. Distance Coefficients d for the Analyses in Table 1

	69-28-7	69-28-5	69-29-5	69-30-2	69-30-6	69-34-7	69-35-7	11-61-35	11-61-45	11-61-65	11-61-7	67-171-1	68-74-16	68-77-1	68-78-1	68-81-10	68-81-11B	69-27-2	69-32-1	70-139-1	70-146-6	178-8
69-28-7	0.0000																					
69-28-5	1.2563	0.0000																				
69-29-5	1.2274	1.6541	0.0000																			
69-30-2	0.6787	1.2591	1.3155	0.0000																		
69-30-6	1.4108	1.2810	1.4615	1.2640	0.0000																	
69-34-7	1.8284	1.6145	2.0675	1.8664	2.1197	0.0000																
69-35-7	1.4886	1.4393	1.5056	1.1801	0.8671	2.0875	0.0000															
11-61-3	0.8856	1.4473	1.1193	0.9568	1.3939	1.8061	1.2051	0.0000														
11-61-4	0.9095	1.5188	0.8260	0.9739	1.3906	1.8957	1.2494	0.6845	0.0000													
11-61-6	1.0675	1.4343	1.0569	0.9936	1.2370	1.9019	1.0651	0.8064	0.6077	0.0000												
11-61-7	1.1584	1.5660	1.3051	0.9862	1.4089	1.9917	1.1279	1.0693	0.7962	0.5316	0.0000											
67-171-1	1.9532	1.5245	2.0687	1.8654	1.8287	2.2672	1.9578	1.8925	2.0530	1.9499	2.1953	0.0000										
68-74-16	1.0640	1.0985	1.2379	0.9435	0.8009	1.8410	0.8344	1.1043	1.0197	0.8343	0.9603	1.8120	0.0000									
68-77-1	1.5225	0.6065	1.8834	1.5405	1.5214	2.0531	1.6532	1.6939	1.7630	1.6933	1.7982	1.7447	1.3929	0.0000								
68-78-1	1.2220	0.9131	1.6879	1.2448	1.1597	1.6573	1.3536	1.4805	1.5481	1.4147	1.5748	1.6197	0.9368	1.3304	0.0000							
68-81-10	1.0787	0.9310	1.7529	1.1799	1.3989	1.5840	1.5300	1.3523	1.5430	1.4522	1.5897	1.6576	1.1236	1.3471	0.6340	0.0000						
68-81-11B	1.8677	1.0556	2.1457	1.8521	1.8564	2.3535	1.8948	1.9329	2.0196	1.9529	2.0356	1.9360	1.7431	0.5089	1.7401	1.7418	0.0000					
69-27-2	1.0867	1.0050	1.7186	1.0992	1.1515	1.8634	1.2034	1.3901	1.4978	1.4078	1.5319	1.7599	0.9325	1.3129	0.5562	0.6847	1.7146	0.0000				
69-32-1	0.8210	0.9140	1.6138	0.9395	1.2769	1.6570	1.4156	1.2885	1.3714	1.3185	1.4014	1.8082	0.9187	1.2437	0.7211	0.5082	1.6587	0.5890	0.0000			
70-139-1	1.2909	0.6655	1.7242	1.3582	1.3342	1.9390	1.4964	1.5288	1.6013	1.4922	1.6483	1.5233	1.1268	0.8325	0.7972	0.8254	1.1827	0.8727	0.8414	0.0000		
70-146-6	1.2651	0.6551	1.7234	1.2334	1.4064	1.4903	1.5202	1.4162	1.5762	1.4758	1.5804	1.5222	1.1910	1.1136	0.9157	0.6460	1.4882	1.0391	0.8268	0.7766	0.0000	
178-8	1.1400	0.8402	1.7044	1.2645	1.3114	1.7828	1.4956	1.4553	1.5749	1.4706	1.6253	1.6426	1.0844	1.1622	0.6782	0.5502	1.5474	0.7044	0.5877	0.4843	0.7425	0.0000

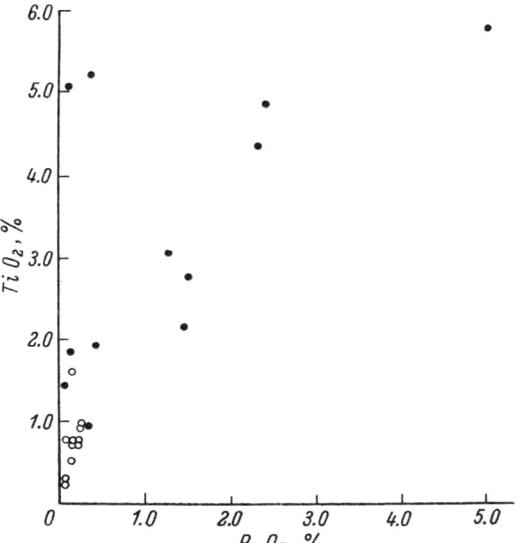

Fig. 3. TiO$_2$—P$_2$O$_5$ scatter diagram for 22 amphibolites.

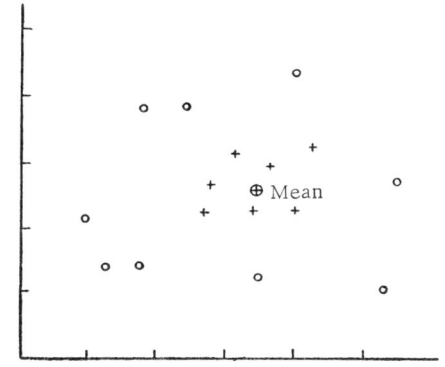

Fig. 4. A possible distribution of two sets of sample points in 2-space. Distances from the mean to inner group points (+) are all less than distances to the outer group (O), but the latter do not form a cluster which is distinct from the inner one.

Examination of the values of d leads to the following appraisal. The upper left-hand 11 rows and columns of the matrix shown in Table 2 give d-values* between rocks recognized in the field to be ortho-amphibolites (A$_0$); the lower right-hand 11 rows and columns similarly measure distances between recognized para-amphibolites (A$_p$); the upper right-hand 11 rows and columns give distances between members belonging to different groups. The ranges of values (abbreviated) are

$A_0 - A_0 \ldots 0.53-2.12,$ midpoint 1.33,
$A_p - A_p \ldots 0.48-1.94$ » 1.21,
$A_0 - A_p \ldots 0.66-2.35$ » 1.51.

These values suggest that the range in each cluster is about the same and, also, that the average distance coefficient within a cluster is less than between clusters. This evidence does not conflict with the possibility that the two clusters might occupy different, nonoverlapping hypervolumes in 21-space.

However, the approach used so far does not permit the determination of whether clusters actually overlap, and this would be useful knowledge. The following analysis will assist in this question.

One condition for no overlap between clusters is that the distance from the mean value of a cluster to any point within the cluster will be less than the distance to any point in a second cluster.† To use this approach it is clearly necessary to decide in advance which group of points will be taken as a cluster.

Taking the groups already determined on geological grounds, the averages may be called the orthomean and the paramean (see Table 1). Distance coefficients from these mean values were first calculated [equation (1)] by normalizing with the total variance of the 22 sample rocks (Table 3). Distances from both means are clearly necessary. For if only the orthomean were used as the basis of comparison, then if all the para-amphibolites were shown to lie further from the orthomean than the ortho-amphibolites, the possibility would remain that the former do not form a cluster, but, rather, are distributed in a hyperspherical shell around the orthomean. This possibility is illustrated in Fig. 4. The use of two mean-value points serves to clarify whether two clusters in fact exist, as well as to test whether they overlap.

---

* The matrix is symmetrical: the lower left-hand values have been omitted to prevent confusion.
† This condition is sufficient but not necessary: its application needs thoughtful consideration.

Table 3. Distance Coefficients from Orthomean and Paramean, Normalized by Using Total Valence

Ortho-amphibolite

	69-28-2	69-28-5	69-29-5	69-30-2	69-30-6	69-34-7	69-35-7	T1-61-3	T1-61-4	T1-61-6	T1-61-7
Orthomean....	0.7414	1.0416	0.9663	0.6697	0.9913	1.5913	0.9212	0.6624	0.6168	0.5838	0.7754
Paramean....	1.0639	0.5014	1.5646	1.0783	1.1324	1.6935	1.2752	1.2827	1.3978	1.2862	1.4433

Para-amphibolite

	67-171-1	68-74-16	68-77-1	68-78-1	68-81-10	68-81-11B	69-27-2	69-32-1	70-139-1	70-146-6	178-8
Orthomean....	1.7449	0.6197	1.3758	1.0691	1.0872	1.7013	1.0426	0.9281	1.1835	1.0893	1.1242
Paramean....	1.4068	0.8843	0.8534	0.6100	0.6043	1.2589	0.6551	0.5954	0.4193	0.6025	0.4895

Table 4. Distance Coefficients from Orthomean and Paramean, Normalized by Using the Variance of Each Group

Ortho-amphibolite

	69-28-2	69-28-5	69-29-5	69-30-2	69-30-6	69-34-7	69-35-7	T1-61-3	T1-61-4	T1-61-6	T1-61-7
Orthomean....	0.9187	1.3435	1.0609	0.7645	0.9650	1.4194	0.9073	0.7288	0.6285	0.5861	0.7762
Paramean....	1.3552	0.8279	1.8871	1.1962	1.2805	1.6153	1.3920	1.5355	1.5940	1.4802	1.6565

Para-amphibolite

	67-171-1	68-74-16	68-77-1	68-78-1	68-81-10	68-81-11B	69-27-2	69-32-1	70-139-1	70-146-6	178-8
Orthomean....	4.0283	2.8440	3.6279	3.4578	3.1286	3.9941	3.1980	2.9521	3.4541	3.3105	3.4537
Paramean....	1.3972	1.4746	0.7619	0.9099	0.6491	1.2248	0.7935	0.7453	0.6510	0.7777	0.5744

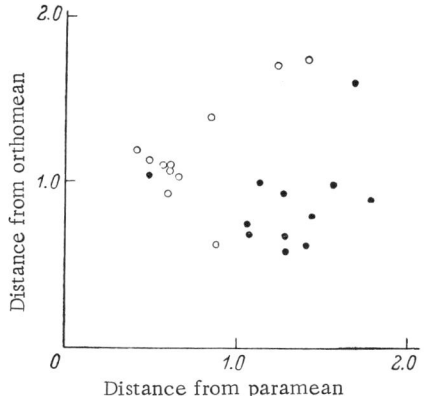

Fig. 5. Distance coefficients for amphibolites, relative to the orthomean and the paramean, normalized by using the total variance of all amphibolites.

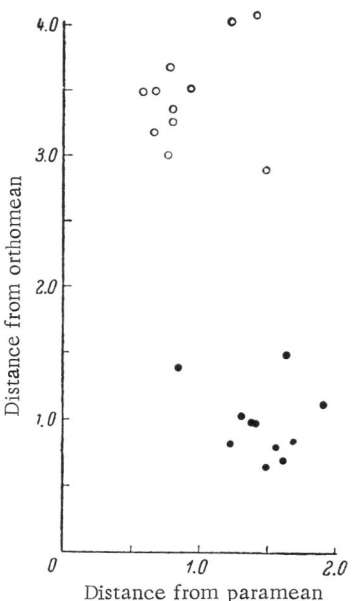

Fig. 6. Distance coefficients for amphibolites, relative to the orthomean and the paramean, normalized by using the variance of each group.

The values of d from Table 3 are shown in Fig. 5 and it is apparent that the two clusters are separated, with the exception of one ortho-amphibolite (No. 69-28-5), which falls in the para-amphibolite field: this rock is in fact a para-amphibolite, originally classified incorrectly (see [Shaw and Kudo, 1965, p. 426]).

Since the separation of the clusters is not perhaps as convincing as might be desired, an alternative approach may be used. This consists again of calculating d-values from the orthomean and paramean, but normalizing the variables separately by using the variance in each group. This leads to the distance coefficients shown in Table 4 and plotted in Fig. 6. Separation of the clusters is much more apparent, and is in fact exaggerated to such a degree that sample No. 69-28-5 falls within the ortho-amphibolite group: this approach is clearly less objective than the former one.

Figures 5 and 6 provide a convenient method of comparing two-cluster analyses. It should, however, be remembered that distance coefficients are vectors, possessing direction as well as length: these figures do not illustrate the directive properties and are in any case only applicable to two-cluster situations.

An Application to Some Silicic Gneisses

The next example illustrates a somewhat less clear-cut situation which brings out some of the limitations of the use of the taxonomic distance measure.

The Apsley paragneiss is a leucocratic plagioclase—quartz—microcline gneiss which contains small amounts of biotite and muscovite. It outcrops extensively in Chandos and adjacent townships in southeastern Ontario and has been recognized [Shaw, 1958; Simony, 1960] as an important stratigraphic unit in this part of the Grenville province of the Canadian Precambrian Shield. High-grade regional metamorphism and extensive tectonic movement have obliterated almost all evidence of the original nature of the sedimentary precursor of this gneissic unit. Similar units have been studied in detail by Engel and Engel [1953] in the Adirondack Mountains of New York State, and there, as at Apsley, it has been difficult to choose between the two possible sedimentary parent rocks, i.e., graywacke or silicic volcanic tuff (dacite). A study of the geochemistry of the Apsley paragneiss has been made by Simony [1960] and two of his analyses are presented in Table 5 for comparison with some representative graywackes, tonalitic granites, and shale.

Table 5. Composition of Some Paragneisses and Similar Rocks
(% and ppm)

Element	A	B	C	D	E	F	G
Ca	1.34	1.99	3.26	4.48	2.53	2.3	2.21
Na	2.94	2.69	1.05	1.33	2.84	3.2	0.96
K	1.60	2.13	1.59	1.51	2.52	2.3	2.66
Ga	13	12	10	18	17	20	19
Cr	5	1	170	180	22	5	90
V	139	21	62	62	88	50	130
Ni	1	1	67	53	15	5	68
Co	5	9	20	27	7	10	19
Zr	167	98	530	610	140	100	160
Sr	213	81	170	350	440	1500	300
Ba	353	750	470	450	420	500	580

Note. A. Apsley sodic paragneiss, average [Simony, 1960]. B. Apsley sodic paragneiss: sample 620712-1 (unpublished). C. Average graywacke [trace elements, Weber and Middleton, 1961, Table 15; major elements are average of all sandstones, Middleton, 1960, Table 3]. D. Normanskill graywacke [trace elements, Weber and Middleton, 1961, Table 12; major elements, unpublished]. E. Average high-Ca granite [Turekian and Wedepohl, 1961, Table 2]. F. Hornblende mica dacite, Lassen Peak, California [Nockolds and Allen, 1953, Table 2, No. 8]. G. Average shale [Turekian and Wedepohl, 1961, Table 2].

Table 6. Distance Coefficients for Quartzo-Feldspathic Rocks

	A	B	C	D	E	F	G
A		1.3204	1.5630	1.8311	0.8737	1.3332	1.4760
B			1.5464	1.8322	1.0667	1.2650	1.4601
C				0.8026	1.4693	1.8079	1.3067
D					1.5722	1.7635	1.3932
E						0.8137	1.0639
F							1.4248
G							

Note. A, B, C, D, E, F, G, same as in Table 5.

Distance coefficients have been computed from this data and are presented in Table 6. Clearly, it would have been preferable to represent each group by a substantial number of individual analyses rather than by an average and a single individual; however, this particular approach was chosen not only to minimize the computation but also to see whether any profitable conclusions could be drawn from the relationships between the cluster midpoints or averages, supplemented by a single randomly chosen individual item.

The situation is undoubtedly complicated by the ill-defined nature of the classification categories. The coherence of a cluster of points representing "high-Ca granites" would be very much a function of how one defines such a granite, and the same is perhaps even more applicable to graywacke and sandstone.

Interpretations may therefore be of dubious significance. Nevertheless, it may be noticed that the largest d-values are between Apsley points and graywackes, and between graywackes and silicic igneous rocks. Secondly, the Apsley paragneisses show a closer resemblance with the silicic igneous rocks than with the graywackes. Thirdly, both the graywackes and the silicic igneous rocks show some resemblance to shale.

It must be remembered that several major elements, in particular Si, Al, Fe, and Mg have been omitted from the computations. Their inclusion would almost certainly have increased the distance between shale and any other point, but would not otherwise have made any important differences, the concentration levels of these elements being similar in the other samples.

It may tentatively be concluded that this analysis suggests a stronger likelihood that the precursor of the Apsley paragneiss was a dacite than that it was a graywacke.

Conclusions

The taxonomic distance measure has a certain value for making qualitative statements concerning geochemical samples which come from multivariate populations. Its chief advantages are computational simplicity and a rather simple physical interpretation which makes it suitable for cautious probing of topological relationships in polydimensional space. Its application, however, to any but the most straightforward classificatory situations is unlikely to be profitable, although it might be worthwhile giving some thought to the use of the angles subtended by the distance vectors at each of the coordinate axes as a means to seek a "preferred orientation" in appropriate circumstances. In many cases, the best conclusion to a study using the distance coefficient would be confirmation of the desirability to proceed with discriminant function analyses.

REFERENCES

Engel, A. E. J., and Engel, C. G., "Grenville series in Northwest Adirondack Mountains, New York. Parts I and II," Geol. Soc. Am. Bull., Vol. 64, pp. 1013-1097 (1953).

Harbaugh, J. W., Balgol Programs for Calculation of Distance Coefficients and Correlation Coefficients Using an IBM 7090 Computer, State Geol. Survey, University of Kansas, Special Distribution Pub. 9 (1964).

Middleton, G. V., "Chemical composition of sandstones," Geol. Soc. Am. Bull., Vol. 71, pp. 1011-1026 (1960).

Nockolds, S. R., and Allen, R., "The geochemistry of some igneous rock series," Geochim. Cosmochim. Acta, Vol. 4, pp. 105-142 (1953).

Shaw, D. M., Chandos Township, Ontario Department of Mines Geol. Report No. 11, pp. 1-28 (1962).

Shaw, D. M., and Kudo, A. M., "A test of the discriminant function in the amphibolite problem," Mineral. Mag. (Tilley Vol.), Vol. 34, No. 268, pp. 423-435 (1965).

Simony, P. S., Origin of the Apsley Paragneiss, M. Sc. Thesis, McMaster University (unpublished) (1960).

Sokal, R. R., and Sneath, P. H., Principles of Numerical Taxonomy. W. H. Freeman and Co., pp. 1-359 (1963).

Turekian, K. K., and Wedepohl, K. H., "Distribution of the elements in some major units of the earth's crust," Geol. Soc. Am. Bull., Vol. 72, pp. 175-192 (1961).

Weber, J. N., and Middleton, G. V., "Geochemistry of turbidites of the Normanskill and Charny Formations, Parts I and II," Geochim. Cosmochim. Acta, Vol. 22, Nos. 2-4, pp. 200-288 (1961).

III. *Paragenetic Analysis*

DISTRIBUTION OF PERCENTAGE VALUES

A. V. Faas

Laboratory of Mathematical Geology
V. A. Steklov Mathematical Institute
Academy of Sciences of the USSR
Leningrad, USSR

O. V. Sarmanov

V. A. Steklov Mathematical Institute
Academy of Sciences of the USSR
Moscow, USSR

Conditions under which observations of percentage values may be used to reconstruct distribution laws of absolute values inaccessible to direct observation have been outlined. In addition, a number of examples are cited in which distribution laws of percentage values are found from known laws of absolute values. These examples are important in testing the hypothesis of absolute-value distribution obtained from observations of percentage values.

Let us take the random vector $\mathbf{X} = (x_1, x_2, \ldots, x_n)$ with nonnegative components that have been transformed to the vector $\mathbf{Y} = (y_1, y_2, \ldots, y_n)$ by the formula

$$y_i = \frac{x_i}{\sum_{j=1}^{n} x_j}, \quad i = 1, 2, \ldots, n.$$

It is assumed that the distribution function of the "percentage" vector $\mathbf{Y}$ is known, and it is necessary to obtain information concerning the original vector $\mathbf{X}$. Reconstruction of the distribution of the vector $\mathbf{X}$ from a known (and available to observation) distribution of the vector $\mathbf{Y}$ is called the percentage problem.

This problem arises, for example, when we investigate the distribution of a component x_1 in the crystal lattice. We assume that each investigated sample of material is homogeneous, but different samples have different compositions. We are interested in the distribution of x_1 within the unit cell. This value is not accessible to direct observation, since the number of cells varying from sample to sample is unknown. However, the percentage of the component x_1 within the cell is equal to the percentage of x_1 in the sample, and this latter value is known.

Another possible application might be the acquisition of information concerning absolute economic indices when the corresponding indices are given only in percentages.

Chayes [1960], Sarmanov [1961], and Vistelius [Sarmanov and Vistelius, 1959] have studied questions connected with the correlation of x_i values. We shall consider the distribution functions of these values.

There is commonly some basis for assuming that one of the components of **X** is statistically independent of the others. This assumption sometimes becomes sufficient for explaining the distribution forms of the other components. If we admit that the law of the independent variable is known, then, as we shall show below, it is possible to reconstruct fully the distribution functions of the other variables from observations of percentage values of y_i.

Hypotheses of relatively *a priori* distributions of x_i are found in geochemical constructions. For example, the hypothesis of Barth, to the effect that the number of oxygens in the unit cell of a crystal lattice is either constant or is regularly distributed (in a narrow segment) is of this kind. It may be commonly assumed that the appearance of 0, 1, 2, . . . ions of a given component conforms to the Poisson distribution.

A number of similar assumptions admit of statistical testing by percentage data.

In addition to the solution of percentage problems such as those described above, when it is necessary to draw conclusions from **Y** relative to **X**, we may examine some "direct" problems of joint distribution of percentage values of y_i from a known joint density $f(x_1, x_2, \ldots, x_n)$ of components of the vector **X**. For some densities, frequently encountered among positive values, corresponding densities of percentage values will be derived. These formulas may be used for goodness-of-fit criteria in testing hypotheses concerning the form of joint density of initial values.

Distribution and Density Functions of Percentage Values

It is assumed that the random values x_i (i = 1, 2, . . . , n) are nonnegative and have distribution densities f_i. We shall use F_i to designate the corresponding distribution functions, and f_i will be the joint density of x_i. For the percentage values of y_i of the density, the distribution function, and the joint density, we shall use the symbols p_i, P_i, and p, respectively.

Then (for more definite application we shall examine y_1 instead of y_i)

$$P_1(z) = P(y_1 < z) = P\left(\frac{x_1}{\sum_{i=1}^{n} x_i} < z\right) = P\left[x_1 < \frac{z}{1-z}(x_2 + \ldots + x_n)\right] =$$

$$= \int_0^\infty dx_n \ldots \int_0^\infty dx_2 \int_0^{\frac{z}{1-z}(x_2 + \ldots + x_n)} f(x_1, x_2, \ldots, x_n) \, dx_1. \tag{1}$$

In case the values of x_i are independent, we obtain

$$P_1(z) = \int_0^\infty \ldots \int_0^\infty f_2(x_2) \ldots f_n(x_n) F_1\left[\frac{z}{1-z}(x_2 + \ldots + x_n)\right] dx_2, \ldots, dx_n. \tag{2}$$

Differentiation of (1) and (2) with respect to z in the general case and in the case of independent values of x_i gives the density of y_1:

$$p_1(z) = \frac{1}{(1-z)^2} \int_0^\infty \ldots \int_0^\infty (x_2 + \ldots + x_n) f\left[\frac{z}{1-z}(x_2 + \ldots + x_n), x_2, \ldots, x_n\right] dx_2 \ldots dx_n, \tag{3}$$

$$p_1(z) = \frac{1}{(1-z)^2} \int_0^\infty \cdots \int_0^\infty (x_2 + \cdots + x_n) f_1\left[\frac{z}{1-z}(x_2 + \cdots + x_n)\right] f_2(x_2) \cdots f_n(x_n) \, dx_2 \cdots dx_n. \tag{4}$$

Since $\frac{x_i}{x_j} = \frac{y_i}{y_j}$, then, by observing the ratio of percentage values $y_{i/j} = \frac{y_i}{y_j}$, we thereby observe the ratio of initial values.

For the value $y_{i/j}$ it is easy to obtain the following expression of its distribution and density functions (for simplification we set i = 1, j = 2):

$$P_{1/2}(z) = P(y_{1/2} < z) = P\left(\frac{x_1}{x_2} < z\right) = \int_0^\infty dx_2 \int_0^{x_2 z} f(x_1, x_2) \, dx_1; \tag{5}$$

if x_1 and x_2 are independent, then

$$P_{1/2}(z) = \int_0^\infty F_1(x_2 z) f_2(x_2) \, dx_2. \tag{6}$$

Differentiation of (5) and (6) with respect to z gives the density of $y_{1/2}$:

$$p_{1/2}(z) = \int_0^\infty x_2 f(x_2 z, x_2) \, dx_2, \tag{7}$$

$$p_{1/2}(z) = \int_0^\infty x_2 f_1(x_2 z) f_2(x_2) \, dx_2. \tag{8}$$

It may be noted that instead of the density of the values of $y_{1/2} = \frac{x_1}{x_2}$ it is sometimes more convenient to use the density of the values of $y_{\frac{1}{1+2}} = \frac{x_1}{x_1 + x_2}$ it is clear that

$$p_{\frac{1}{1+2}}(z) = \frac{1}{(1-z)^2} \int_0^\infty x_2 f\left(\frac{z}{1-z} x_2, x_2\right) dx_2, \tag{9}$$

where $0 \leq z \leq 1$.

We may note further that the densities $p_{1/2}(z)$ and $p_{\frac{1}{1+2}}(z)$ are related by the following:

$$p_{1/2}(z) = \frac{1}{(1+z)^2} p_{\frac{1}{1+2}}\left(\frac{z}{1+z}\right), \tag{9'}$$

where $0 \leq z < \infty$.

Reconstruction of Laws of x_i

Let us assume that among the values of x_i one, such as x_1, is independent of the rest.

1. If, among the values of $y_{i/1}$ and $y_{1/j}$, we find one that is log-normal, such as $y_{1/k}$, then both values x_1 and x_k will also have log-normal distribution.

Actually, $\ln y_{1/k} = \ln x_1 + (-\ln x_k)$, where $\ln x_1$ and $(-\ln x_k)$ are independent, and their sum has normal distribution. Therefore, according to Cramer's theorem, $\ln x_1$ and $(-\ln x_k)$ have normal distribution, but x_i and x_k are log-normal.

In this case, a single hypothesis concerning the independence of two components of the vector **X** permits us to reconstruct the distribution laws from percentage values of $y_{1/k}$.

2. Let us say that the discrete value of x is distributed according to the Poisson law with the parameter $\lambda > 0$, when

$$P(x = ar + b) = \frac{e^{-\lambda}\lambda^r}{r!}, \qquad (10)$$

in which r = 0, 1, 2, ..., and a and b are arbitrary constants.

If $\ln y_{1/k}$ has Poisson distribution, then $\ln x_1$ and $-\ln x_k$ also have Poisson distribution if x_1 and x_k are independent (proof of this follows from the theorem of Raikov [1937]).

3. Let us now assume that x_1 is independent of other values of x_i, and that its distribution function is known. We shall demonstrate that, in this case, it is possible to reconstruct the distribution functions of all the components $x_2, x_3, \ldots, x_n$, from observations of percentage values y_i.

In the equation $\ln y_{1/2} = \ln x_1 - \ln x_2$, let us set $\ln y_{1/2} = u$, $\ln x_1 = v_1$, and $-\ln x_2 = v_2$; let $\varphi_u(t)$, $\varphi_{v_1}(t)$, $\varphi_{v_2}(t)$ be characteristic functions of the random values u, v_1, and v_2. By virtue of the independence of v_1 and v_2, $\varphi_u(t) = \varphi_{v_1}(t)\varphi_{v_2}(t)$, whence

$$\varphi_{v_2}(t) = \frac{\varphi_u(t)}{\varphi_{v_1}(t)}, \qquad (11)$$

but $\varphi_u(t)$ is known to us from observations of $y_{1/2}$, and $\varphi_{v_1}(t)$ from our assumption.

By knowing the characteristic function $v_2 = -\ln x_2$, it is possible to find the distribution function of $\ln x_2$ (by means of Fourier reconversion), and, lastly, the distribution function of the value $x_2 = e^{-v_2}$ itself.

This method, possible in principle, of reconstructing the distribution law of x_1, is of fundamental theoretical significance. Other possible methods of reconstructing the distribution of x_i may be noted.

4. Let us assume (in addition) that we know the densities $p_{1/2}(z)$ and $f_1(z)$ have derivatives of zero. Let $f_1(0) = a_0, \ldots, f_1^{(k)}(0) = a_k$, $p_{1/2}(0) = b_0, \ldots, p_{1/2}^{(k)}(0) = b_k$. After setting z = 0 in (8), we find $b_0 = a_0 \int_0^\infty x_2 f_2(x_2) dx_2 = a_0 m_1$, where m_1 is the first moment of x_2. Hence, $m_1 = b_0/a_0$.

By differentiating (8) k times with respect to z and then substituting z = 0, we obtain similarly

$$m_{k+1} = \frac{b_k}{a_k}, \quad k = 0, 1, 2, \ldots \qquad (12)$$

Thus, with the assumptions we have made, it is possible to find the first k + 1 moments of the value x_2, which in many cases is sufficient for complete characterization of the distribution function.

5. Let us assume that x_1 has a truncated distribution, so that $0 \le x_1 \le a$, in which $f_1(uv) = \varphi_1(u)\varphi_2(v)$, where φ_1 and φ_2 are known functions; thus,

$$f_1(x_2 z) = \begin{cases} \varphi_1(x_2)\varphi_2(z), & \text{when } x_2 z \le a, \\ 0, & \text{when } x_2 z > a. \end{cases}$$

Such, for example, is the density $f_1(z) = l z^\alpha$. The hypothesis of Barth is an example of a similar assumption with $\alpha = 0$.

In this case, equation (8) assumes the form:

$$p_{1/2}(z) = \varphi_2(z) \int_0^{a/z} x f_2(x) \varphi_1(x) \, dx.$$

Setting $a/z = u$, let us designate $\varphi_3(u) = \dfrac{p_{1/2}\left(\frac{a}{u}\right)}{\varphi_2\left(\frac{a}{u}\right)}$, then

$$\varphi_3(u) = \int_0^u x f_2(x) \varphi_1(x) \, dx$$

or

$$\varphi_3'(u) = u f_2(u) \varphi_1(u);$$

since $p_{1/2}(z)$, $\varphi_1(z)$, $\varphi_2(z)$, and consequently $\varphi_3(u)$ are also known to us, then

$$f_2(u) = \frac{\varphi_3'(u)}{u \varphi_1(u)}. \tag{13}$$

Let us examine the case, important in geochemical problems, of the particular case of x_1 distribution uniform at $[0, a]$. Here,

$$p_{1/2}(z) = \frac{1}{a} \int_0^{a/z} x f_2(x) \, dx, \tag{14}$$

and for the density $f_2(z)$ we obtain the following expression:

$$f_2(z) = \frac{a}{z} \left\{ p_{1/2}\left(\frac{a}{z}\right) \right\}_z'. \tag{15}$$

From (14) it follows that $p_{1/2}(z)$ declines steadily with increase in z; therefore, if x_1 has uniform distribution in the interval $(0, a)$, for the entire series of values $y_{1/2}, y_{1/3}, \ldots, y_{1/n}$ we should then have a series of steadily decreasing densities. This should be discernible to the eye when examining percentage data.

6. Let us assume, finally, that x_1 has gamma distribution:

$$f_1(z) = \frac{z^\alpha e^{-z}}{\Gamma(\alpha + 1)} \quad (\alpha > -1),$$

and in this case, equation (8) assumes the form

$$\Gamma(\alpha + 1) p_{1/2}(z) z^{-\alpha} = \int_z^\infty x^{\alpha+1} f_2(x) e^{-zx} dx. \tag{16}$$

From the known "image" (left part), we find the corresponding "original" in the Laplace transform: $\varphi(x) \doteq \Gamma(\alpha + 1) p_{1/2}(z) z^{-\alpha}$. Hence, we finally obtain

$$f_2(x) = \varphi(x) x^{-\alpha - 1}. \tag{17}$$

This ends the brief survey of methods of reconstructing the distribution law of x_i for a known distribution law of x_1 when x_1 is independent of x_i.

Examples of Distributions of Percentage Values

We shall now assume that we know the joint density of the components of the vector **X**, and we shall denote a series of distributions of the corresponding percentage values.

1. Let x_i be uniformly distributed in the space bounded by the hyperplane $\sum_{i=1}^{n} \beta_i x_i = \alpha$ and the coordinates of the planes, so that the variation range of the variables is given by the system of inequalities

$$\left. \begin{array}{l} \sum_{i=1}^{n} \beta_i x_i \leq \alpha, \\ x_i \geq 0, \ i = 1, 2, \ldots, n, \end{array} \right\} \tag{18}$$

where all $\beta_i > 0$.

Then, from (1), we find the following expression for the distribution function of y_1:

$$P_1(z) = 1 - \frac{(1-z)^{n-1} \prod_{i=2}^{n} \beta_i}{\prod_{i=2}^{n} [z \beta_1 + (1-z) \beta_i]}; \tag{19}$$

equations (6) and (7) give, respectively,

$$P_{1/2}(z) = 1 - \frac{\beta_2}{\beta_1 z + \beta_2}, \quad p_{1/2}(z) = \frac{\beta_1 \beta_2}{(\beta_1 z + \beta_2)^2}. \tag{20}$$

The hypothesis concerning initial density may be tested by the relation

$$\frac{p_{i}(0)}{j} \frac{p_{j}(0)}{i} = 1, \tag{20'}$$

which is a simple consequence of (20). We may note that when n = 2 and $\beta_1 = \beta_2$, the percentage value of y_1 has uniform distribution in the segment (0, 1).

Note 1. If the values of x_i are uniformly distributed in the segments $0 \leq x_i \leq \beta_i$ (i = 1, 2, ..., n) and are independent, the percentage values of y_i then have a more complex distribution. For example, when n = 2 for the density of y_i, from equation (4) we obtain the following expression:

$$p_1(z) = \begin{cases} \dfrac{\beta_2}{2\beta_1 (1-z)^2}, & \text{for } 0 \leq z \leq \dfrac{\beta_1}{\beta_1 + \beta_2}, \\ \dfrac{\beta_1}{2\beta_2 z^2}, & \text{for } \dfrac{\beta_1}{\beta_1 + \beta_2} \leq z \leq 1, \end{cases} \tag{20''}$$

Note 2. Another initial density gives distributions completely identical to (19) and (20); this corresponds to independent values of x_i having the exponential distribution

$$f(x_1, x_2, \ldots, x_n) = e^{-\sum_{i=1}^{n} \beta_i x_i} \prod_{i=1}^{n} \beta_i.$$

2. Let us assume that the positive values of x_i have the joint density

$$f(x_1, x_2, \ldots, x_n) = \frac{\prod_{i=1}^{n} x_i^{\alpha_i} e^{-x_i}}{\prod_{i=1}^{n} \Gamma(\alpha_i + 1)}, \quad \alpha_i > -1, \tag{21}$$

i.e., that the components of the vector **X** have gamma distribution. Then

$$p_1(z) = z^{p-1}(1-z)^{q-1} \frac{\Gamma(p+q)}{\Gamma(p)\Gamma(q)}, \tag{22}$$

where $p = \alpha_1 + 1$, $q = \sum_{i=2}^{n} \alpha_i + n - 1$, i.e., y_1 has beta distribution.

3. Let us examine further the density $p_{\frac{1}{1+2}}(z)$ of values $\frac{x_1}{x_1 + x_2}$, when x_1 and x_2 are dependent and have joint gamma distribution. Then

$$f(x_1, x_2) = \frac{(x_1 x_2)^{\alpha} e^{-\frac{x_1 + x_2}{1-R}}}{(1-R)^{1+\alpha}\Gamma(\alpha+1)} \sum_{k=0}^{\infty} \frac{\left[\frac{x_1 x_2 R}{(1-R)^2}\right]^k}{k!\,\Gamma(\alpha+k+1)}, \tag{23}$$

where $0 \leq R < 1$; $\alpha > -1$; R is the correlation coefficient between x_1 and x_2 [Sarmanov, 1960].

By using equation (9) we obtain

$$p_{1/2}(z) = \frac{(1-R)^{\alpha+1}}{\Gamma(\alpha+1)} z^{\alpha}(1-z)^{\alpha} \sum_{k=0}^{\infty} \frac{\Gamma(2\alpha+2k+2)}{\Gamma(\alpha+k+1)} \frac{[Rz(1-z)]^k}{k!} = \frac{(1-R)^{\alpha+1} z^{\alpha}(1-z)^{\alpha}}{\beta(\alpha+1, \alpha+1)[1-4Rz(1-z)]^{\alpha+\frac{3}{2}}}. \tag{24}$$

When R = 0 we obtain beta distribution, but when R = α = 0, the distribution is uniform at (0, 1).

4. Let us assume that $f(x_1, x_2, \ldots, x_n)$ is n-dimensional normal density

$$f(x_1, x_2, \ldots, x_n) = \frac{1}{(2\pi)^{\frac{n}{2}} \sqrt{\Delta}} e^{-\frac{\sum_{i,j=1}^{n} \Delta_{ij}(\xi_i - \bar{\xi}_i)(\xi_j - \bar{\xi}_j)}{2\Delta}}, \tag{25}$$

where **A** = $\{\alpha_{ij}\}$ is the covariance matrix, and Δ_{ij} is the cofactor of the determinant Δ.

In view of the nonnegative character of the values examined by us, equation (25) may be only an approximate expression for density of the component when the average values $\bar{\xi}_i$ are positive and considerably greater than the standard deviation.

For the given case

$$P_1(z) = \Phi\left(\frac{z-a}{\sqrt{\alpha z^2 + \beta z + \gamma}}\right), \tag{26}$$

where

$$a=\frac{\xi_1}{\sum_{i=1}^{n}\xi_i}; \quad \alpha=\frac{\sum_{i,j}a_{ij}}{\left(\sum_{i=1}^{n}\xi_i\right)^2}; \quad \beta=\frac{-2\sum_{i=1}^{n}a_{1i}}{\left(\sum_{i=1}^{n}\xi_i\right)^2}; \quad \gamma=\frac{a_{11}}{\left(\sum_{i=1}^{n}\xi_i\right)^2};$$

$$\Phi(u)=\frac{1}{\sqrt{2\pi}}\int_{-\infty}^{u}e^{-\frac{z^2}{2}}dz.$$

Thus, in the case of n-dimensional normal correlation, y_1 has the distribution of S. N. Bernstein [1926] with a dispersion function of the second degree.

It is possible that biased distributions of percentage values, observed in geochemistry, may arise at times merely through percentage calculations of normal distribution values.

REFERENCES

Bernstein, S. N., "Sur les courbes des distributions probabilités," Mathem. Z., Vol. 24, pp. 199-211 (1926).

Chayes, F., "On correlation between variables of constant sum," J. Geophys. Res., Vol. 65, No. 12, pp. 4185-4193 (1960).

Raikov, D. A., "Decomposition of the Poisson law," Dokl. Akad. Nauk SSSR, Vol. 14, No. 1, pp. 9-12 (1937).

Sarmanov, O. V., "Pseudonormal correlation and various inferences concerning it," Dokl. Akad. Nauk SSSR, Vol. 132, No. 2, pp. 299-302 (1960).

Sarmanov, O. V., "False correlation between random values," Tr. Mat. Inst. Akad. Nauk SSSR, Vol. 64, pp. 174-184 (1961).

Sarmanov, O. V., and Vistelius, A. B., "Correlation between percentage values," Dokl. Akad. Nauk SSSR, Vol. 126, No. 1, pp. 22-25 (1959).

NIOBIUM IN METAGRANITES OF THE POLAR URALS

E. P. Kalinin, M. V. Fishman and B. A. Goldin

Geological Institute
Komi Branch of the Academy of Sciences of the USSR
Syktyvkar, USSR

This paper is devoted to a study of the correlation between modal and chemical compositions of granitoid rocks and their niobium content. For studying the correlation, linear correlation analysis has been employed, using conclusions from the Sarmanov—Vistelius theorem to avoid false correlation because of percentage calculations.

Data for this study were gathered by the authors during 1960-61 when making geologic investigations in the Polar Urals. Chemical analyses of the rocks and some niobium analyses were made by L. P. Pavlov, L. A. Raznitsyn, and É. Babushkina in the chemical laboratory of the Komi Branch of the Academy of Sciences of the USSR. Most of the niobium analyses were made by M. I. Zaboeva at the Department of Analytical Chemistry of the Ural State University at Sverdlovsk. Quantitative mineral analyses were made on an Andin integrating stage, and roundness of grains was determined by the method of Chayes [1963]. The standard error of the mean for the principal three rock-forming minerals does not exceed 2% in any of the investigated samples.

Mathematical processing of the data was carried out by means of the Mathematical Geology Group at the Leningrad Division of the V. A. Steklov Mathematical Institute of the Academy of Sciences of the USSR. The authors express their thanks to all the above-mentioned workers. They also thank A. B. Vistelius for guidance and consultation in the mathematical methods used in the present paper, and A. Shumilova and A. V. Faas for aid in making the computations.

In the region of the Polar Urals, within the Lyapinskii anticlinorium, magmatic rocks are widespread, both acidic and basic series, among which several magmatic complexes of different ages may be distinguished. One group consists of pre-Ordovician Baikalian rocks, the second of post-Ordovician Caledonian-Variscan rocks [Fishman, 1962, 1963].

Secondary injections of magma and post-magmatic solutions along what were essentially tectonic zones for long periods gave rise to complex multiphase masses widely exhibiting hybridism and superimposed metamorphism.

Granitic rocks represent the most widespread type among the intrusive masses. Existing data indicate that two phases of different ages may be definitely distinguished, and possibly more, among these intrusions. Normal and subalkalic granitic rocks are distinguished on the basis of chemical and mineral composition [Fishman, 1962].

The subalkalic granitic rocks, as also the normal, are found in groups of different ages. Among these, both magmatic and metasomatic varieties (metagranites)* may be recognized.

The metasomatic granitic rocks were formed by transformation of magmatic granitic rocks (normal and subalkalic) by means of high-temperature (pneumatolytic-hydrothermal) post-magmatic solutions [Beus, 1962].

Metasomatically transformed subalkalic granites, widespread in the southern part of the Pechora Urals, are localized in discontinuous bands or irregularly formed zones ranging in size from a few square meters to 1500 m^2 (headwaters of the Southern Perchuk-Elya River) extending along tectonic zones of schist development, common among magmatic rocks at the contacts with basic rocks.

Metasomatically transformed normal granites are most widely distributed in the central part of the Polar Urals. They either form small isolated tabular bodies with exposed areas of 0.5-1.0 km^2 among quartz—mica and chlorite schists, hyperbasites, and gabbroic rocks, or they represent endogene contact zones of large granitic masses at contacts with basic rocks. The age of the metagranites of both groups is computed to be chiefly 350-390 million years [Fishman, 1962].

Metasomatically transformed granites (metagranites) are distinguished by variable grain size, structure, and composition. They contain porphyroblastic microcline and quartz and are rich in fine, scaly green and colorless muscovite or lepidomelane. Dark violet fluorite and, more rarely, calcite, barite, and tourmaline are found in disseminated grains or thin veinlets along shear planes or in isolated patches. An enrichment of accessory minerals is observed: magnetite, pyrite, and, more rarely, fergusonite, columbite, cyrtolite, thorite, samarskite, allanite, and still other minerals. At least two varieties of potassium—sodium feldspars may be distinguished in the metagranites: magmatic and metasomatic.

Magmatic feldspars are microcline and sodium orthoclase, up to 5 mm across, argillized and replaced by albite and quartz. Different types of perthite are commonly present (string, film, lamellar, patch, veined), replacement perthite predominating. The second type of potassium—sodium feldspar forms irregular grains of microcline up to 1.0 mm across. Perthite is comparatively rare. Albite of perthitic intergrowths commonly exhibits polysynthetic twinning. The microcline contains relics of plagioclase. In paragenetic association with this potassium feldspar, cyrtolite is frequently observed, more rarely magnetite, ilmenite, and lepidomelane ($Ng \approx Nm = 1.701 \pm 0.002$).

Among the plagioclases, at least three genetic varieties have been recognized: one magmatic and at least two metasomatic.

The magmatic plagioclase — albite and albite—oligoclase — occurs in irregularly formed grains up to 0.3 mm across, sericitized and replaced by epidote, zoisite, garnet, microcline, and quartz.

The first metasomatic variety is represented by checkerboard albite (Ab_{94-97}), forming pseudomorphs after potassium feldspar of the variety described above. The second meta-

---

* Metagranites are metasomatically altered granites, in many places approaching apogranites in aspect because of the intensity of the metasomatic processes [Beus et al., 1962].

Table 1. Chemical Analysis (in wt.%) and Quantitative Mineral

O	SiO₂ / Si/O	TiO₂ / Ti/O	Al₂O₃ / Al/O	Fe₂O₃ / Fe³⁺/O	FeO / Fe²⁺/O	MnO / Mn/O	MgO / Mg/O	CaO / Ca/O	Na₂O / Na/O	K₂O / K/O	Nb₂O₅ / Nb/O
49.13	76.2	0.05	13.82	0.33	0.09	—	0.14	0.90	3.29	4.23	0.008
	724	0.6	149	4.7	1.4	—	1.6	13.0	49.6	71.4	0.12
48.56	73.92	0.04	14.91	0.70	0.48	0.01	0.06	0.26	4.17	3.82	0.048
	711	0.4	163	10.1	7.8	0.2	0.8	3.9	63.6	65.3	0.72
48.82	74.46	0.02	14.66	0.73	0.50	0.01	0.08	0.28	4.34	3.97	0.007
	712	0.2	159	10.4	8.0	0.2	1.0	4.1	66.0	67.6	0.10
48.55	74.04	0.03	14.39	0.97	0.42	0.01	0.14	0.22	4.28	4.26	0.008
	713	0.4	157	14.0	6.8	0.2	1.6	3.3	65.5	72.9	0.12
49.01	75.98	0.19	13.03	0.33	0.71	0.02	0.19	0.94	3.73	4.41	0.004
	724	2.4	141	4.7	11.2	0.4	2.2	13.7	56.5	74.7	0.06
49.19	77.64	0.07	12.18	0.69	0.01	0.02	0.16	0.30	3.66	4.51	0.010
	737	0.8	131	9.8	0.2	0.4	2.0	4.5	55.3	76.0	0.14
48.50	75.90	0.21	12.22	0.71	0.13	0.01	0.14	0.44	3.25	5.63	0.006
	731	2.7	134	10.3	2.0	0.2	1.6	6.4	49.7	96.3	0.08
49.00	77.06	0.14	11.84	0.71	0.23	0.02	0.11	0.44	4.45	4.23	0.014
	734	1.6	128	10.2	3.7	0.4	1.4	6.3	67.3	71.6	0.20
49.39	78.72	0.12	11.44	0.37	0.11	0.02	Tr.	0.66	3.67	4.40	0.010
	744	1.4	123	5.3	1.8	0.4	—	9.5	55.1	73.9	0.14
48.85	74.20	0.06	14.04	0.23	0.03	0.02	0.52	1.60	4.69	4.14	0.008
	709	0.8	153	3.3	0.6	0.4	6.3	23.3	71.2	69.4	0.12
49.23	76.48	0.08	13.14	0.31	0.03	0.03	0.64	0.62	3.68	4.77	0.018
	726	1.0	142	4.5	0.6	0.6	7.9	8.9	55.5	80.4	0.26
47.92	69.28	0.21	16.05	0.17	0.46	0.08	1.34	3.10	3.98	4.45	0.012
	675	2.7	178	2.5	7.5	1.3	16.9	46.1	61.6	77.0	0.17
49.16	77.52	0.16	11.76	0.95	0.33	0.03	0.05	0.76	3.39	4.72	0.014
	736	2.0	127	13.4	5.3	0.6	0.8	11.2	51.3	79.7	0.20
49.09	76.64	—	12.41	1.12	0.26	—	0.15	0.52	4.14	4.36	0.014
	729	—	134	16.1	4.0	—	1.8	7.5	62.5	73.7	0.20
49.19	78.22	0.16	11.35	0.98	0.20	0.01	0.05	0.36	3.63	4.25	0.012
	743	2.0	122	14.0	3.3	0.2	0.6	5.3	54.7	71.8	0.16
49.01	76.82	0.16	12.27	0.90	0.17	0.03	0.06	0.47	4.01	4.35	0.010
	732	2.0	133	12.9	2.7	0.6	0.8	6.9	60.8	71.6	0.14
48.86	76.76	0.14	12.07	0.47	0.30	0.01	0.59	Tr.	4.22	4.13	0.013
	734	1.8	131	6.8	4.7	0.2	7.4	—	64.1	74.7	0.18
49.15	78.06	0.13	11.15	0.89	0.26	0.04	0.19	0.80	3.55	4.12	0.015
	742	1.6	120	12.6	4.0	0.6	2.4	11.6	53.5	69.6	0.22
49.32	78.30	0.18	10.70	1.76	0.18	Tr.	0.30	0.48	3.93	3.90	0.006
	741	22	115	24.9	3.0	—	3.6	6.9	59.2	65.7	0.08
49.02	75.90	0.04	13.36	0.84	0.11	0.02	0.22	0.56	3.97	4.43	0.017
	723	0.6	144	12.0	1.8	0.4	2.7	8.2	60.4	75.1	0.24
48.81	76.66	0.16	11.78	1.02	0.48	0.08	0.06	0.28	3.66	5.16	0.010
	733	2.0	128	14.8	7.7	1.2	0.8	4.1	55.7	87.7	0.14
49.03	75.60	0.18	13.16	0.89	0.45	0.03	Tr.	0.72	4.31	4.70	0.011
	720	2.2	142	12.8	7.3	0.6	—	10.4	65.3	79.5	0.16
48.92	76.10	0.07	12.88	1.01	0.07	0.01	0.36	0.60	4.56	2.84	0.008
	726	0.8	145	14.5	1.2	0.2	4.5	8.8	69.1	48.2	0.12
48.75	76.32	0.14	11.17	1.45	0.72	0.05	0.29	1.42	3.77	3.90	0.007
	731	1.6	121	20.7	11.6	0.8	3.5	20.7	57.4	66.5	0.10
48.80	76.12	0.09	12.34	0.86	0.36	0.02	0.03	0.20	4.75	4.39	0.019
	728	1.0	134	12.3	5.7	0.4	0.4	2.9	72.1	74.6	0.27
49.26	77.74	0.16	11.68	0.91	0.48	0.03	Tr.	0.62	3.66	4.58	0.008
	737	2.0	126	13.0	7.7	0.6	—	8.9	55.2	77.1	0.12
49.85	80.46	0.16	10.69	0.69	0.14	Tr.	0.02	0.46	3.12	4.13	0.010
	754	2.0	114	9.6	2.2	—	0.2	6.6	46.5	68.8	0.14
49.28	76.00	0.03	13.51	1.09	0.03	0.03	0.08	0.46	4.68	4.15	0.010
	720	0.4	145	15.4	0.6	0.6	1.0	6.7	70.4	69.8	0.12

Note. The ratios element/O and mineral/O have been multiplied by 1000. Here and elsewhere in this article: Q) opaque accessory minerals (zircon, sphene, allanite, apatite); Op) opaque ore minerals (magnetite, ilmenite, ilmenorutile);

Composition (in vol.%) of Metagranites of the Polar Urals

Q / Q/O	K	Pl Pl/O	Bi Bi/O	Mu Mu/O	Sm Sm/O	Ac Ac/O	Op Op/O	F F/O	Coarseness of grains acc. to Chayes	Total measured area, mm²	No. of thin sections
39	23	27	—	10.9	—	—	0.02	0.1	80	400	1
794	468	550	—	224	—	—	0.41	2.0			
42	16	29	—	13	—	—	—	—	46	1200	3
865	329	597	—	268	—	—	—	—			
39	19	25	—	15	0.3	0.4	0.3	1	59	780	2
799	389	512	—	307	6.1	8.2	6.1	20.4			
32	22	34	—	10	1.5	—	—	0.5	43	1206	3
659	453	700	—	206	30.8	—	—	10.3			
37	29	23	2	7	1	1	—	—	49	1086	3
755	592	469	40.8	143	20.4	20.4	—	—			
47	27	24	2	—	—	—	—	—	37	1380	4
955	549	488	40.7	—	—	—	—	—			
40	38	20	—	—	1.3	0.3	0.4	—	45	1140	3
825	784	412	—	—	26.8	6.2	8.2	—			
47	35	15.5	—	2	—	—	0.5	0.5	73	800	2
959	714	316	—	40.8	—	—	10.2	10.2			
31	50	17	—	1	—	—	1	—	65	970	2
628	1012	344	—	20.2	—	—	20.2	—			
53	24	18	—	2	1	—	1	1	50	1200	3
1085	491	368	—	40.9	20.4	—	20.4	20.4			
34	37	23	1	2	0.9	0.5	1.5	0.1	35	2450	6
691	751	467	20.3	40.6	18.3	10.2	30.5	2.0			
36	37	15	4	4	1	2	1	—	60	740	2
751	772	313	83.5	83.5	20.8	41.7	20.8	—			
40	37	18	—	3.5	—	—	0.5	1	55	830	2
814	753	366	—	71.2	—	—	10.2	20.3			
31	20	42	—	6.5	—	—	0.5	—	50	1020	3
631	407	856	—	132.0	—	—	10.2	—			
38	48	8	2	1.5	—	—	2	0.5	57	960	2
773	976	163	40.6	30.5	—	—	40.7	10.2			
44	31	21	1	0.5	—	0.5	2	—	62	820	2
898	633	428	20.4	10.2	—	10.2	40.8	—			
42	23	31	1	2	—	1	—	—	42	1460	3
860	471	634	20.4	40.9	—	20.4	—	—			
56	21	12	0.5	6	—	1	1	2.5	48	1180	3
1139	427	244	10.2	122.0	—	20.3	20.3	50.9			3
52	32	14	—	2	—	—	—	—	51	985	2
1054	648	284	—	40.6	—	—	—	—			
42	41	12	1.5	—	—	—	2	1.5	56	1080	3
857	836	245	30.6	—	—	—	40.8	30.6			
36	40	23	0.4	—	0.4	—	0.2	—	48	1200	3
738	820	471	8.2	—	8.2	—	4.1	—			
46	34	14	—	4	0.5	0.6	0.3	0.6	52	1150	3
938	693	286	—	81.6	10.2	12.2	6.1	12.2			
35	31	31	—	2	—	0.2	0.4	0.4	38	1300	4
715	634	634	—	40.9	—	4.1	8.2	8.2			
19	41	38	0.7	0.6	—	—	0.2	0.5	53	1100	3
390	841	779	14.4	12.3	—	—	4.1	10.3			
40	28	30	—	1.3	—	—	0.5	0.2	61	1230	3
820	574	615	—	26.6	—	—	10.2	4.1			
52	26	16	0.4	3.6	—	0.4	0.6	1.0	58	1080	3
1056	528	325	8.1	73.1	—	8.1	12.2	20.3			

quartz; K) essentially potassium feldspar; Pl) plagioclase; Bi) biotite; Mu) muscovite; Sm) secondary minerals; Ac) non-F) fluorite.

Table 2. Matrix of Normal (Paired) Correlation Coefficients
(of Niobium and Minerals)

	Nb/O	F/O	Op/O	Ac/O	Mu/O	Bi/O	Pl/O	K/O
Q/O	+0.066	**+0.416**	+0.070	+0.110	−0.028	−0.088	*−0.595*	*−0.354*
K/O	+0.103	−0.135	**+0.462**	−0.066	*−0.674*	+0.284	*−0.441*	
Pl/O	−0.103	*−0.387*	*−0.506*	−0.201	+0.286	−0.269		
Bi/O	+0.087	−0.129	+0.341	**+0.654**	−0.203			
Mu/O	−0.262	+0.106	*−0.384*	+0.078				
Ac/O	+0.048	+0.039	+0.089					
Op/O	**+0.489**	+0.293						
F/O	+0.276							

Note. Levels of 99% significance, r > 0.478; of 95% significance, r > 0.374. Here and elsewhere in the article: figures in boldface indicate significant positive correlation coefficients; figures in italics give significant negative correlation coefficients.

somatic variety is lathlike albite (Ab_{92-96}) of elongated prismatic form. It generally forms simple twins, rarely polysynthetic twins.

The accessory minerals cyrtolite, magnetite (groundmass), garnet (almandite, rarely grossularite), hastingsite (Ng = 1.682 ± 0.002, Np = 1.671 ± 0.002, 2V = −80°, c∧Z = 10°), and fluorite are associated with the metasomatic albite. Accessory fergusonite and samarskite are also found in this association.

Analysis of the evolution of composition and properties of the rock-forming minerals during the formation of the metagranites, and also of the interrelations among them, permits discrimination of several stages of the metasomatic process, sequentially giving way to the next (sometimes with some superimposition). The earliest stage is dominantly of potassium metasomatism, giving way to the stages of sodium or iron—sodium metasomatism, and, finally, of fluorine—silicon metasomatism [Apel'tsin, 1966].

Correlation of Niobium Concentration

with the Mineral Composition of Metagranites

Table 1 shows 28 chemical and 26 quantitative mineral analyses of metagranites examined in this paper.

Linear correlation analysis has been used to study correlations among the elements indicated in the table [Vistelius, 1956, 1962]. To avoid false correlation because of percentage calculations, conclusions from the Vistelius—Sarmanov theorem have been used [Sarmanov and Vistelius, 1959].

Since the quantitative relations of elements in rocks are such that approximately 160 ions of oxygen are required for 100 cations (the "standard cell" of Barth [1962]), we may consider the oxygen content to be a constant value. The weight percentage of oxygen, taken as a constant value, is the total content of oxygen in the sample. Correlation was examined for particular values, obtained by division of the indicated values and multiplied by 1000 for convenience in computation. These ratios are shown in Table 1 in the column of each analysis, in the second row, and they are designated in the table and elsewhere in this paper by element/O (such as Si/O and Nb/O) or by mineral/O (such as Q/O and Pl/O).

The matrix of the normal correlation coefficients is given in Table 2. To examine the degree of correlation between each component and all members of the paragenetic association, we furnish the multiple correlation coefficients (R):

$$\frac{Q}{O} \ldots \ldots 0.998, \quad \frac{Ac}{O} \ldots \ldots 0.809,$$

$$\frac{K}{O} \ldots \ldots 0.999, \quad \frac{Ot}{O} \ldots \ldots 0.883,$$

$$\frac{Pl}{O} \ldots \ldots 0.998, \quad \frac{E}{O} \ldots \ldots 0.829,$$

$$\frac{Bi}{O} \ldots \ldots 0.906, \quad \frac{Nb}{O} \ldots \ldots 0.630.$$

$$\frac{Mu}{O} \ldots \ldots 0.993,$$

The normal correlation coefficients define the strength of the correlation between pairs of values without isolating the force of correlation of the other components [Vistelius, 1956]. Partial correlation coefficients were therefore computed for all combinations of minerals and niobium (Table 3), except for quartz, biotite, and the nonopaque accessories, which gave such small values with niobium that it made no sense to perform laborious computations. The partial correlation coefficients define the strength of the correlation between two computations at fixed contents of the remaining components, i.e., they indicate the correlation between individual pairs of components in a "pure form."

From Table 2 it may be seen that a positive correlation between concentrations of niobium and opaque accessories exists ($r = +0.489$ at 99% significance). This fact confirms the general view that in granitic rocks niobium is normally disseminated in titanium-bearing, ferruginous, and titanium minerals, such as magnetite, ilmenite, hematite, and others. The presence of niobium as an impurity in the ore (opaque) accessory minerals is due to the similarity of crystallochemical properties of titanium, iron, and other elements. This permits them to substitute isomorphously for each other and sometimes to form minerals of complex composition. During the late-magmatic stage of mineralization, niobium also enters substantially into the accessory minerals (columbite, fergusonite, and others). These conclusions, made possible by the analysis of a positive correlation between niobium and the opaque accessory minerals, find support in petrographic and mineralogic study of the granites.

Analysis of the partial correlation coefficients (Table 3) in still greater measure confirms the existence of a positive correlation between niobium and the opaque accessories ($\rho = +0.52$, $r = +0.49$ at 99% significance). The remaining correlations are insignificant.

At first glance it seems strange that no meaningful correlation exists between niobium and biotite, since it is known from the literature that 83% of niobium is associated with biotite in granites and only 17% with the accessory minerals [Lyakhovich, 1964]. We should note a peculiarity of ore-bearing granitic rocks, emphasized by Lyakhovich, that "in a number of cases the rocks become impoverished in just that rock-forming mineral that is the chief concentrator of the ore element" [Lyakhovich, 1964, p. 79]. In the present example, the granitic rocks are leucocratic varieties. This is a consequence of the superimposition of post-magmatic processes, which caused partial or complete replacement of biotite, extraction of niobium from this mineral, and its redeposition with the formation of independent accessory minerals.

It is known that albitized varieties of granitic rocks are richest in niobium. And, although a correlation between niobium mineralization and albitization in these rocks has been established beyond doubt, the insignificant partial and normal correlation coefficients between niobium and albite indicate that there is no direct dependence of degree of albitization on niobium concentration. This is true also of fluorite, which is frequently found in paragenetic association with niobium-bearing accessory minerals.

Table 3. Successive Elimination of the Effect of Components of the Investigated System on the Value of the Correlation Coefficient

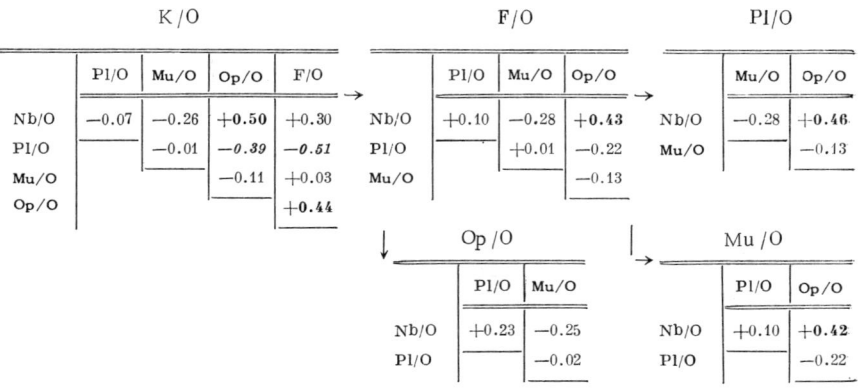

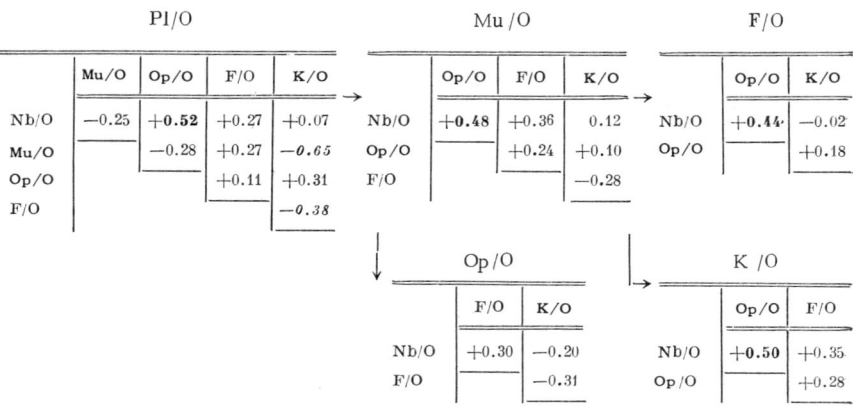

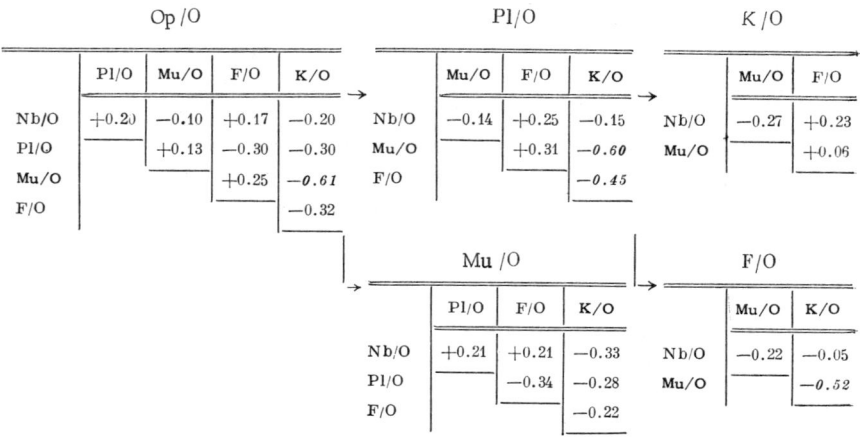

Table 3 (continued)

Mu/O

	Pl/O	Op/O	F/O	K/O
Nb/O	−0.03	+0.43	+0.33	−0.16
Pl/O		−0.46	−0.44	−0.36
Op/O			+0.37	+0.31
F/O				−0.10

K/O

	Pl/O	Op/O	F/O
Nb/O	−0.10	+0.51	+0.33
Pl/O		−0.40	−0.52
Op/O			+0.43

Op/O

	Pl/O	F/O
Nb/O	+0.13	+0.14
Pl/O		−0.42

F/O

	Pl/O	Mu/O	Op/O	K/O
Nb/O	+0.02	−0.31	+0.46	+0.14
Pl/O		+0.36	−0.46	−0.54
Mu/O			−0.43	−0.66
Op/O				+0.53

Mu/O

	Pl/O	Op/O	K/O
Nb/O	+0.15	+0.38	−0.09
Pl/O		−0.37	−0.43
Op/O			0.37

Op/O

	Pl/O	K/O
Nb/O	+0.35	−0.28
Pl/O		−0.34

Thus, as a result of our investigations, it is maintained that the chemical and crystallo-chemical properties of niobium are significant factors in its distribution, dissemination, and concentration. Because of these properties, niobium does not enter the crystal structure of the chief rock-forming minerals (potassium feldspar, quartz, and others), which explains the complete absence of correlation or the low significance (90% and lower) of correlation between niobium and these components. This is confirmed as well by the fact that the multiple correlation coefficient for niobium is lower for these minerals than for the others.

As a result of crystallization of the rock-forming minerals, making up 80–95% of the entire rock, the concentration of niobium is substantially increased in the later stages, sometimes the final stage, of the mineral-forming process. In this case, niobium may become an isomorphous admixture in accessory minerals of iron and titanium or it may form independent accessory minerals, as indicated by the positive correlation between niobium and the opaque accessories.

Analysis of the interrelations of the rock-forming and accessory minerals among themselves (from an examination of the matrices of multiple and partial correlation coefficients, Tables 2 and 3) permits us to establish the following types of correlation.

1. Significant negative correlation between Q and Pl, Q and K, K and Mu, Pl and K, Pl and F, F and K, Pl and Op (t > 2.78, 99% level of significance) and also between K and Op (t > 2.06, 95% level of significance). The correlation between these component-antagonists is almost unaffected by their correlation with other components, or the effect merely strengthens the negative correlation, as between Op and Pl. In the investigated rocks, therefore, these tend to have no connection.

2. Significant positive correlation between Bi and Ac, K and Op (t > 2.78, 99% level of significance) and also between Q and F, Op and F (t > 2.06, 95% level of significance).

These actual correlations reflect a definite sequence in the process of mineral formation.

The negative correlations indicate replacement (displacement) of one rock component by another, later in time. The actual interrelations observed by us in thin sections may be expressed by very definite quantitative values. Thus, late formation of quartz and fluorite after feldspar may be observed, formation of albite and muscovite after potassium feldspar, etc.

The positive correlations indicate that one component accompanies another, occasioning its large concentration. Thus, fluoritization of granites is invariably accompanied by development of quartz and ore minerals. All nonopaque accessory minerals (zircon, sphene, allanite, apatite) occur chiefly in biotite segregations.

Table 4. Matrix of Normal (Paired) Correlation Coefficients
(Niobium and Metals)

	Nb/O	K/O	Na/O	Ca/O	Mg/O	Mn/O	Fe²⁺/O	Fe³⁺/O	Al/O	Ti/O
Si/O	+0.057	+0.018	−0.488	−0.628	−0.633	−0.370	−0.263	+0.368	−0.941	+0.172
Ti/O	−0.169	+0.419	−0.439	+0.300	+0.184	+0.379	+0.262	−0.015	−0.355	
Al/O	−0.074	−0.081	+0.460	+0.423	+0.483	+0.168	+0.140	−0.467		
Fe³⁺/O	−0.135	−0.221	−0.075	−0.341	−0.394	−0.101	+0.153			
Fe²⁺/O	−0.238	−0.072	+0.048	+0.194	−0.002	+0.355				
Mn/O	+0.175	+0.334	+0.029	+0.543	+0.397					
Mg/O	+0.145	−0.027	+0.153	+0.734						
Ca/O	−0.059	−0.014	−0.009							
Na/O	+0.118	−0.397								
K/O	+0.146									

Note. 99% level of significance, r > 0.463; 95% level of significance, r > 0.361.

Table 5. Successive Elimination of the Effect of Components of the Investigated System on the Values of Partial Correlation Coefficients

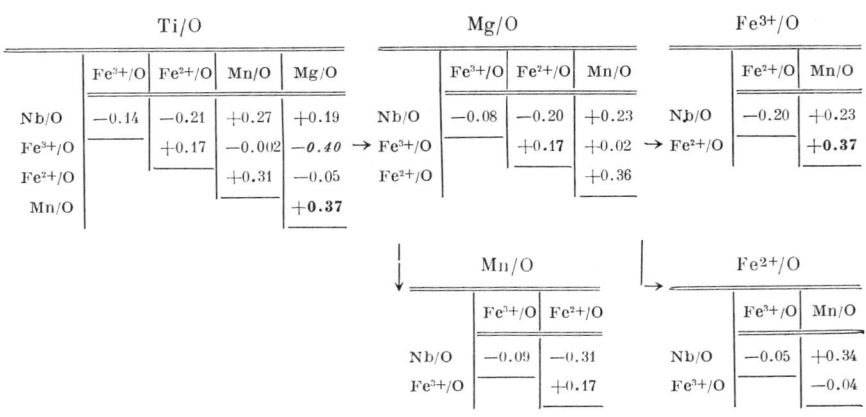

Table 5 (continued)

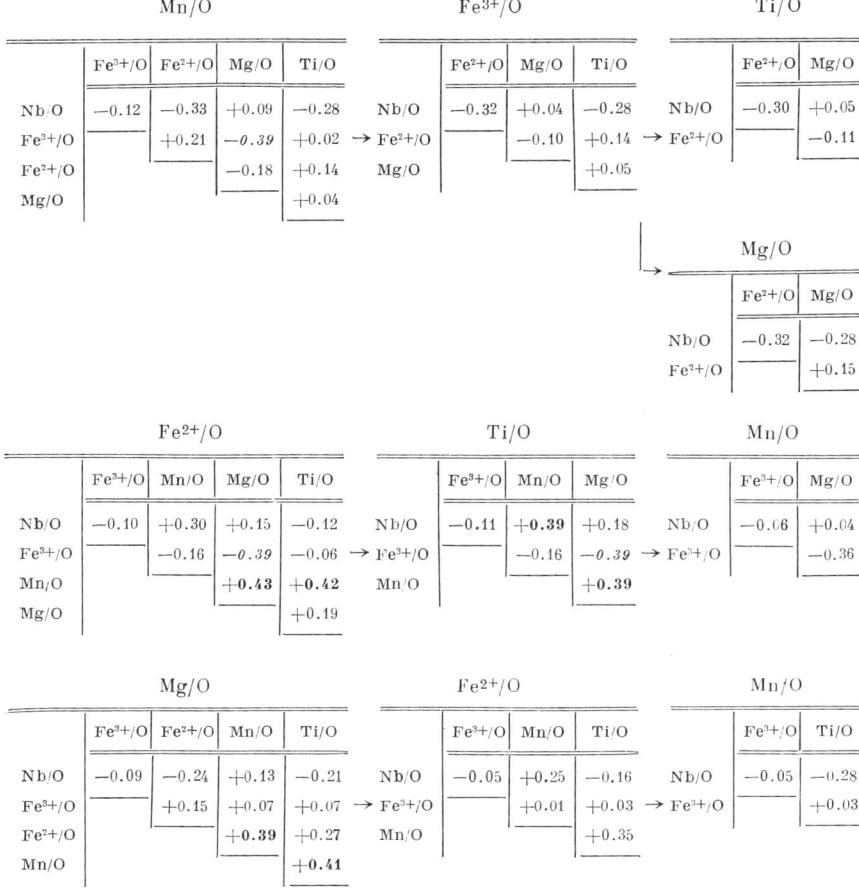

Correlation of Niobium Concentration with the Chemical Composition of Metagranites

A. E. Fersman proposed a geochemical star diagram for niobium in order to represent the aspects of isomorphism of this element more completely. From his scheme it may be seen that the strongest correlation of niobium is found with Ti, Ta, and Zr, moderate correlation with Mn, Fe, W, Mo, Th, and Sn, and the weakest (Nb is rarest) with rare earths and hafnium.

We have studied these correlations and have given quantitative evaluations for limited material. From the elements listed above we have data at our disposal only for Ti, Fe, and Mn.

The matrix of normal correlation coefficients (Table 4) has not yielded a single significant correlation between niobium and the rock-forming elements of rocks.

The multiple correlation coefficients are given below. A study of these shows that low values characterize those elements (Mn, Ti, Fe^{2+}) that have the closest geochemical affinity for niobium and that participate with it in heterovalent isomorphism. These elements, not having sufficiently strong normal correlation with all members of the paragenetic association, are characterized by significant correlation with each other, as may be seen from an analysis of the partial correlation coefficients. The multiple correlation coefficients (R) are: Si/O... 0.999, Ti/O... 0.810, Al/O... 0.998, Fe^{3+}/O... 0.950, Fe^{2+}/O... 0.868, Mn/O... 0.751, Mg/O... 0.936, Ca/O... 0.988, Na/O... 0.970, K/O... 0.960, Nb/O... 0.623.

Table 6. Successive Elimination of the Effect of Components of the Investigated System on the Values of Partial Correlation Coefficients

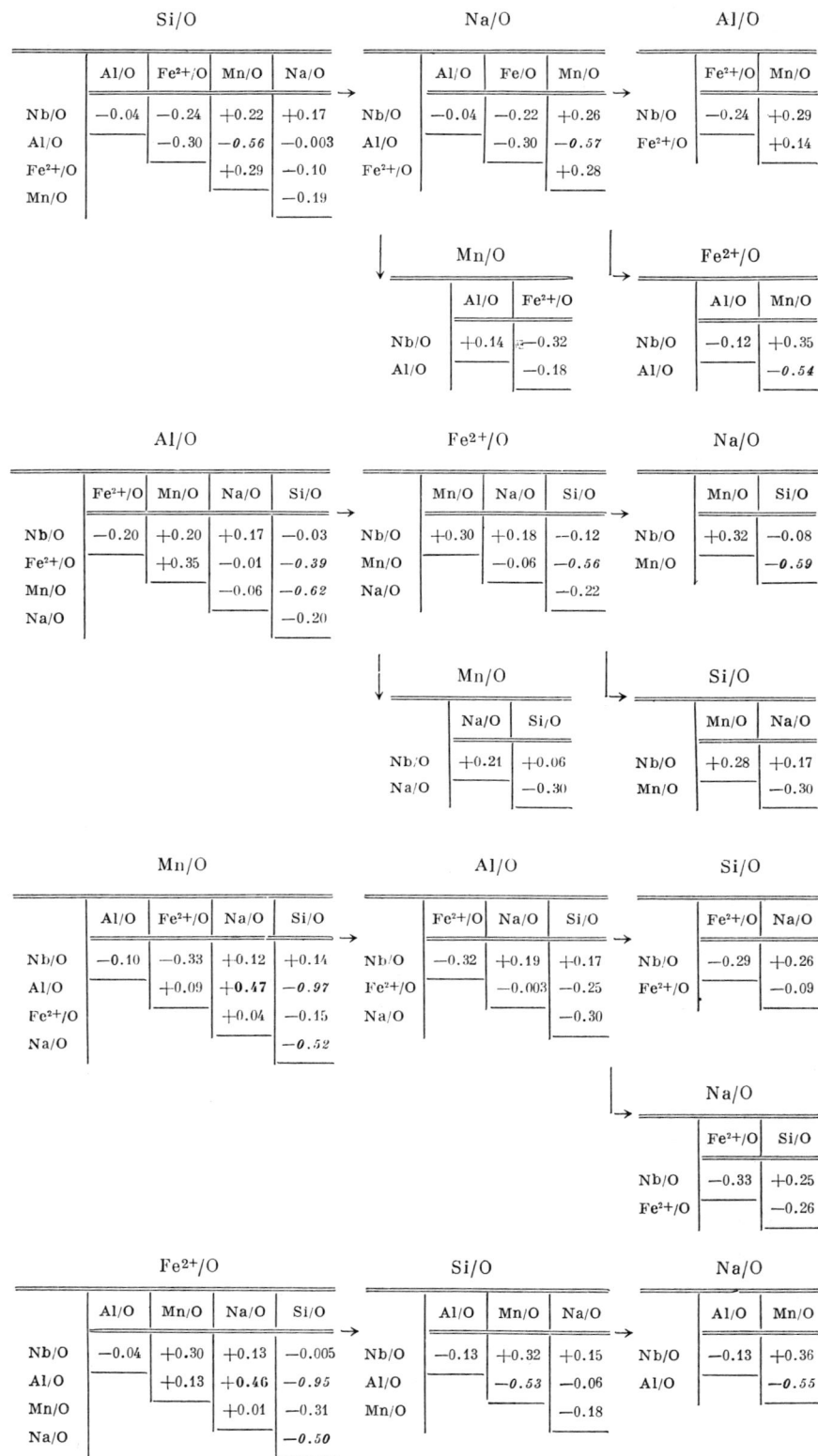

Table 6 (continued)

	Na/O					Fe²⁺/O				Mn/O	
	Al/O	Fe²⁺/O	Mn/O	Si/O		Al/O	Mn/O	Si/O		Al/O	Si/O
Nb/O	−0.15	−0.25	+0.18	+0.14	Nb/O	−0.13	+0.30	+0.08	Nb/O	−0.18	+0.21
Al/O		+0.13	+0.18	−0.93	Al/O		+0.14	−0.93	Al/O		−0.95
Fe²⁺/O			+0.36	−0.28	Mn/O			−0.35			
Mn/O				−0.41							

Table 5 shows the computed partial correlation coefficients of all possible combinations of the elements Ti, Fe^{2+}, Fe^{3+}, Mn, Mg, and Nb (those that have the greatest paired correlation with Nb) in order to obtain significant partial correlation coefficients.

In addition, to provide a more complete picture of the linear paragenetic associations of the principal rock-forming elements, the partial correlation coefficients have been computed for all possible combinations of elements with the more significant paired correlations (Table 6).

From an analysis of the data in Table 5, a positive correlation between niobium and manganese has been established ($\rho_{Nb/O, Mn/O, Ti/O, Fe^{2+}/O} = +0.39$, 95% level of significance). The correlation between niobium and ferrous iron and titanium is below the 95% level of significance.

Correlation between niobium and manganese may be explained in the following way. Niobium apparently replaces ferric iron in iron-bearing minerals. In ferric oxides the compensation of valence is probably effected in the following manner: $3Fe^{3+} \leftarrow Nb^{5+} + 2(Mn, Mg)^{2+}$. This scheme confirms the existence of a positive correlation between niobium and manganese (the more complete the replacement process, the greater the necessity of manganese occurring along with niobium). If the replacement process proceeds to the formation of independent accessory minerals of columbite type, this will more likely be manganocolumbite ($MnNb_2O_6$) than ferrocolumbite ($FeNb_2O_6$) in the investigated rocks, and this again is reflected in the nature of the correlation between Nb and Mn.

From analysis of the interrelations between the rock-forming elements (Tables 4-6), the following types of correlations appear.

1. Significant negative correlation between Al and Si, Mn and Si, Mg and Si, Ca and Si, Na and Si, Al and Mn, Al and Fe^{3+} (t > 2.763, 99% level of significance) and also between Fe^{3+} and Mg, K and Na, Na and Ti, Fe^{2+} and Si (t > 2.048, 95% level of significance).

2. Significant positive correlation between Ca and Mg, Ca and Mn, Mg and Al, Na and Al (t > 2.763, 99% level of significance) and also between K and Ti, Ca and Al, Mg and Mn, Mn and Ti, Mn and Fe^{2+}, and Fe^{3+} and Si (t > 2.048, 95% level of significance).

In most cases the partial correlation coefficients do not contradict the paired coefficients, but, on the contrary, the significance of the correlation increases (of Na and Al, for example, and Mn and Si, and others).

It is characteristic that silicon is antagonistic to most elements. This may be due to the fact that Si is the only cation of which an increase or decrease of content in the rock takes place by corresponding decrease or increase in total content of the salic components (quartz, potassium feldspar, albite) regardless of which salic mineral undergoes this change.

In the present example, this correlation may be explained more concretely, since the observed development of later quartz in the granites during the final stages of metasomatism takes place throughout the entire mass of the rock and proves to be chiefly the replacement of feldspars. This is confirmed by the strong negative correlation between Si and Al ($r_{Si/O,\ Al/O}= -0.94$, $\rho_{Si/O,\ Al/O \cdot Mn/O} = -0.97$), since it is clear that the principal bearers of aluminum are those very feldspars. Confirmation is also found in the previously directly established negative correlation between quartz and both plagioclase and potassium feldspar (Table 2).

Analysis of the remaining components also shows close interrelation between the elements and their principal mineral hosts. Thus, the antagonism between plagioclase and potassium feldspar is confirmed by the negative correlation between Na and K; the positive correlation between opaque accessories with potassium feldspar in part confirms the positive correlation between Ti and K, etc.

The analysis thus shows that the distribution of niobium is due mainly to peculiarities of the chemical and crystallochemical properties of the element. Because of these, niobium does not enter the crystal structure of the principal rock-forming minerals but is bound chiefly to the opaque accessory minerals. In these minerals it is observed either in isomorphous admixtures or in independent minerals (columbite and others). The correlation between niobium concentration in the investigated granites and the opaque accessory minerals and manganese is demonstrated by correlation analysis.

From analysis of the correlation between minerals and elements in the investigated group of rocks, it is possible to recognize a definite stagelike character in the mineral-forming process and to identify basic pairs of antagonists and associates among both elements and minerals. The two types of correlation (between elements and minerals) mutually complement and control one another.

The conclusions we have reached are in complete agreement with direct petrographic observations and mineral studies of the metagranites.

REFERENCES

Apel'tsin, F. R., "Evolution of the composition of micas as a criterion of rare-metal mineralization in granitized schists," Geol. Mestorozhd. Redkikh Elementov, No. 30, pp. 144-159 (1966).

Barth, T. F. W., Theoretical Petrology, John Wiley and Sons, New York (1952).

Beus, A. A., Severov, É. A., Sitnin, A. A., and Subbotin, K. D., Albitized and Greisenized Granites (Apogranites), Izd. Akad. Nauk SSSR, Moscow (1962).

Chayes, F., Petrographic Modal Analysis, Wiley, New York (1956).

Fishman, M. V., "New data on the age of the granitic rocks in the Polar Urals," Dokl. Akad. Nauk SSSR, Vol. 145, No. 2, pp. 400-403 (1962).

Fishman, M. V., "Some aspects of magmatism in the Polar Urals," Tr. Inst. Geol. Komi Fil. Akad. Nauk SSSR, No. 4, pp. 25-32 (1963).

Fishman, M. V., and Goldin, B. A., "Albitized and greisenized granites in the central part of the Polar Urals," Tr. Inst. Geol. Komi Fil. Akad. Nauk SSSR, No. 3, pp. 125-138 (1962).

Lyakhovich, V. V., "An aspect of ore-bearing granitic rocks," Geol. Rudn. Mestorozhd., Vol. 6, No. 2, p. 79 (1964).

Sarmanov, O. V., and Vistelius, A. B., "Correlation of percentage values," Dokl. Akad. Nauk SSSR, Vol. 126, No. 1, pp. 22-25 (1959).

Vistelius, A. B., "Problems of studying correlation in mineralogy and petrography," Zap. Vses. Mineralog. Obshchestva, Pt. 85, No. 1, pp. 58-75 (1956).

Vistelius, A. B., "Phosphorus in the granitoidal rocks of Tien Shan," Geokhimiya, No. 2, pp. 116-134 (1962).

LINEAR PARAGENETIC ASSOCIATIONS IN ROCKS AND MINERALS OF THE LADOGA FORMATION

Yu. V. Nagaitsev
Institute of the Earth's Crust
A. A. Zhdanov Leningrad State University
Leningrad, USSR

Yu. V. Podol'skii
Laboratory of Aerial Methods
Ministry of Geology of the USSR
Leningrad, USSR

Results of studying the strength of linear correlation between elements in minerals (biotite, garnet) and rocks (mica schist and gneiss) of the Ladoga Formation in Karelia are discussed. The correlations, as it will be shown, permit us to recognize some paragenetic associations of elements and to provide a statistical basis for previously stated conclusions that the contents of individual components (Fe", Mg, Ca, and other elements) in minerals depend on the composition of the rock and the grade of regional metamorphism.

Investigations have been made of biotite schists and gneisses of the Ladoga Formation, which is widely developed in the northern and northwestern Ladoga region. The following metamorphic zones have been distinguished on the basis of mineral associations in the rocks of the Ladoga region [Nagaitsev, 1965]:

1. Staurolite—andalusite (corresponding to the high-temperature part of the epidote—amphibolite facies and the low-temperature part of the amphibolite facies);

2. Sillimanite—muscovite (medium-temperature part of the amphibolite facies); and

3. Regional migmatization with: a) sillimanite—potassium feldspar (high-temperature part of the amphibolite facies); and, b) hypersthene (granulite facies) subzones.

Micaceous rocks of the staurolite—andalusite zone are microgranular and very fine-grained quartzitic and biotite—quartz schists, thinly interlaminated with biotite schists containing muscovite, staurolite, andalusite, and, rarely, garnet and cordierite. In some of the schists relics of sedimentary textures are preserved (pelitic, silty). Plagioclase, preserved in small quantities, is generally An_{9-20}. Garnet, which forms very small porphyroblasts up to 2-3 mm in diameter, has a pyrope—spessartite—almandite or spessartite—pyrope—almandite composition (with a pyrope content of 7.0-11.9%). The refractive index of the garnet ranges from 1.810 to 1.814. Biotite is found in very small brown plates with an intermediate index of 1.639-1.645 at an iron content of $f = 50.3$-52.6 mol.%.

For the sillimanite—muscovite zone, medium- and coarse-grained biotite—quartz and biotite schists are characteristic, commonly containing garnet, sillimanite, and muscovite. The plagioclase content is noticeably higher (up to 10-15%), and the type is oligoclase or oligoclase—andesine (An_{15-30}). Garnet, which forms large porphyroblasts up to 2-5 cm in diameter, has a spessartite—pyrope—almandite or pyrope—almandite composition (with a pyrope content of 14.3-17.4%). The refractive index of the garnet is 1.805-1.810. Biotite contains somewhat less iron (f = 46.0-50.5 mol.%) than the biotites of the staurolite—andalusite zone. The intermediate index of refraction is 1.640-1.644.

Rocks in the zone of regional migmatization have one feature in common: they consist of biotite and garnet—biotite gneisses, migmatized in lit-par-lit fashion. In the sillimanite—potassium feldspar subzone, they contain separate layers of gneiss with cordierite, sillimanite, and potassium feldspar, and, in the hypersthene subzone, they contain layers of gneiss with hypersthene, apparently replacing cummingtonite gneisses of the sillimanite—potassium feldspar subzone. The plagioclase in the gneisses is oligoclase—andesine or andesine (An_{30-40}). The garnet is pyrope—almandite with a pyrope content of 17.9-31.7%. Gneisses characterizing the hypersthene subzone always contain garnet with higher pyrope content than the sillimanite—cordierite gneisses of the sillimanite—potassium feldspar subzone. The refractive index of the garnet ranges from 1.788 to 1.805. The composition of biotite from gneisses in the zone of regional migmatization varies considerably, as shown by the intermediate refractive index (ranging from 1.639 to 1.666) and chemical analyses. The iron content of the biotite ranges from 42.4-60.5 mol.%. The TiO_2 content in the biotite is appreciably higher (averaging 3.5%) than in the biotites of the staurolite—andalusite (about 1.75%) and the sillimanite—muscovite (about 1.90%) zones.

From the data cited it thus becomes apparent that increase in regional metamorphism is accompanied by crystallization of garnets with higher pyrope content and biotite with higher titanium content. The dependence of these mineral compositions (especially biotite) on the chemistry of the rock, however, has not yet been sufficiently investigated. Only the relation of the calcium components in garnet to the CaO content in the rock has been noted [Nagaitsev, 1965].

To obtain a statistical basis for systematic relations in compositional changes of minerals, we have investigated the strength of linear correlation between the individual chemical elements of garnet and biotite and their contents in the rocks.

Investigation Data

The initial material used in making the investigation consists of complete silicate analyses of the micaceous rocks of the Ladoga Formation (27 analyses) and of the garnets and biotites in these rocks, taken from the different metamorphic zones [Lebedev et al., 1964]. The analyzed rocks and minerals had the following distribution among the metamorphic zones: 2 from the staurolite—andalusite zone, 5 from the sillimanite—muscovite zone, and 20 from the zone of regional migmatization.

Since rocks of similar composition were investigated, rhythmically bedded sand—clay rocks metamorphosed in different degree, one naturally expects gradual transitions in composition of the indicated minerals from one metamorphic zone to the next. In view of this, the analyses of the different metamorphic zones were combined into a single group when computing correlation coefficients. Such combination into groups leads to possible extinction of correlations characteristic of individual mineral associations, such as the fundamental correlation between Fe^{2+} in garnet—plagioclase—quartz rocks and Fe^{2+} in garnet. The most stable correlations, such as that between Ca in rocks and in garnet, are manifest even when combining rocks with different mineral associations into a single group.

Table 1

		Biotite										Garnet								Rock								
		K/O	Na/O	Ca/O	Mg/O	Mn/O	Fe^{2+}/O	Fe^{3+}/O	Al/O	Ti/O	Si/O	Ca/O	Mg/O	Mn/O	Fe^{2+}/O	Fe^{3+}/O	Al/O	Ti/O	Si/O	K/O	Na/O	Ca/O	Mg/O	Mn/O	Fe^{2+}/O	Fe^{3+}/O	Al/O	Ti/O
Rock	Si/O	0.10	0.05	−0.27	−0.33	0.20	−0.04	−0.15	−0.13	0.11	−0.45	−0.10	0.15	−0.13	0.24	0.06	0.51	−0.45	−0.50	−0.32	−0.34	−0.33	−0.61	−0.46	−0.54	−0.48	−0.90	−0.37
	Ti/O	−0.07	−0.13	0.22	0.23	0.09	0.24	0.16	0.00	0.03	0.14	0.17	0.06	−0.06	−0.05	−0.31	−0.32	0.01	0.29	0.13	−0.32	0.38	0.33	−0.25	0.11	0.54	0.15	
	Al/O	−0.16	−0.03	0.13	−0.07	−0.10	−0.11	0.24	0.18	−0.13	0.40	0.05	−0.24	0.27	−0.17	0.00	−0.48	0.55	0.42	0.32	0.41	0.22	0.42	0.44	0.34	0.26		
	Fe^{3+}/O	−0.12	0.27	0.22	0.33	−0.03	0.23	−0.06	−0.04	−0.07	0.43	0.29	−0.26	0.17	−0.12	−0.25	−0.23	−0.05	0.37	0.00	−0.24	0.36	0.30	0.11	0.34			
	Fe^{2+}/O	0.22	−0.10	−0.04	0.42	−0.16	0.33	−0.29	−0.19	0.14	0.26	0.15	0.21	−0.34	−0.29	−0.14	−0.05	0.06	0.21	−0.13	−0.19	0.24	0.35	0.73				
	Mn/O	0.25	−0.22	−0.02	0.28	−0.08	0.22	−0.35	−0.24	0.25	0.17	0.10	0.16	−0.22	−0.34	0.11	−0.06	0.42	0.12	−0.28	0.13	0.31	0.06					
	Mg/O	−0.25	0.19	0.29	0.47	−0.54	−0.36	0.21	0.39	−0.32	0.29	−0.44	−0.08	0.15	−0.15	−0.12	−0.30	−0.02	0.52	0.52	0.09	−0.31						
	Ca/O	0.22	−0.49	0.27	0.21	0.39	0.66	−0.17	−0.52	0.38	0.15	0.65	0.00	−0.26	0.06	−0.08	−0.31	0.34	0.01	−0.48	0.12							
	Na/O	−0.05	−0.19	0.40	0.00	−0.10	−0.05	0.17	−0.24	−0.18	0.06	−0.06	−0.12	0.13	0.06	0.29	−0.21	0.17	−0.03	0.13								
	K/O	0.00	0.06	−0.10	0.08	−0.47	−0.52	0.52	0.27	−0.09	0.10	−0.32	−0.04	0.13	0.03	−0.12	0.03	0.06	0.15									
Garnet	Si/O	−0.13	−0.04	0.22	0.33	−0.37	−0.27	−0.17	0.22	−0.12	0.15	−0.15	−0.04	0.11	−0.56	−0.32	−0.55	−0.02										
	Ti/O	0.08	−0.25	−0.21	−0.20	0.17	0.16	0.02	−0.07	−0.06	0.00	0.22	−0.04	−0.02	−0.25	0.34	−0.21											
	Al/O	0.48	0.01	−0.19	−0.14	0.08	0.10	−0.05	−0.14	0.34	−0.12	−0.20	0.35	−0.34	0.06	−0.18												
	Fe^{3+}/O	−0.11	−0.07	−0.26	−0.12	0.17	0.01	−0.10	0.01	−0.05	−0.05	0.06	0.14	−0.02	−0.05													
	Fe^{2+}/O	−0.39	0.42	−0.04	−0.45	0.25	0.15	0.32	0.21	−0.31	0.03	0.08	−0.50	0.39														
	Mn/O	−0.74	0.73	0.12	−0.42	0.10	−0.24	0.22	0.51	−0.73	0.33	−0.06	−0.73															
	Mg/O	0.66	−0.56	−0.01	0.42	−0.19	−0.08	−0.15	−0.35	0.56	−0.09	−0.40																
	Ca/O	0.00	−0.21	−0.01	0.03	0.33	0.54	0.01	−0.43	0.14	−0.04																	
Biotite	Si/O	−0.08	0.32	0.15	0.27	0.03	0.10	0.25	0.00	−0.01																		
	Ti/O	0.76	−0.63	−0.11	0.40	−0.05	0.16	0.04	−0.77																			
	Al/O	−0.41	0.43	0.05	−0.43	−0.02	0.10	−0.11																				
	Fe^{3+}/O	−0.25	0.19	0.05	0.04	−0.21	−0.40																					
	Fe^{2+}/O	0.29	−0.24	0.05	−0.08	0.50																						
	Mn/O	−0.05	0.03	0.08	−0.54																							
	Mg/O	0.23	−0.35	−0.20																								
	Ca/O	−0.25	−0.07																									
	Na/O	−0.66																										

Note. 99%-level of significance, $r \geq 0.55$; 95%-level of significance, $r \geq 0.39$.

In order to free ourselves from percentage calculations and, even more, to reconstruct the true relations between elements in keeping with the Vistelius—Sarmanov theorem [Sarmanov and Vistelius, 1959], correlation coefficients were computed between weight ratios Si/O, Ti/O, Al/O, etc., not between percentage contents of the chief oxides in the rocks and minerals. The analyses were processed and the correlation coefficients computed by the method developed by Vistelius [1948, 1956].

Apart from the strength of the correlations between elements within rocks, garnets, and biotites, in order to shed light on the dependence of mineral compositions on rock composition and grade of metamorphism, paired correlation coefficients were calculated for elements in rock and garnet, elements in rock and biotite, elements in garnet and biotite. The matrix of the normal (paired) correlation coefficients is shown in Table 1.

Interpretation of Correlation Strength

From an analysis of the computed correlation coefficients, definite correlations have been noted between the corresponding pairs of elements, and these permit us to draw several conclusions.

1. Correlations between Elements in a Rock

Silicon is found in antipathetic relations with Al, Fe^{3+}, Fe^{2+}, Mn, and Mg. Only positive correlation is observed between elements of the femic group, or the correlation coefficients are not significant. In addition, a positive correlation coefficient is observed between Al and Na, and Mg and K.

Such relations indicate that the amount of SiO_2 in the investigated rocks is defined primarily by the quartz content. The presence of paragenetic associations of elements in the femic group indicates that the contents of these elements depend on the content of dark minerals, chiefly biotite ($r_{Mg/O, K/O} = +0.52$) and garnet. Positive correlation between Na and Al is due to the presence of plagioclase with an appreciable content of the alkali molecule in the rock. The amount of plagioclase in the rock depends on the quartz content, as attested to by the strong negative correlation between Si and Al.

No documented correlation has been observed between Ca and Si or elements of the femic group. With K, however, the correlation coefficient for Ca is negative. This means that the more biotite and potassium feldspar in the rock, these being the chief potassium-bearing minerals in the investigated rocks, the less calcium. It is natural to suppose that most of the CaO is concentrated in plagioclase, and the content of this mineral in the rock is inversely proportional to the amount of biotite and potassium feldspar.

These conclusions are all widely accepted, and they are in complete agreement with chemical analyses and quantitative mineral counts in thin sections (Table 2).

Table 2

Mineral	90A	13-1	92	188B	163	26	13	168
Plagioclase, vol.%	57.4	49.2	39.1	35.8	32.2	21.7	21.3	13.9
Biotite + potassium feldspar, vol. %	19.7	23.2	41.6	21.3	30.6	37.9	69.7	83.3

Note. Numbers at heads of columns indicate analysis number.

Table 3

	1	2	3	4	5	6	7
T_i, at. quant.	21	23	23	19	51	48	51
K_{Mg}	0.429	0.466	0.463	0.482	0.471	0.429	0.431
K_{Mn}	0.0022	0.0	0.0019	0.0	0.0018	0.0	0.0015
K_{Fe}	0.523	0.492	0.490	0.482	0.437	0.481	0.489
K_{Fe}	0.783	0.796	0.766	0.770	0.760	0.739	0.842
K_{Mg}	0.092	0.130	0.147	0.185	0.231	0.244	0.142
K_{Mn}	0.125	0.074	0.086	0.045	0.009	0.017	0.016
K_{Ca}	0.055	0.097	0.033	0.068	0.047	0.071	0.177

	8	9	10	11	12	13	14
T_i, at. quant.	36	39	26	24	47	35	39
K_{Mg}	0.455	0.414	0.473	0.494	0.481	0.515	0.466
K_{Mn}	0.0018	0.0	0.0	0.0	0.0	0.0	0.0018
K_{Fe}	0.515	0.478	0.461	0.434	0.421	0.481	0.481
K_{Fe}	0.759	0.789	0.778	0.770	0.739	0.670	0.748
K_{Mg}	0.226	0.197	0.180	0.161	0.242	0.320	0.212
K_{Mn}	0.015	0.014	0.042	0.069	0.019	0.010	0.040
K_{Ca}	0.067	0.040	0.058	0.054	0.051	0.036	0.050

	15	16	17	18	19	20	21
T_i, at. quant.	41	50	42	36	39	54	38
K_{Mg}	0.522	0.488	0.462	0.437	0.495	0.469	0.447
K_{Mn}	0.0	0.0	0.0018	0.0018	0.0	0.0	0.0019
K_{Fe}	0.409	0.425	0.462	0.497	0.433	0.435	0.480
K_{Fe}	0.730	0.778	0.778	0.772	0.743	0.730	0.720
K_{Mg}	0.288	0.248	0.196	0.258	0.231	0.238	0.237
K_{Mn}	0.030	0.022	0.026	0.020	0.026	0.033	0.036
K_{Ca}	0.065	0.045	0.084	0.038	0.055	0.044	0.041

	22	23	24	25	26	27	Mean	Standard dev.
T_i, at. quant.	46	40	25	43	40	55	38	10.47
K_{Mg}	0.474	0.367	0.459	0.511	0.368	0.500	0.461	0.038
K_{Mn}	0.0	0.0036	0.0006	0.0005	0.0026	0.0	0.0009	0.0011
K_{Fe}	0.446	0.556	0.493	0.414	0.557	0.407	0.470	0.041
K_{Fe}	0.753	0.733	0.781	0.724	0.787	0.697	0.750	0.036
K_{Mg}	0.244	0.249	0.180	0.254	0.173	0.286	0.214	0.051
K_{Mn}	0.023	0.018	0.039	0.022	0.029	0.017	0.034	0.026
K_{Ca}	0.035	0.040	0.033	0.050	0.121	0.034	0.059	0.031

Note. Numbers at heads of columns indicate analysis number; upper group of atomic ratios belongs to biotite, the lower to garnet.

2. Correlation between Elements in Garnet

In studying the correlation between elements in garnet, special interest was found in the correlation of elements essential to the composition of the mineral. These elements in the garnet are Fe^{2+}, Mg, Ca, and Mn.

As the correlation coefficients show, Mg has a negative correlation with Ca and with the group of Fe^{2+} and Mn. The correlation coefficients between Ca and Fe^{2+} and between Ca and Mn are meaningless.

Such correlation between the bivalent minerals in the octahedral positions of the garnet lattice indicates that the increase in pyrope component takes place by decrease in content of both grossularite—andradite and spessartite—almandite components. However, increase in grossularite—andradite (or spessartite—almandite) component in garnets may occur only at the expense of the pyrope content, since Mg shows a negative correlation with Fe^{2+} and Mn and with Ca, but the correlation coefficients between Fe^{2+} and Ca and between Mn and Ca are meaningless. A positive correlation between Fe^{2+} and Mn reflects the fact that garnets of high iron content are characterized by high Mn content.

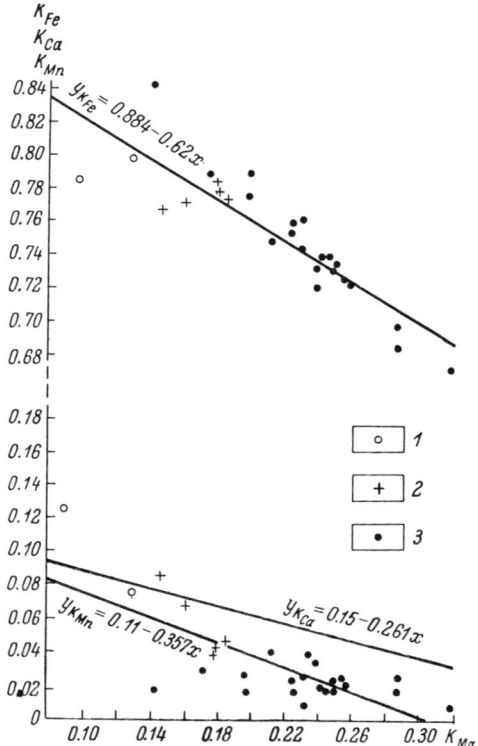

Fig. 1. Relations of the contents of Fe, Ca, and Mn to Mg content, in garnet rocks of the Ladoga Formation from different metamorphic zones. 1) Staurolite—andalusite metamorphic zone; 2) sillimanite—muscovite zone; 3) zone of regional migmatization.

For graphic proof of the existence of these relations between the elements in garnet, we computed for each garnet analysis the atomic ratios (K_{el}) of the elements Mg, Mn, Fe (because of the insignificant Fe^{3+} content, total iron was considered), and Ca, these ratios being measures of concentration. From the positions of the indicated elements and the interatomic distances Ca—O and (Fe, Mg, Mn)—O in the garnet lattice [Menzer, 1929], these ratios were computed by means of the following formulas:

$$K_{Mg} = \frac{Mg}{Fe + Mg + Mn}, \quad K_{Fe} = \frac{Fe}{Fe + Mg + Mn}, \quad K_{Mn} = \frac{Mn}{Fe + Mg + Mn}, \quad K_{Ca} = \frac{Ca}{Fe + Mg + Mn + Ca},$$

and they are shown in Table 3.

The correlation coefficients between the indicated ratios were computed (Table 4), and the parameters of the regression lines of K_{Fe} for K_{Mg}, K_{Mn} for K_{Mg}, and K_{Ca} for K_{Mg} were calculated [Hald, 1961]. The regression lines and the points corresponding to the relations of K_{Fe} and K_{Mg} and of K_{Mn} and K_{Mg} in the investigated garnets from the different metamorphic zones are shown in Fig. 1. From a study of this figure we may see that, with increase in grade of metamorphism, the concentration of Mg in the garnet increases and the contents of Fe, Mn, and Ca simultaneously decline. This fully supports the conclusion made above concerning the relations among the components of garnet.

The presence of positive correlation between Ca and Fe in the garnet follows directly from the investigated correlation coefficients between the atomic ratios, once again confirming the reliability of the computed ratios.

Table 4

	$K_{Ca}(Gr)$	$K_{Mn}(Gr)$	$K_{Mg}(Gr)$	$K_{Fe}(Gr)$	$K_{Fe}(Bt)$	$K_{Mn}(Bt)$	$K_{Mg}(Bt)$
Ti(Bt) ..	−0.09	−0.71	0.57	−0.30	−0.43	−0.06	0.02
$K_{Mg}(Bt)$ ..	−0.30	−0.03	0.37	−0.46	−0.91	−0.70	
$K_{Mn}(Bt)$ ..	0.17	0.12	−0.23	0.21	−0.66		
$K_{Fe}(Bt)$ ..	0.28	0.30	−0.58	0.57			
$K_{Fe}(Gr)$ ..	0.61	0.33	−0.88				
$K_{Mg}(Gr)$ ..	−0.43	−0.74					
$K_{Mn}(Gr)$ ..	−0.03						

Note. 99% level of significance, r ≥ 0.55; 95% level of significance, r ≥ 0.39.

Table 5

Component	Staurolite—andalusite zone (av. of 2 anal.)	Sillimanite—muscovite (av. of 5 anal.)	Zone of regional migmatization (av. of 20 anal.)
Almandite	75.2	73.6	68.0
Spessartite	9.2	6.1	2.5
Pyrope	8.2	15.3	24.8
Grossularite—andradite	7.4	5.0	4.7

Garnet compositions computed for the different zones of metamorphism are given in Table 5.

3. Correlation between Elements in Biotite

In examining the matrix of the correlation coefficients between the elements in biotite, one may clearly discern two groups of associated elements. The first group contains Ti, K, and Mg; the second, Na and Al. The elements of these two groups are antipathetic in their linear relations.

The positions of Fe^{3+}, Ca, Mn, and Fe^{2+} are somewhat peculiar; these components exhibit no significant correlations. In considering the presence of positive correlations between Mg and Ca and between Mn and Fe^{2+} and negative correlation coefficients between Fe^{3+} and Fe^{2+} and between Mn and Mg, however, these elements may be divided into two subgroups. The first, containing Fe^{3+} and Ca, belongs to the group of associated elements containing Ti, K, Mg; the second, represented by Mn and Fe^{2+}, belongs to the group containing Na and Al.

This division of elements in biotite by the nature of their correlations into two antipathetic groups indicates a systematic change in the contents of the mineral-forming elements in biotite.

From the chemical analyses of biotite, the atomic ratios for Mg, Fe (total), and Mn have also been computed, using the formulas

$$K_{Mg} = \frac{Mg}{Fe + Mg + Mn + Ti}, \quad K_{Fe} = \frac{Fe}{Fe + Mg + Mn + Ti}, \quad K_{Mn} = \frac{Mn}{Fe + Mg + Mn + Ti}.$$

The structural positions of Ti in biotite are not accurately known [Jacob, 1924], and we have therefore computed the atomic quantity of this element by means of its concentration in

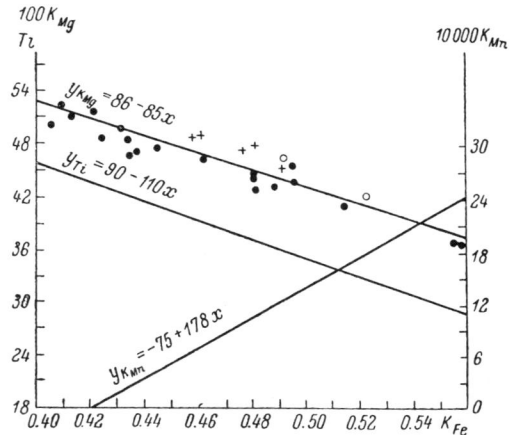

Fig. 2. Relations of the coefficients of Fe, Mg, Mn, and Ti contents in biotites of the Ladoga Formation from different zones of metamorphism. Symbols as in Fig. 1.

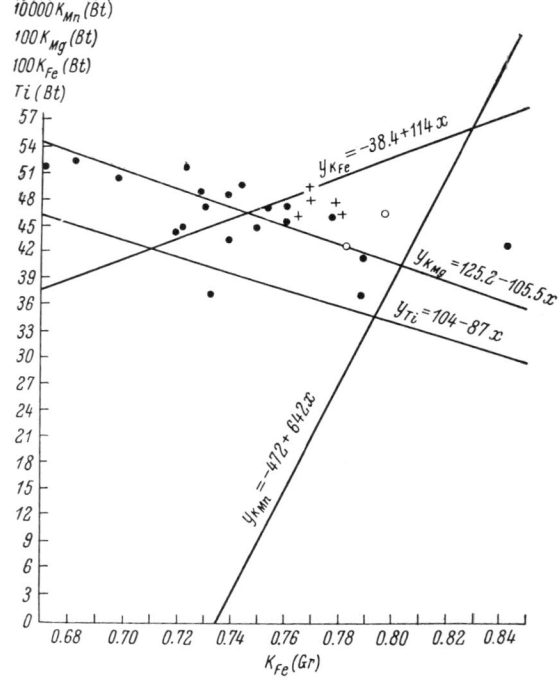

Fig. 3. Dependence of Fe, Mg, and Ti contents in biotite on the Fe content of garnet in rocks of the Ladoga Formation from different zones of metamorphism. Symbols as in Fig. 1.

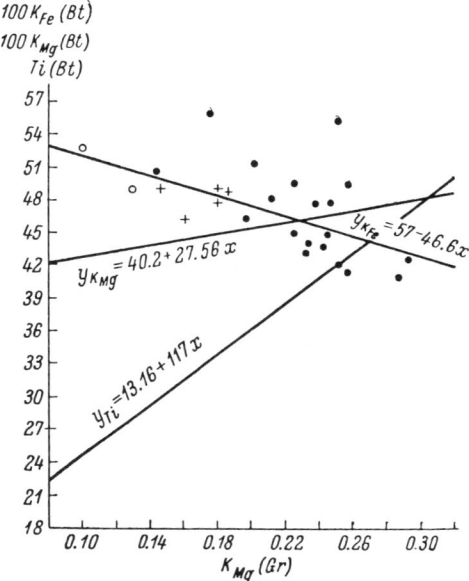

Fig. 4. Dependence of Fe, Mg, and Ti contents of biotite on the Mg content of garnet in rocks of the Ladoga Formation from different zones of metamorphism. Symbols as in Fig. 1.

the mineral. The computed atomic ratios are given in Table 3. The computed correlation coefficients between these ratios are shown in Table 4. The regression equations of K_{Mg} for K_{Fe}, Ti for K_{Fe}, K_{mn} for K_{Fe} and the points corresponding to the relations of K_{Mg} and K_{Fe} in biotite are shown in Fig. 2.

Analysis of these data indicates that the division of the elements of biotite into two associated groups is correct. In addition, this analysis establishes the fact that the contents of Mg and Ti increase and the contents of Fe and Mn decrease with increase in grade of metamorphism.

4. Correlation of Elements in Garnet and Biotite with Elements of the Rock

The computed correlation coefficients between elements of the rock and of the minerals indicate that the Ca content in garnet and the Mg content in biotite depend directly on the amounts of these elements in the rock. Therefore, we may naturally assume that in garnets of essentially calcium rocks (rich in plagioclase) the relative content of the grossularite—

andradite component should increase. In these rocks, on the basis that Mg and K of the rock have positive correlation and K and Ca negative correlation, the MgO content should decline and, consequently, biotite in these rocks should be impoverished in Mg because of an increase in the Fe^{2+} content (in the biotite Fe^{2+} and Mg are in different associated groups). This arrangement is reflected also in the correlation coefficients: Ca of the rock has a strong positive correlation with Fe^{2+} of biotite.

Fe^{2+}, Mn, and Mg of garnets and Ti and K of biotites have no correlation with elements of the rock, and a change in the contents of these elements in the minerals clearly depends but weakly on variations in the chemical composition of the rock.

One can hardly avoid noticing the negative correlation coefficient of the Fe^{2+} of biotite with K in the rock, although it is not entirely clear what process gives rise to this correlation. The crystallization of biotite impoverished in Fe and, consequently, enriched in Mg, may be associated both with regional metamorphism (Fig. 2) and with subsequent alteration of the rocks, during which late biotite formed, also poor in Fe and rich in Mg.

The remaining significant correlations between elements of the rock and of the minerals are less interesting, since they follow for the most part from correlations of elements within the rock, garnet, and biotite. An example of such correlation is that mentioned above between Ca of the rock with Fe^{2+} of biotite. In addition, many of the correlation coefficients not considered represent relations between element impurities (such as Ca, Na, and Mn in mica) of minerals and elements of the rock, a change in the content of which, in a mineral, is weakly reflected in variations among the principal mineral-forming elements.

5. Correlation between the Elements of Garnet and Biotite

The following significant correlations between elements in the investigated minerals have been noted.

1. Ca in garnet is related positively to Fe^{2+} and negatively to Al in biotite.

2. Mg in garnet has positive correlation coefficients with Ti, Mg, and K and negative with Na in biotite.

3. Mn in garnet is characterized by negative correlation with Ti, Mg, and K and positive with Na and Al in biotite.

4. Fe^{2+} in garnet is related negatively to Mg and K and positively to Na in biotite.

5. Al in garnet is positively related to K in biotite.

In analyzing the relations among the elements of garnet and biotite, and in considering the relations previously examined, we may note definite paragenetic associations in the compositions of garnet and biotite.

1. Garnet with high contents of Fe^{2+} and Mn (Ca) must be associated with ferruginous (Fe^{2+}) biotite.

2. Garnet with substantial amounts of the pyrope component, i.e., relatively rich in Mg, must be found in paragenetic association with biotite rich in Ti, Mg, and K.

It is rather difficult to explain these paragenetic associations merely by the chemical composition of the rock, because Fe^{2+}, Mn, and Mg in garnet and K and Ti in biotite have no proved correlation with elements of the rock. We are therefore forced to conclude that metamorphism has been a decisive factor in the formation of the associations. The first paragenetic association must correspond to the compositions of minerals from the zone of low-grade metamorphism, the second to compositions of minerals in the zone of regional migmatization.

The correlation coefficients between elements essential to the compositions of the minerals (Table 4) and the regression equations that we have computed for these elements (Figs. 3 and 4) fully support these systematic relations.

Conclusions

1. The presence of correlations between the principal rock-forming elements attests to systematic change in the mineralogical composition of rocks depending on variations in their chemical composition.

2. The garnet composition of the investigated rocks of the Ladoga Formation is determined both by the metamorphic grade and by the chemical composition of the rock. With increasing metamorphic grade, the garnet becomes impoverished in Fe^{2+} and Mn and the amount of Mg increases. The calcium content of the garnet depends strongly on the chemical composition of the rock, but a tendency toward diminution in CaO content is observed in garnet in the zone of regional migmatization.

3. The principal mineral-forming elements in biotite are separated into two antipathetic groups: Ti, Mg, and K (Fe^{3+}, Ca), and Na and Al (Fe^{2+}, Mn). The content of elements of the first group increases with increase in metamorphic grade. Furthermore, an increase in the Mg content in biotite may be explained by the general increase in MgO content of the rock.

4. From an analysis of the correlation coefficients we note very definite paragenetic associations in the compositions of biotite and garnet, corresponding to definite metamorphic grades.

The correlation coefficients were computed in the Laboratory of Mathematical Geology of the V. A. Steklov Order of Lenin Mathematical Institute.

REFERENCES

Albee, A., "Distribution of Fe, Mg, and Mn between garnet and biotite in natural mineral assemblages," J. Geol., Vol. 75, No. 1, pp. 155-163 (1965).
Hald, A., Statistical Theory with Engineering Applications, John Wiley and Sons, New York (1952).
Jacob, J., "Beiträge zur chemischen Konstitution der Glimmer, I. Die Schwedischen Manganophylle," Z. Krist., Vol. 61, pp. 155-163 (1924).
Kretz, R., "Chemical study of garnet, biotite, and hornblende from gneisses of southwestern Quebec, with emphasis on distribution of elements in coexisting minerals," J. Geol., Vol. 67, No. 4, pp. 371-402 (1959).
Lebedev, V. I., "Garnet and biotite from rocks of the Ladoga Formation and their petrologic significance," in: Problems of Magmatism and Metamorphism, Vol. 2, Izd. Leningr. Gos. Univ. (1964).
Lebedev, V. I., Nagaitsev, Yu. V., Pototskaya, V. E., Prudnikov, E. D., Shapkina, Yu. S., and Yurova, G. M., "Data on a mineralogical study of the metamorphic rocks in the northwestern Ladoga region," in: Mineralogy and Geochemistry, No. 1, Izd. Leningr. Gos. Univ. (1964).
Menzer, G., "Die Kristallstruktur der Granate," Z. Krist., Vol. 69, pp. 300-396 (1929).
Nagaitsev, Yu. V., "Characteristics of the metamorphic zones of the Ladoga Formation" Vest. Leningr. Gos. Univ., Ser. Geol. i Geogr., Issue 3, No. 18, pp. 25-33 (1965).
Sarmanov, O. B., and Vistelius, A. B., "Correlation between percentage values," Dokl. Akad. Nauk SSSR, Vol. 126, No. 1, pp. 22-25 (1959).

Vistelius, A. B., "The measure of correlation between paragenetic members and methods of studying it," Zap. Vses. Mineralog. Obshchestva, Pt. 77, No. 2, pp. 147-158 (1948).

Vistelius, A. B., "Problems of studying correlation in mineralogy and petrography," Zap. Vses. Mineralog. Obshchestva, Pt. 85, No. 1, pp. 58-74 (1956).

PROCESSES OF MAGMATIC DIFFERENTIATION IN CONNECTION WITH PARAGENETIC FEATURES AMONG ROCK-FORMING ELEMENTS IN NATURAL GLASS

V. V. Gruza

All-Union Scientific Research Institute of Geology
Leningrad, USSR

From an analysis of correlation coefficients defining the strength of correlation between rock-forming elements in natural acidic glass, it has been concluded that these rocks are characterized by micro-inhomogeneities due to differentiation of ions in homogeneous melts. The peculiarities of correlations between elements in acid crystalline rocks of a number of intrusive complexes correspond to systematic patterns of correlation among elements in glasses, and these may be satisfactorily explained not only by crystallization differentiation but also by differentiation of ions in homogeneous melts.

Study of sequential separation of phases in systems similar to those in nature has led to the conclusion, shared by many petrologists, that fractionation is the single most important process giving rise to variation in igneous rocks [Bowen, 1956]. In addition, a number of investigators justifiably contend that no one process is universally valid, and they assert that an essential factor during magmatic differentiation is the separation of molecules or ions of various types in homogeneous melts through the effect of gravity [Barth, 1961; Brewer, 1951; Kennedy, 1957; Kuz'min, 1964; Kadik, 1963; et al.] (molecular gravitational differentiation according to Grigor'ev [1946]) or through the effect of the temperature gradient that arises between the melt and the country rock [Wahl, 1946; Zlobin, 1960; et al.] (temperature-diffusion differentiation according to Grigor'ev). In connection with establishment of the ionic nature of silicate melts [Bockris et al., 1956; Esin, 1957], one should consider the possible role of ion differentiation in homogeneous melts, which, as a final result, may well lead to liquation.

The possibility and trend of ion differentiation in a magma may be clarified by studying the features of the joint distribution of elements in natural glasses constituting any single body, since the distribution of rock-forming elements in glasses corresponds to their distribution in the melt. This problem may be solved most successfully by using the methods of correlation analysis, which have been repeatedly used by Vistelius [1948, 1956, 1963, and other papers] for studying paragenetic relations of elements in different natural occurrences.

Table 1. Matrix of Paired Correlation Coefficients Determining the Force of Correlation between Petrogenetic Elements in the Acidic Glasses of the Mukhor-Tala Extrusive

	K	Na	Mg	Ca	Fe^{2+}	Fe^{3+}	Al	Ti
Si	+0.57	—0.07	—0.27	—0.08	—0.26	—0.14	—0.56	—0.31
Ti	—0.09	—0.15	—0.17	—0.42	—0.05	—0.09	+0.64	
Al	—0.25	—0.17	—0.12	—0.35	—0.09	—0.06		
Fe^{3+}	—0.35	—0.12	—0.05	—0.40	—0.32			
Fe^{2+}	—0.24	—0.05	—0.25	—0.13				
Ca	—0.16	—0.34	—0.03					
Mg	—0.34	—0.29						
Na	—0.24							

Note. 99% level of significance for r at $|r| \geq 0.38$; 95% level of significance for r at $|r| \geq 0.30$; n = 42.

Table 2. Matrix of Paired Correlation Coefficients Defining the Force of Correlation between Petrogenetic Elements in Rocks of the Moderately Acidic Group

	K	Na	Mg	Ca	Fe^{2+}	Fe^{3+}	Al	Ti
Si	+0.07	—0.35	—0.50	—0.55	—0.40	—0.11	—0.82	—0.63
Ti	—0.20	+0.26	+0.44	+0.40	+0.53	+0.19	+0.28	
Al	+0.02	+0.35	+0.13	+0.17	+0.11	—0.13		
Fe^{3+}	—0.02	—0.10	+0.21	+0.20	—0.04			
Fe^{2+}	—0.06	+0.26	+0.21	+0.21				
Ca	—0.29	+0.04	+0.68					
Mg	—0.28	—0.06						
Na	—0.57							

Note. 99% level of significance for r at $|r| \geq 0.31$; 95% level of significance for r at $|r| \geq 0.24$; n = 66.

Table 3. Matrix of Paired Correlation Coefficients Defining the Force of Correlation between Petrogenetic Elements in Rocks of the Potassium Granite Group

	K	Na	Mg	Ca	Fe^{2+}	Fe^{3+}	Al	Ti
Si	+0.26	—0.20	—0.76	—0.70	—0.51	—0.06	—0.74	—0.65
Ti	—0.19	+0.03	+0.67	+0.59	+0.60	+0.32	+0.37	
Al	—0.30	+0.27	+0.51	+0.47	+0.31	—0.19		
Fe^{3+}	—0.11	+0.19	+0.19	—0.04	—0.03			
Fe^{2+}	—0.005	—0.09	+0.47	+0.35				
Ca	—0.31	—0.11	+0.57					
Mg	—0.30	+0.17						
Na	—0.55							

Note. 99% level of significance for r at $|r| \geq 0.28$; 95% level of significance for r at $|r| \geq 0.22$; n = 83.

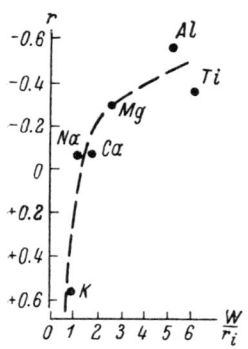

Fig. 1. Force of correlation between elements and silicon in glasses of the Mukhor-Tala extrusive. Values of ionic potentials are plotted on the abscissa, correlation coefficients on the ordinate axis.

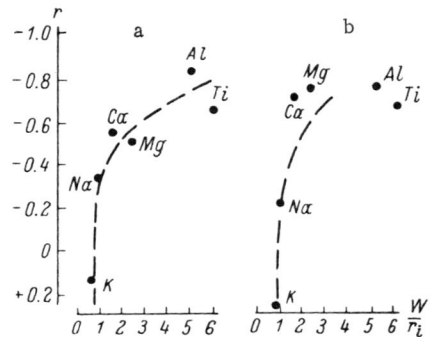

Fig. 2. Force of correlation between elements and silicon in the moderately acidic granitic rocks (a) and the group of potassium granites (b) in the central part of the Altai-Sayan region. Symbols as in Fig. 1.

Data for this study came from 42 chemical analyses of glasses, chiefly perlites, of the Mukhor-Tala extrusive, published in a number of reports [Volyanyuk, 1962; Manuilova, Nasedkin, Petrov, and Polinkovskaya, 1961; and Nasedkin, 1963]. In view of the fact that the perlites were formed by hydration of obsidian containing almost no juvenile water [Ross and Smith, 1961; Ross, 1964; Manuilova, Varshal, and Maier, 1962], the results of the chemical analyses are recalculated for water-free material. After recalculation, it was discovered that all glasses of the examined extrusion are very similar and, in their chemical composition, they represent a closed system according to Chayes [1960], i.e., in all the analyses the sums of the equal-volume comparative contents of components are the same. In further calculations, therefore, values expressed in weight percentages of elements have been used.

As seen from Table 1, the concentrations of elements in these rocks have been differentiated. If we exclude ferric and ferrous oxides from our consideration, the concentrations of which are commonly oxidation—reduction reactions, the petrogenetic elements of the Mukhor-Tala glasses may be divided into three groups by the character of the correlations. The first group includes silicon and potassium, which are positively related (r_{Si-K} = +0.57, 99% level of significance). The second contains titanium and aluminum, also positively related (r_{Ti-Al} = +0.64, 99% level of significance) but inversely related to silicon (r_{Si-Al} = −0.56, 99% significance level; r_{Si-Ti} = −0.31, 95% significance level). The third group contains sodium, magnesium, and calcium, which exhibit no positive correlations with other elements.

The rather sharp antipathetic relations between silicon and aluminum in the glasses is due to the fact that the ions of these elements in acidic melts fulfill similar functions, participating in the formation of complex aluminosilicate anions, in which they replace each other [Esin, 1957]. An increase in the silicon content in melts leads to an increase in the role of complex anions of strong acids, whereas an increase in the aluminum content indicates an increase in the role of anions of weak acids. The association of potassium with silicon and of titanium with aluminum is systematic, since a general rule is followed [Shcherbina, 1964], involving the tendency of anions of strong acids to associate with cations of elements of high alkalinity (potassium and sodium), whereas anions of weak acids associate with cations of weakly alkaline elements (titanium). In keeping with this rule, antipathetic relations among elements also increase in the glasses relative to silica according to decrease in their alkaline properties (Fig. 1). The nature of the joint distribution of elements in natural glasses of the Mukhor-Tala extrusive is thus defined by the energy relations of the ions of these elements in the complex ions into which they enter. On the basis of the data we have

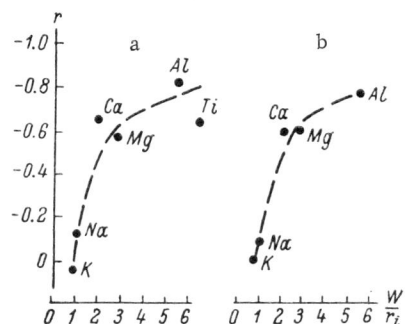

Fig. 3. Force of correlation between elements and silicon in granites of Central Kazakhstan. a) From D. N. Ivanov's data; b) from V. E. Gendler's data.

cited, it is naturally impossible to evaluate the factors controlling the separation of ions in a homogeneous melt from which they have formed.

It is interesting to compare our results, obtained by studying the natural glasses of the Mukhor-Tala extrusive, with data on aspects of correlation among petrogenetic elements in acidic crystalline rocks. For this purpose we have computed the correlation coefficients defining the force of correlation between elements in rocks of the granitoidal complexes of the Altai-Sayan fold region: Upper Cambrian moderately acidic granitic rocks and Devonian potassium granites.

Rock masses of the moderately acidic granitic rocks are found within the eastern slopes of the Kuznetsk Alatau (Ulen'-Tuim, Uibat, Tygertysh, and Askiz) and are composed of a wide range of calc-alkalic and alkalic rocks. Biotite granites, very persistent in their petrographic character, occur in the central parts of these masses. Rocks in the marginal zones exhibit well-defined hybridism. The shapes of the rock masses have not been determined with assurance, but there are grounds for believing that they are isolated exposures of a single large batholith [Ken, 1964].

The masses of Devonian potassium granites are found in the Western Sayan, where they are generally distinguished in composition from the Dzhoi intrusive complex [Ivanova, 1961; Orlov, 1961]. They are also found in the northern part of the Kuznetsk Alatau. The massifs are composed of alaskitic and biotitic granites, persistent in composition.

To study the correlation coefficients, 66 chemical analyses of biotite granites from the moderately acidic group and 83 chemical analyses of rocks from the potassium granite group were used. A large number of these analyses have been published [Bazhenov, 1934; Ziv, 1931; Musatov, 1961; Orlov, 1962; Polyakov and Teleshov, 1960; Stankevich, 1963]. A number of analyses were made of samples from the author's collection.

In order to avoid errors introduced by percentage calculations, the analytical data were first transformed according to established procedure [Sarmanov and Vistelius, 1959; Vistelius and Sarmanov, 1961].

Judging from the data given in Tables 2 and 3, the correlations between elements in rocks of the investigated groups are also rather systematic, and the same tendency of change in correlation force between elements and silicon as that found in the acidic glasses of the Mukhor-Tala extrusive appears: the higher the ionic potential of the elements the greater the antagonistic relations between these elements and silicon (Fig. 2). This systematic relationship between elements has also been found not only in rocks of the Altai-Sayan region but also in the granites of Central Kazakhstan, studied by Gendler [1964] and Ivanov [1963]. The relationship is shown graphically (Fig. 3) on the basis of data taken from these investigators. As in the silicic glasses, this systematic pattern reflects the association in rocks, of complex anions of strong acids with cations of the most alkaline elements, and of anions of weak acids with cations of the least alkaline elements. Such relations between petrogenic elements may arise in homogeneous melts, and there is no necessity of seeking an explanation for chemical peculiarities of the investigated granites [Ivanov, 1963], of considering the crystalline phases.

Study of the peculiarities of the joint distribution of petrogenic elements in natural glasses of the Mukhor-Tala extrusive thus shows that the concentrations of these elements in the glasses have been differentiated, that the differences are due to separation of ions in a homogeneous melt. Patterns of joint distribution of elements in the glasses have been controlled by the energy relations of ions of the elements.

Correlation peculiarities between elements in acid crystalline rocks from a number of intrusive complexes correspond to the systematic patterns of correlation between elements in the glasses, and the relations may be explained successfully by differentiation of ions in homogeneous melts as well as by crystallization differentiation.

It should be brought to notice that the idea of studying peculiarities of elemental distribution in glasses belongs to A. B. Vistelius.

REFERENCES

Bazhenov, I. K., "Western Sayan," in: Sketches of the Geology of Siberia, Izd. Akad. Nauk SSSR, Leningrad (1934).

Barth, T. F. W., "Composition and evolution of magma in the southern part of the mid-Atlantic Ridge," in: Physicochemical Problems of the Formation of Rocks and Ores, Vol. 1, Izd. Akad. Nauk SSSR, Moscow (1961).

Bockris, J. O. U., Tomplinson, Y. W., and White, I. Z., "The structure of the liquid silicates," Trans. Faraday Soc., Vol. 52, p. 299 (1956).

Bowen, N. L., Igneous Rocks, Dover, New York (1956).

Brewer, L., "Equilibrium distribution of the elements in the earth's gravitational field," J. Geol., Vol. 59, p. 490 (1951).

Chayes, F., "On correlation between variables of constant sum," J. Geophys. Res., Vol. 65, No. 12 (1960).

Esin, O. A., "The structure of liquid silicates," Usp. Khim., Vol. 26, No. 12 (1957).

Esin, O. A., and Gel'd, P. V., "Structural features of glass-forming and liquid silicates," in: The Structure of Glass, Izd. Akad. Nauk SSSR, Moscow (1953).

Gendler, V. E., "Chemical composition of post-Lower Cambrian granites in the Tarbagatai Range," Author's abstract in Byul. Mosk. Obshchestva Ispytatelei Prirody, Nov. Ser., Vol. 69, Otd. Geol., Vol. 39, No. 3 (1964).

Grigor'ev, D. P., "Experience in systematics and terminology of elemental processes of magmatic differentiation," in: Academician D. S. Belyankin on His 70th Birthday, Izd. Akad. Nauk SSSR, Moscow (1946).

Ivanov, D. N., "Linear paragenetic relations of the principal rock-forming elements in granites of Central Kazakhstan," Dokl. Akad. Nauk SSSR, Vol. 150, No. 2, p. 392 (1963).

Ivanova, T. N., Principal Systematic Features in Development of Intrusive Magmatism in the Eastern Part of the Altai-Sayan Region, Byull. Vses. Nauchn.-Issled. Inst., No. 3, Gosgeoltekhizdat, Moscow (1961).

Kadik, A. A., "Evaluation of the possible role of gravitation during differentiation of magmas in a homogeneous state," in: Chemistry of the Earth's Crust, Vol. 1, Izd. Akad. Nauk SSSR, Moscow (1963).

Ken, A. N., Geologic Formations in the Central Part of the Altai-Sayan Fold Region and Their Ore Content, Tr. Vses. Nauchn.-Issled. Geol. Inst., Nov. Ser., Vol. 103, Izd. Nedra, Leningrad (1964).

Kennedy, G. C., "Some aspects of the role of water in rock melts," in: The Crust of the Earth, (eds.: Rapport, S. and Wright, H.), paperback, Signet Books, New American Library, New York (1955).

Kuz'min, A. M., "The hypsometric law and its value in solving problems of magma development," Geol. Geofiz., No. 4, p. 108 (1964).

Manuilova, N. S., Varshal, B. G., and Maier, A. A., Investigation of the Structure and Some Physicochemical Properties of Perlites, Sb. Respubl. Nauchn.-Issled. Inst. Mestnykh Stroitel'nykh Materialov, No. 25 (1962).

Manuilova, N. S., Nasedkin, V. V., Petrov, V. P., and Polinkovskaya, A. I., "Petrography and practical value of the perlites of Mukhor-Tala (Buryat ASSR)," in: The Petrography and Mineralogy of Deposits of Perlite, Ceramic Raw Material, and Mica, Tr. Inst. Geol. Rydnykh Mestorozh., Petrog., Mineralog., Geokhim., No. 48, Izd. Akad. Nauk SSSR, Moscow (1961).

Musatov, D. I., Intrusive Magmatism on the Eastern Slope of the Kuznetsk Alatau, Material on the Geology and Mineral Deposits of the Krasnoyarsk Krai, No. 1, Krasnoyarsk (1961).

Nasedkin, V. V., Water-Bearing Glass of Acidic Composition, Its Origin and Alteration, Tr. Inst. Geol. Rydnykh Mestorozh., Petrog., Mineralog., Geokhim., No. 98, Izd. Akad. Nauk SSSR, Moscow (1963).

Orlov, D. M., A Brief Outline of Magmatism in Western Sayan, Tr. Vses. Nauchn.-Issled. Geol. Inst., Vol. 58, Gosgeoltekhizdat, Moscow (1961).

Orlov, D. M., Differentiated Masses of Granitic Rocks of the Dzhoi Complex of the Western Sayan, Tr. Veses. Nauchn.-Issled. Geol. Inst., Vol. 73, Gosgeoltekhizdat, Moscow (1962).

Polyakov, G. V., and Teleshov, A. E., "Magmatic complexes in the region of the Teya Group of iron-ore deposits (Kuznetsk Alatau)," in: Principal Ideas of M. A. Usov in Geology, Izd. Akad. Nauk KazSSR, Alma-Ata (1960).

Ross, C. S., and Smith, R. L., "Water and other volatiles in volcanic glasses," Am. Mineralogist, Vol. 40, No. 11, p. 1071 (1955).

Ross, C. S., "Volatiles in volcanic glasses and their stability relations," Am. Mineralogist, Vol. 49, No. 3-4, p. 258 (1964).

Sarmanov, O. V., and Vistelius, A. B., "Correlation between percentage values," Dokl. Akad. Nauk SSSR, Vol. 126, No. 1, p. 22 (1959).

Shcherbina, V. V., "The geochemistry of silicate melts," Zap. Vses. Mineralog. Obshchestva, Pt. 93, No. 5, p. 537 (1964).

Stankevich, E. K., Geology of the Eastern Part of the Tygertysh Pluton (Kuznetsk Alatau), Tr. Vses. Nauchn.-Issled. Geol. Inst., Vol. 98, Petrography Collection No. 5, Gosgeoltekhizdat, Moscow (1963).

Vistelius, A. B., "The measure of correlation between paragenetic members and methods of studying it," Zap. Vses. Mineralog. Obshchestva, Pt. 77, No. 2, p. 147 (1948).

Vistelius, A. B., "Problems of studying correlation in mineralogy and petrography," Zap. Vses. Mineralog. Obshchestva, Pt. 85, No. 1, p. 58 (1956).

Vistelius, A. B.,"Problems of mathematical geology," Geol. Geofiz., No. 7, p. 3 (1963).

Vistelius, A. B., and Sarmanov, O. V., "On the correlation between percentage values, major component correlation in ferromagnesium micas," J. Geol., Vol. 69, No. 2, p. 145 (1961).

Volyanyuk, N. Ya., Petrography of Rocks of the Volcanic Glass Deposits of Mukhor-Tala and Western Transbaikalia, Zap. Vost.-Sib. Otd. Vses. Mineralog. Obshchestva, No. 3, Irkutsk (1962).

Wahl, W., "Thermal diffusion-convection as a cause of magmatic differentiation," Am. J. Sci., New Haven, Vol. 244, No. 6, p. 417 (1946).

Ziv, E. F., Scheelite-Bearing Skarns of the Eastern Slope of the Kuznetsk Alatau, Tr. Vses. Nauchn.-Issled. Inst. Mineral. Syr'ya, No. 145, GGU NKTP SSSR, Moscow-Leningrad (1939).

Zlobin, B. I., "Some questions on differentiation of alkalic magma as exemplified by the intrusion in the Sandyk Mountains," in: Magmatism and Related Mineral Deposits, Gosgeoltekhizdat, Moscow (1960).

IV. Analysis of Geologic Sections

A STOCHASTIC MODEL OF STRATIFICATION (THE CASE OF UNLIMITED INTERSTRATAL EROSION)

T. S. Rivlina

*Laboratory of Mathematical Geology
V. A. Steklov Mathematical Institute
Academy of Sciences of the USSR
Leningrad, USSR*

A stochastic model of stratification is examined in this paper. A Markov character of the stratification process is assumed. Expressions are given for computing transitional probabilities from one bed to another when the number of eroded beds from interstratal erosion is not restricted. The derived formulas may be used for constructing models of the process on an electronic computer and also for restoring beds eliminated from the section because of Markov erosion.

The problem of the origin of bedding in thick sedimentary formations has been the repeated subject of numerous investigations. For a long time study has been made of individual beds with detailed consideration of structure. During this time, the problem concerning the process causing the formation of bedding as a whole has not been seriously raised. Only Vistelius [1949b], in studying the thickness of sections of the Productive sequence on the Apsheron Peninsula and the flysch of northeastern Caucasus and Kakhetiya (the latter from descriptions of N. V. Vassoevich), remarked on the necessity of reconstructing the type of stratification process, i.e., the general mechanism causing bedded structure as a whole and not of a single bed, since the objective of reconstructing conditions under which a single bed formed is not precisely the problem of type of process causing stratification. The study of flysch sections by Vistelius has shown that the sequence of beds in these sections is not inconsistent with the view that the sequence has Markov characteristics [Vistelius, 1959a].

At approximately the same time the question concerning the mechanism of stratification was formulated, the ideas of Kuenen [1948], previously advanced, received wide attention. According to this author, sequences of terrigenous flysch form by simple roiling of the sediments and subsequent differentiation during settling in conformity with Stokes' law. These ideas have been accepted by a wide circle of geologists, and tests of their validity in the most varied parts of the earth have shown that, fundamentally, Kuenen's views explain observed facts. This has led us to accept the idea that thick beds may result from roiling and that the process may be repeated many times, i.e., this indirectly confirms the view that flysch may be the realization of a Markov process.

In studying the problem of stratification, a model was first developed in which, during interstratal erosion, only erosion of a single bed was allowed [Vistelius and Feigel'son, 1965]. Further investigations have shown that this restriction is not necessary. A solution of the problem, with an unrestricted number of eroded layers, is given in the present paper.

I take this opportunity to thank R. A. Zaidman for consultation and valuable advice, facilitating solution of the problem, and I also thank A. M. Kagan for his useful remarks.

Formulation of the Problem

In the process of stratification, accumulation of sand, silt, and clay occur and interstratal erosion takes place. As a mathematical model of this phenomenon, let us examine a steady-state Markov process having four constants — S, A, C, E (S representing accumulation of sand, A of silt, C of clay, and E representing interstratal erosion) — infinite in both directions, such that the final probability $P_E < 1/2$. It is given by a matrix of transitional probabilities

$$\begin{matrix} P_{SS} & P_{SA} & P_{SC} & P_{SE} \\ P_{AS} & P_{AA} & P_{AC} & P_{AE} \\ P_{CS} & P_{CA} & P_{CC} & P_{CE} \\ P_{ES} & P_{EA} & P_{EC} & P_{EE} \end{matrix}$$

In practice, when studying a section, the state of erosion is not observed, and we go from the above-described process, which we designate by $X(t)$, $t = \ldots, -1, 0, 1, 2 \ldots$, to another, $Y(t)$, $t = \ldots, -1, 0, 1, 2 \ldots$ having three constants, S, A, C. The transition from $X(t)$ to $Y(t)$ is effected by the following rule: $IE = \wedge$, where $I = S, A, C$; $\wedge$ is an empty event, being satisfied by the associative property. For example, $\ldots ASECSEEC \ldots = \ldots A(SE)[C(SE)E]C \ldots = \ldots AC \ldots$ In the transition from $X(t)$ to $Y(t)$ a chain of events $X(t)$ is "cemented" into a single event. It may be said that the Markov process is not disturbed by this. The matrix of transitional probabilities $Y(t)$ will be

$$\begin{matrix} \varphi_{SS} & \varphi_{SA} & \varphi_{SC} \\ \varphi_{AS} & \varphi_{AA} & \varphi_{AC} \\ \varphi_{CS} & \varphi_{CA} & \varphi_{CC} \end{matrix}$$

It is necessary to set up equations relating φ_{IJ} (where $I, J = S, A, C$) to P_{IJ} (where $I, J = S, A, C, E$).

Let U_n^I be the probability that there is a chain of states in the process $X(t)$ with a length $2n+1$ with an initial state I, in which, after I, there follows regularly n accumulations and n erosions in any order, but such that the number of erosions never exceeds the number of accumulations. All accumulated layers, except the first, are eroded. The probability that, when cutting off the process $X(t)$ at any point, the bed J at the n-th site is not eroded is expressed by $_n\Phi_j$. In view of the steady-state process, $_n\Phi_j$ does not depend on n:

$$_n\Phi_J = \Phi_J.$$

If, in realization of the process $X(t)$ we observe a succession of beds I and J, one directly after the other, then, according to the transition from $X(t)$ to $Y(t)$, the formation of this combination of beds might have occurred in the following way: either bed J was deposited immediately after accumulation of I and was not subsequently eroded, or, after deposition of I, there took place any number of erosions (n) and accumulations (n) in any order, but such that the number of erosions never exceeded the number of accumulations, after which the bed J was deposited and was not subsequently eroded.

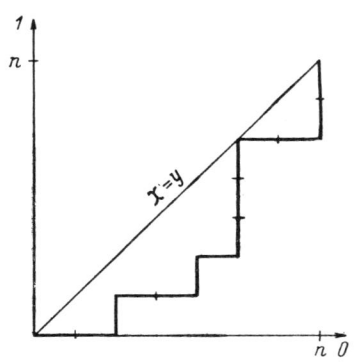

Fig. 1. Graphic representation of the stratification process.

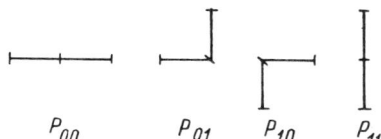

Fig. 2. Types of transitions in the path from one state to another, with corresponding probabilities.

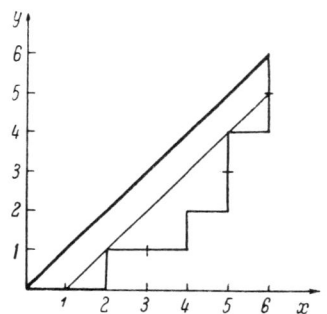

Fig. 3. Correspondence between the classes $\mathfrak{A}_{n-1, m}$ and $\mathfrak{B}_{n, m}$. The paths belong to $\mathfrak{B}_{6,4}$ and $\mathfrak{A}_{5,4}$.

We may then set down a system of equations

$$\varphi_{IJ} = \frac{1}{\Phi_I}\left(P_{IJ}\Phi_J + \sum_{n=1}^{\infty} U_n^I P_{EJ}\Phi_J\right), \quad (1)$$

$$I, J = S, A, C,$$

where

$$\Phi_J = 1 - \left(P_{JE} + \sum_{n=1}^{\infty} U_n^I P_{EE}\right).$$

The problem now consists of finding U_n^I.

The Case of Two Constants

Let us consider the processes of accumulation and erosion as a random walk.

For simplicity we will assume that we have two states: accumulation and erosion. These states will be designated 0 and 1. In Fig. 1, using a system of Cartesian coordinates (x, y), the horizontal scale unit is the accumulation interval and the vertical unit is the erosion interval. The origin corresponds to the initial state.

In order to return to the original state after 2n intervals, we must make n steps along the horizontal and n steps along the vertical, during which the line x = y may be touched but not crossed, otherwise the original bed is eroded (by virtue of the geometric properties of the triangle). We assume the first step to be always accumulation, the succeeding step to be erosion.

Let there be in all m horizontal and m vertical degrees of freedom. Then n, the number of steps along the horizontal, is broken down into m positive terms: $n = k_1 + k_2 + \ldots + k_m$, where $k_i > 0$ is the number of steps at the i-th horizontal degree of freedom. Similarly, n is the number of steps along the vertical, is represented in the form $n = l_1 + l_2 + \ldots + l_m$, where $l_i > 0$ is the number of steps at the i-th vertical degree of freedom. Four types of transitions are possible on the described path (Fig. 2).

The number of transitions of the type "00" on such a path is equal to $k_1 - 1 + k_2 - 1 + \ldots + k_m - 1 = n - m$, of type "11," $- l_1 - 1 + \ldots + l_m - 1 = n - m$, of type "01," m, and of type "10," m − 1. The probability of one such path is

$$(P_{00}P_{11})^{n-m} P_{01}^m P_{10}^{m-1}.$$

Let us designate by $\mathfrak{A}_{n,m}$ the set of all possible paths with n steps and m degrees of freedom along vertical and horizontal axes that do not go beyond the line x = y.

Let $A_{n,m}$ be the number of elements $\mathfrak{A}_{n,m}$, then

$$U_n = \sum_{m=1}^{n} A_{n,m} (P_{00}P_{11})^{n-m} P_{01}^{m} P_{10}^{m-1}. \qquad (2)$$

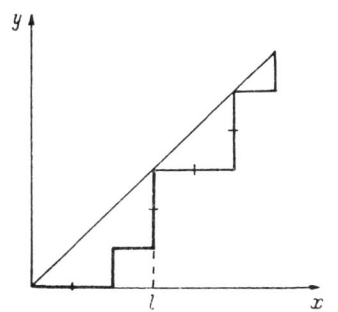

Fig. 4. Separation of the path into two parts. The first belongs to $\mathfrak{B}_{l,r}$, the second to $\mathfrak{A}_{n-l, m-r}$.

Let us assume that $\mathfrak{B}_{n,m}$ is the subset of paths of $\mathfrak{A}_{n,m}$, that have no points in common with the line x = y earlier than 2n steps, and $B_{n,m}$ is the number of elements $\mathfrak{B}_{n,m}$.

There is one-to-one correspondence between $\mathfrak{A}_{n-1, m}$ and $\mathfrak{B}_{n,m}$, demonstrated by the manner of adding to any path of $\mathfrak{A}_{n-1,m}$ the first horizontal and last vertical degrees of freedom so that the resulting path belongs to $\mathfrak{B}_{n,m}$. On the other hand, by rejection of the first horizontal and last vertical degrees of freedom from any path of $\mathfrak{B}_{n,m}$ we obtain the path of $\mathfrak{A}_{n-1, m}$ (Fig. 3). Hence, $B_{n,m} = A_{n-1,m}$.

Let us examine the path of $\mathfrak{A}_{n,m}$. Let the line x = y first touch at l steps along the horizontal. The path is then broken into two parts, the first belonging to $\mathfrak{B}_{l,r}$, the second to $\mathfrak{A}_{n-l, m-r}$ (Fig. 4).

The following recurrent relation is valid:

$$A_{n,m} = \sum_{l,r} B_{l,r} A_{n-l, m-r} = B_{1,1} A_{n-1, m-1} + \sum_{l,r} A_{l-1, r} A_{n-l, m-r},$$

where the range of summation is $l \le n$, $1 \le r \le m$, $r \le l-1$, $l \ge 2$, $m-r \le n-l$, and $B_{1,1} = 1$.

We then have

$$A_{n,m} = \sum_{\substack{1 \le r \le m \\ 2 \le l \le n \\ 1 \le l-r \le n-m}} A_{l-1, r} A_{n-l, m-r} + A_{n-1, m-1} \qquad (3)$$

under the initial conditions $A_{n,n} = 1$, $n \ge 0$; $A_{n,0} = 0$, $n \ge 1$; $A_{n,1} = 1$, $n \ge 1$;

Let us now introduce a generating function for $A_{n,m}$:

$$U(x, y) = A_{0,0} + \sum_{n=1}^{\infty} \sum_{m=1}^{n} A_{n,m} x^n y^m. \qquad (4)$$

Consequently,

$$U(x, y)[U(x, y) - 1] = \sum_{n=2}^{\infty} \sum_{m=1}^{n-1} \sum_{\substack{1 \le r \le m \\ 2 \le l \le n \\ 1 \le l-r \le n-m}} A_{l-1, r} A_{n-l, m-r} x^{n-1} y^m. \qquad (5)$$

Formula (4) may be rewritten in the form

$$U(x, y) = 1 + \sum_{n=2}^{\infty} \sum_{m=2}^{n} A_{n-1, m-1} x^{n-1} y^{m-1}, \qquad (6)$$

then

$$y(U-1) = \sum_{n=2}^{\infty} \sum_{m=2}^{n} A_{n-1, m-1} x^{n-1} y^m, \tag{7}$$

$$\frac{U-1}{x} = \sum_{n=1}^{\infty} \sum_{m=1}^{n} A_{n, m} x^{n-1} y^m. \tag{8}$$

By means of (3) let us introduce an equation for U(x, y). In order to do this it is necessary that the summation in (5), (7), and (8) be made within the same limits: n = 2, 3, . . . , m = 2, 3, . . . , n - 1.

In (5) we subtract a member when m = 1: if m = 1, then r = 1; when n − l > 0, $A_{n-l,0} = 0$; therefore,

$$\sum_{n=2}^{\infty} A_{n-1, 1} A_{0, 0} x^{n-1} y = y \sum_{n=2}^{\infty} x^{n-1} = \frac{xy}{1-x}$$

and

$$U(U-1) - \frac{xy}{1-x} = \sum_{n=2}^{\infty} \sum_{m=2}^{n-1} \sum_{l, r} A_{l-1, r} A_{n-l, m-r} x^{n-1} y^m. \tag{9}$$

In (7) we subtract a member when m = n; specifically,

$$\sum_{n=2}^{\infty} A_{n-1, n-1} x^{n-1} y^n = y \sum_{n=2}^{\infty} (xy)^{n-1} = \frac{xy^2}{1-xy},$$

then

$$y(U-1) - \frac{xy^2}{1-xy} = \sum_{n=2}^{\infty} \sum_{m=2}^{n-1} A_{n-1, m-1} x^{n-1} y^m. \tag{10}$$

In (8) we subtract a member when m = 1, n = 1, 2, . . . :

$$\sum_{n=1}^{\infty} A_{n, 1} x^{n-1} y = \frac{y}{1-x}$$

and when m = n,

$$\sum_{n=2}^{\infty} A_{n, n} x^{n-1} y^n = \frac{xy^2}{1-xy},$$

then

$$\frac{U-1}{x} - \frac{y}{1-x} - \frac{xy^2}{1-xy} = \sum_{n=2}^{\infty} \sum_{m=2}^{n-1} A_{n, m} x^{n-1} y^m. \tag{11}$$

According to (3), (9), (10), and (11),

$$\frac{U-1}{x} - \frac{y}{1-x} - \frac{xy^2}{1-xy} = U(U-1) - \frac{xy}{1-x} + y(U-1) - \frac{xy^2}{1-xy},$$

whence, keeping in mind the initial condition $A_{0,0} = 1$, we obtain

$$U(x, y) = \frac{1 - xy + x - \sqrt{1 + x^2y^2 + x^2 - 2x - 2xy - 2x^2y}}{2x}. \tag{12}$$

In comparing (2) and (4), we see that

$$\sum_{n=1}^{\infty} U_n = \frac{1}{P_{10}} \sum_{n=1}^{\infty} \sum_{m=1}^{n} A_{n,m} (P_{00}P_{11})^n \left(\frac{P_{01}P_{10}}{P_{00}P_{11}}\right)^m = \frac{1}{P_{10}} \left\{ U\left[(P_{00}P_{11}), \left(\frac{P_{01}P_{10}}{P_{00}P_{11}}\right)\right] - 1 \right\}, \tag{13}$$

where $U(x, y)$ is given in explicit form by (12).

We have thus found the probability $\sum_{n=1}^{\infty} U_n$ in the case of two states. In like manner, we determine all the values in equation (1).

The General Case

Let us now examine the case when the state of accumulation has been differentiated: we shall assume that there are three states of accumulation, S, A, C (in practice one may study the process with any number of accumulation states).

If k_i is the number of steps at the i-th horizontal degree of freedom, then k_i is divided into t_i positive members: $k_i = k_{i1} + k_{i2} + \ldots + k_{it_i}$, where $k_{ij} > 0$ is the number of steps in the j-th succession of beds of identical composition at the i-th degree of freedom, and t_i is the number of sequences of beds of identical composition at the i-th degree of freedom, $1 \leq t_i \leq k_i$ (Fig. 5).

The sum of the products of probabilities of horizontal degrees of freedom of all paths with initial states I = S, A, C, E, and determined by the number of degrees of freedom m, is then

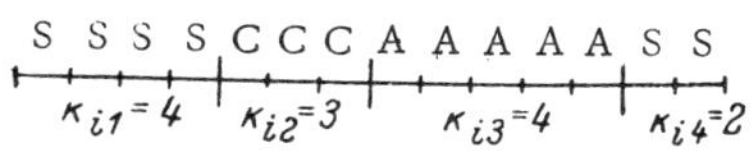

$$\sum_{\substack{m \\ \sum_{i=1}^{m} k_i = n}} \left\{ \left(\prod_{i=2}^{m} \sum_{t_i=1}^{k_i} \sum_{\substack{\{\varepsilon_j\} \\ \varepsilon_j = S,A,C \\ \varepsilon_j \neq \varepsilon_{j+1}}} P_{E\varepsilon_1} P_{\varepsilon_1\varepsilon_2} \ldots P_{\varepsilon_{t_i-1}\varepsilon_{t_i}} P_{\varepsilon_{t_i}E} \sum_{\substack{t_i \\ \sum_{j=1} k_{ij}=k_i}} P_{\varepsilon_1\varepsilon_1}^{k_{i1}-1} \ldots \right. \right.$$

$$\left. \ldots P_{\varepsilon_{t_i}\varepsilon_{t_i}}^{k_{it_i}-1} \right) \sum_{t_1=1}^{k_1} \sum_{\substack{\{\varepsilon_j\} \\ \varepsilon_j = S,A,C \\ \varepsilon_j \neq \varepsilon_{j+1}}} P_{I\varepsilon_1} P_{\varepsilon_1\varepsilon_2} \ldots P_{\varepsilon_{t_1-1}\varepsilon_{t_1}} P_{\varepsilon_{t_1}E} \sum_{\substack{t_1 \\ \sum_{j=1} k_{1j}=k_1}} P_{\varepsilon_1\varepsilon_1}^{k_{11}-1} \ldots$$

$$\left. \ldots P_{\varepsilon_{t_1}\varepsilon_{t_1}}^{k_{1t_1}-1} \right\}. \tag{14}$$

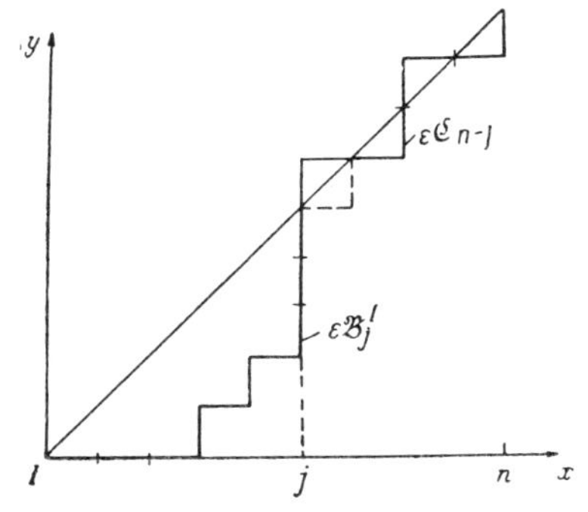

Fig. 5. An example of subdividing the horizontal degree of freedom.

Fig. 6. Subdividing the path into classes. All paths belong to $\mathfrak{D}_n^I$.

Let us introduce the arbitrary function

$$y_i = \sum_{n=1}^{\infty} P_{ii}^{n-1} x^n = \frac{x}{1 - P_{ii}x}, \quad i = S, A, C,$$

the matrix

$$\Pi(x) = \|\pi_{ij}\|, \text{ where } \pi_{ij} = \begin{cases} P_{ij}y_j, & i \neq j \\ 0, & i = j, \end{cases}$$

$$i, j = S, A, C$$

and the matrix

$$Z_I = \| P_{IA} y_A, \ P_{IS} y_S, \ P_{IC} y_C \|, \ I = S,A,C,E$$

$$T = \begin{Vmatrix} P_{AE} \\ P_{SE} \\ P_{CE} \end{Vmatrix}.$$

Then

$$Z_I S^{t_i-1} T = \sum_{\substack{\{\varepsilon_j\} \\ \varepsilon_j = S,A,C, \\ \varepsilon_j \neq \varepsilon_{j+1}}} P_{I\varepsilon_1} y_{\varepsilon_1} P_{\varepsilon_1 \varepsilon_2} y_{\varepsilon_2} \cdots P_{\varepsilon_{t_i-1} \varepsilon_{t_i}} y_{\varepsilon_{t_i}} P_{\varepsilon_{t_i} E}.$$

We shall designate

$$w_I(x) = \sum_{t_i=1}^{\infty} Z_I S^{t_i-1} T = Z_I (F-S)^{-1} T. \tag{15}$$

If ω_I is expanded into a series for stages of x, then the coefficient at x^{k_i} will be

$$\sum_{t_i=1}^{k_i} \sum_{\substack{t_i \\ \sum_{j=1} k_{ij}=k_i}} \sum_{\substack{\{\varepsilon_j\} \\ \varepsilon_j = S,A,C \\ \varepsilon_j \neq \varepsilon_{j+1}}} P_{I\varepsilon_1} P_{\varepsilon_1 \varepsilon_1}^{k_{i_1}-1} P_{\varepsilon_1 \varepsilon_2} P_{\varepsilon_2 \varepsilon_2}^{k_{i_2}-1} \cdots P_{\varepsilon_{t_i} \varepsilon_{t_i}}^{k_i t_i -1} P_{\varepsilon_{t_i} E}.$$

Then, according to (14), the sum of the products of probabilities of horizontal degrees of freedom of all paths with m degrees of freedom will be equal to the coefficient at x^n in the expanded expression $\omega_I(x)[\omega_E(x)]^{m-1}$.

We shall now introduce a generating function for the probability of the vertical degree of freedom

$$\mu(y) = \sum_{s=1}^{\infty} P_{EE}^{s-1} y^s = \frac{y}{1 - P_{EE} y},$$

then for m degrees of freedom,

$$[\mu(y)]^m = \frac{y^m}{(1 - P_{EE} y)^m} = y^m \sum_{s=0}^{\infty} C_{m+s-1}^{s} P_{EE}^{s} y^s = \sum_{n=m}^{\infty} C_{n-1}^{n-m} P_{EE}^{n-m} y^n. \tag{16}$$

The probability of all paths with m degrees of freedom will be the coefficient at $x^n y^n$ in the expanded expression

$$\omega_I(x) [\omega_E(x)]^{m-1} [\mu(y)]^m. \tag{17}$$

Let us examine some classes of paths (Fig. 6).

Let $\mathfrak{C}_n$ be the set of all possible paths with n steps along the vertical and n along the horizontal, with the last vertical step and with the initial state E (any number of degrees of freedom). C_n is the probability that the path belongs to $\mathfrak{C}_n$; $\mathfrak{D}_n^I$ is the set of all possible paths with n steps along the vertical and n along the horizontal, with the first horizontal and the last vertical step and with the initial state I = S, A, C, E. D_n^I is the probability that the path belongs to $\mathfrak{D}_n^I$; Finally, $\mathfrak{B}_n^I$ is the subset of all paths of $\mathfrak{D}_n^I$, which has no common points with the line x = y before 2n steps. B_n^I is the probability that the path belongs to

The following recurrent relation is then valid:

$$D_n^I = \sum_{j=1}^{n} B_j^I C_{n-j}. \tag{18}$$

Let us find the generating function for the probabilities of paths of all classes.

Class $\mathfrak{D}_n^I$

If (17) is summed for m, the number of degrees of freedom, then, on expanding into a series, the coefficient at $x^n y^n$ will be precisely the probability D_n^I. The coefficient at y^n may be selected, according to (16):

$$\sum_{m=1}^{\infty} C_{n-1}^{m-1} P_{EE}^{n-m} [\omega_E(x)]^{m-1} \omega_I(x) = \omega_I(x) [\omega_E(x) + P_{EE}]^{n-1}.$$

Hence, by the generating function for D_n^I, the member $f_0^I(x)$ will be free along x in the series

$$\sum_{n=1}^{\infty} \frac{\omega_I(x) [\omega_E(x) + P_{EE}]^{n-1}}{x^n} z^n = \frac{\omega_I \frac{z}{x}}{1 - \frac{z}{x}(\omega_E + P_{EE})} = \sum_{j=-\infty}^{\infty} x^j f_j^I(z).$$

Class $\mathfrak{C}_n$

Let us set $\mathfrak{F}_n = \mathfrak{C}_n \setminus \mathfrak{D}_n^P$.

F_n is the probability that the path belongs to $\mathfrak{F}_n$, and is equal to the coefficient at x^n in the expression expanded into a series

$$P_{EE} \sum_{m=1}^{n-1} C_{n-1}^{n-m-1} P_{EE}^{n-m-1} \omega_E^m = P_{EE} \left(\sum_{m=0}^{n-1} C_{n-1}^m P_{EE}^{n-m-1} \omega_E^m - P_{EE}^{n-1} \right) = P_{EE} [(\omega_E + P_{EE})^{n-1} - P_{EE}^{n-1}].$$

By the generating function for C_n, the member $g_0(z)$ will be free along x in the series

$$\sum_{n=1}^{\infty} \frac{\omega_E (\omega_E + P_{EE})^{n-1}}{x^n} z^n + \sum_{n=1}^{\infty} \frac{P_{EE}[(\omega_E + P_{EE})^{n-1} - P_{EE}^{n-1}]}{x^n} z^n =$$

$$= \frac{\omega_E \frac{z}{x}}{1 - \frac{z}{x}(\omega_E + P_{EE})} + P_{EE} \left[\frac{\frac{z}{x}}{1 - \frac{z}{x}(\omega_E + P_{EE})} - \frac{\frac{z}{x}}{1 - P_{EE} \frac{z}{x}} \right] = \sum_{j=-\infty}^{\infty} x^j g_j(z).$$

Let $v^I(z) = \sum_{j=1}^{\infty} B_j^I z^j$. Then, according to (18),

$$f_0^I(z) = v^I(z) g_0(z),$$
$$v^I(z) = \frac{f_0^I(z)}{g_0(z)}. \tag{19}$$

On expanding (19) into a series for z degrees of freedom, the coefficient at z^n is the probability B_n^I. Then

$$v^I(1) = \sum_{n=1}^{\infty} B_n^I = \frac{f_0^I(1)}{g_0(1)}.$$

Similarly, as pointed out earlier, there is one-to-one correspondence between $\mathfrak{B}_n^I$ and $\mathfrak{A}_{n-1}^I$, where $\mathfrak{A}_{n-1}^I$ is the set of those paths of $\mathfrak{D}_{n-1}^I$, that do not extend beyond the line x = y. The probability that the path belongs to $\mathfrak{A}_{n-1}^I$, is then the desired U_{n-1}^I and

$$B_n^I = \sum_{K=S,A,C} P_{IK} U_{n-1}^K P_{EE}.$$

We may then find, $\sum_{n=1}^{\infty} U_n^I$, I = S, A, C from the system of equations

$$v^I(1) = \sum_{n=1}^{\infty} B_n^I = \sum_{K=S,A,C} P_{IK} \sum_{n=2}^{\infty} U_{n-1}^K \cdot P_{EE} + \sum_{K=S,A,C} P_{IK} P_{KE}$$

$$I = S, A, C,$$

By having the system of equations (1), it is possible to model the process of stratification. In practice, the realization of the process corresponding to the proposed model may be obtained on an electronic computer by the Monte Carlo method.

The method of studying the process of stratification discussed in this paper permits us to examine problems associated with a number of geologic periods with complex accumulation and erosion, corresponding to the subdivision into degrees of freedom. A method is given for computing generating functions for probabilities of paths along which, in contrast to those considered in this paper, the number of accumulations and erosions may not be the same. By means of such generating functions it is possible to examine a number of other problems, such as finding the average number of steps necessary for transition from one eroded layer to another, not necessarily the next, and having different statistical characteristics.

REFERENCES

Kuenen, P. N., "Turbidity currents of high density," Intern. Geol. Cong. Rep., 18th session, Great Britain, Pt. 13, p. 44 (1948).

Vistelius, A. B., "The mechanism of stratification," Dokl. Akad. Nauk SSSR, Vol. 65, No. 2, p. 191 (1949a).

Vistelius, A. B., "The mechanism of correlation of stratification," Dokl. Akad. Nauk SSSR, Vol. 65, No. 4, p. 535 (1949b).

Vistelius, A. B., and Feigel'son, T. S., "Theory of formation of sedimentary beds," Dokl. Akad. Nauk SSSR, Vol. 164, No. 1, p. 158 (1965).

THE VERTICAL AND LATERAL VARIATION OF A CARBONIFEROUS LIMESTONE AREA NEAR SLIGO (IRELAND)

W. Schwarzacher

Queen's University
Belfast, Northern Ireland

The causes of vertical and lateral facies variation in Lower Carboniferous limestone are examined.

One of the outstanding aims of stratigraphic geology is the understanding of facies variation both in its areal and vertical development.

The available data are the lithologic and paleontological observations which have to be interpreted in terms of changing depositional conditions. Ideally we would like to explain the lithologic character at a given locality and position in the section as the result of an explainable interaction of environmental factors that are time dependent and operate in an area.

There are two major difficulties in the analysis of geologic data. First, geologic processes are so complicated that a great number of variables behave in an unpredictable way as random components. Secondly, there is no absolute time scale in the stratigraphic record that is accurate enough for the study of the more detailed time variation of facies. Both circumstances seem to lead to a definite limit in interpretation, and further progress can only be made by making assumptions about the nature of the variables.

Our procedure is, therefore, to examine first the data (geologic observations), then consider what assumptions we can make from a geologist's point of view about the nature of the lithologic variables, and, finally, reconsider facies variation as a result of environmental changes.

The Basic Data

The area which has been examined is a relatively small part of the Viséan (Lower Carboniferous) basin north and northeast of Sligo (northwest Ireland). Here we have concentrated on the Glencar Limestone, which is 100-120 m thick and which forms the middle part of a more extensive limestone and shale sequence [Schwarzacher, 1964]. As it was intended that the lateral as well as vertical variation in lithology be studied, every exposed section in this group was measured. However, conditions of exposure are such that only three localities with complete sections of the Glencar Limestone were found and only 14 localities in which the middle part of it is sufficiently exposed. Most conclusions about lateral variation are based on these 14 localities only.

The Glencar Limestone shows a very pronounced grouping of limestone bands. Usually, five to seven massive limestone beds occur closely together and are separated from the next group by a layer of shale. The thickness of such a group, which will be called an oscillation, is 2-3 m. We have chosen the top limestone band of such groups as boundaries to define our smallest stratigraphic units and have correlated them from locality to locality. In its complete development, the Glencar Limestone is built up from 50 such oscillations, but correlation is limited to groups 7-28. This is because the lower parts are not sufficiently exposed, and in the higher parts the oscillating nature of the sediments is not so marked. Correlation thus becomes doubtful. Groups 20-28 in particular make excellent marker horizons and can be identified in the field in all localities. Additional evidence as to the correlation is provided by a horizon of dolomite and chert in group 28 and very persistent coral beds in groups 7, 8, 10, 11, 20, 24, and 26.

The measured sections have been drawn on a scale of 1:20, recording limestone or shale by different symbols. To express the lithology in numerical terms, a limestone percentage for each 20-cm interval was calculated to the nearest 5%; this corresponds to the accuracy of measurement, which was to the nearest centimeter. The choice of the interval used for finding the limestone percentage is somewhat arbitrary and it determines to a certain extent the smoothness of the data. If we assume two series based on, say, a 10-cm interval x_{10} and a 20-cm interval x_{20} of the same data, we find the following relations.

Mean x_{20} = mean x_{10}, and approximately, for a long series:

$$\text{var } x_{20} \approx \frac{1+2r}{2} \text{ var } x_{10},$$

where r_1 is the first serial correlation coefficient of series x_{10}. In a random series of data, doubling the interval reduces the variance by one half; when (as is always the case with our data) serial correlation is present, the effect of smoothing is less. The limestone percentages based on a 20-cm interval are called the basic data. It must be clearly understood that this way of representing lithological variation in intervals without overlap does not introduce artificial serial correlation, but interpolation for smaller intervals will do so.

The Vertical Variation

The analysis of the vertical sequence attempts to resolve the lithologic variation into a number of components. This is done either for purely descriptive purposes such as an aid to stratigraphic correlation, or, in the hope that geologically meaningful variations can be isolated. The separation into components in itself involves some interpretation of the geologic data. This can be demonstrated by the following analysis. Let $y_{(z)}$ be the lithologic composition at a certain position z in the section; then we can write

$$y_{(z)} = u_{(z)} + \varepsilon_{(z)}. \tag{1}$$

We split the lithologic variation into a systematic component $u_{(z)}$ plus a random variable $\varepsilon_{(z)}$. The separation can only be carried out if we assume that $u_{(z)}$ is an analytical function which at least locally can be represented by a polynomial. Fitting of such polynomials is carried out most easily by applying Spencer's 21-point formula [Vistelius, 1961]. The formula

Table 1

Section	Basic values	Spencer	Resid.	2 cov, Spencer, residual
0	1025	420	525	80
1	1209	401	655	155
4	1424	575	664	185
14	1149	382	641	126

represents third-order polynomials accurately and is also a very good approximation for the higher orders.

We have attempted to estimate the order of magnitude of the two components $u_{(z)}$ and $\varepsilon_{(z)}$ using the following methods. First, a direct method is used to calculate the residuals from the Spencer smoothed basic data. We should find that, corresponding to (1).

$$\mathrm{var}_r y_{(z)} = \mathrm{var}\, u_{(z)} + \mathrm{var}\, \varepsilon_{(z)} + 2\,\mathrm{cov}\, u_{(z)}\varepsilon_{(z)}.$$

If $u_{(z)}$ is the systematic part and $\varepsilon_{(z)}$ is a random component, then the covariance term should vanish. In fact, there is a small correlation between the residuals and the Spencer smoothed values. We have therefore not completely succeeded in separating the random component. Data are given for four sections in Table 1.

The calculations suggest that the random component is of the order of 50% or more of the total variation.

Accepting that $u_{(z)}$ is a polynomial of order r, we can form the first, second, up to the r-th difference of the series, thus eliminating the polynomial term but increasing the variance of any random term which may be contained in the series. It can be shown that

$$\mathrm{var}\left(\Delta^r \varepsilon_{(z)}\right) = C_{2r}^r \, \mathrm{var}\, \varepsilon_{(z)}.$$

Calculated values of $\mathrm{var}(\Delta^r_{\varepsilon_z})$, divided by C_{2r}^r should eventually settle down to a value which is an estimate of $\mathrm{var}\, \varepsilon_{(z)}$. The somewhat tedious analysis showed that $\mathrm{var}\, \varepsilon_{(z)}$ is roughly 500, again 50% of var y, and that the sampling fluctuations of $\mathrm{var}(\Delta^r_{\varepsilon_z})$ lose significance after the fifth-order polynomial has been eliminated.

A third method which leads to a similar, not very precise result, is approximation of the basic data by a trigonometric expansion of the form

$$u_{(z)} = A_0 + A_1 \cos z + \ldots + A_i \cos iz + B_1 \sin z + \ldots + B_i \sin iz, \qquad (2)$$

from which series we can form the mean square deviation, a value which converges toward zero with increasing i. The convergence of a random variable should be much slower than that of any systematic part of the series. However, the random element will be incorporated even in the low-order harmonic coefficients and, since $\mathrm{var}\, \varepsilon_{(z)}$ is large in our case, we cannot expect too much from the method. For example, in section 0 (n = 440), a decrease in convergence of the mean square deviation occurs at roughly 550, corresponding to i = 45; that is at a wavelength of approximately 2 m, the random fluctuations take over and any systematic variation shorter than 2 m will be completely masked by the random element.

We come to the following conclusions: first, if we accept the model of a vertical variation consistent with equation (1), we may state that the random element is responsible for about 50% of the total variation. Secondly, because the random element is large, we cannot separate it completely from the systematic component. Finally, the systematic component can be approximated by polynomials of fairly low order and Spencer's method of smoothing seems an adequate procedure.

Plots of Spencer smoothed values are very useful for the general description of the series and for stratigraphic correlation. The series is not stationary, however, and attempts were made to remove a trend such that the mean of the series became zero throughout. This was done by harmonic smoothing; Spencer values were approximated by a trigonometric ex-

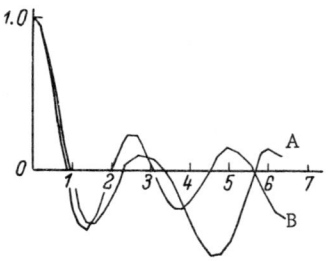

Fig. 1. Correlograms of Section 4. A) lower half (oscillations 6-20); B) upper half (oscillations 21-34).

pansion (2), which was terminated in such a way that it only took in variations with a wavelength exceeding 20 m. Depending on the length of the sections this involved the calculation of a harmonic series up to the fourth order for the longest sections. The trend was subtracted from the Spencer values and we may now write the model of vertical variation slightly differently as

$$y_{(z)} = u_{(z)} + T_{(z)} + \varepsilon_{(z)},$$

where $T_{(z)}$ stands for trend, which in this case is a rather artifical but simple analytical function.

A more detailed examination of $u_{(z)}$ shows that despite the removal of $T_{(z)}$ we have not succeeded in making the series stationary. If we use the method of Kheiskanen [1964], calculating correlograms for subsections, we find that the autocorrelation function changes when we progress along the section. For example, section 4, which comprises the lower and middle Glencar Limestone, gives two quite different correlograms for the lower and upper half of the sequence (see Fig. 1); similar results are found in all longer sections. This is undoubtedly due to a change in oscillation wavelength, which is not entirely random. Consequently, we find an increased damping of correlograms with increasing section length, which we cannot attribute entirely to random fluctuations in wavelength and amplitude of the oscillations. It would be difficult but, we believe, possible, to find a theoretical random process which fits the observed correlograms, but it seems very questionable if this would bring us nearer to a geological interpretation of the vertical variation.

The Problem of a Time Scale

The problem of vertical variation is obviously somehow linked to a variation of environments as a function of time. In such a detailed investigation we cannot equate thickness directly with time, and under the most favorable conditions we may only hope that the vertical dimension corresponds as an average to time. There are, however, many geological observations which suggest that sedimentation velocities vary considerably with different lithologies. A well-known relationship, for instance, is that of grain size and bed thickness in many sandstones [Pettijohn, 1949], which indicates that deposition is more rapid with increasing grain size. Petrographic evidence in the rocks under examination seems to suggest that deposition of the limestone was much more rapid than deposition of the shale. The latter contains many erosional breaks, and, particularly near the top and bottom of individual oscillations, layers of transported fossils, indicating prolonged reworking at that time of the development, occur.

If we make the vertical scale z dependent on the rock type $y_{(t)}$ laid down at a certain time t, we can examine how such a relationship will affect the vertical record, provided we assume a certain time sequence for $y_{(t)}$. Consider, for example, an environmental factor which produces a lithology $y_{(t)}$ in such a way that the time sequence of lithologies is given by

$$y_{(t)} = a - b \cos t.$$

In this, $y_{(t)}$ could stand for limestone percentage, and a and b are constants which can be chosen such that $y_{(t)}$ either always remains positive (continuous deposition) or such that erosion takes place. We now find the vertical position.

$$z_{(t)} = \int_0^t y_{(t)} dt = at - b \sin t.$$

Table 2

Oscilla-	Limestone			Shale		
tion	a_0	a_1	a_2	a_0	a_1	a_2
20	166.6705	3.7615	—4.2303	147.2537	—6.0258	0.9660
21	183.7727	1.5210	—5.6318	88.4764	—1.1218	—0.6212
22	102.9709	2.6270	—1.8759	50.6067	—1.0089	0.4207
23	114.3578	1.3614	—1.8407	68.0365	—0.8107	—2.1246
24	182.9279	—0.9940	—4.9032	58.2487	—1.3502	—0.9833
25	118.7208	2.1299	—2.7955	6.2096	0.9230	0.1139
26	161.7377	—0.0611	—4.3257	50.1275	—1.9706	—0.8655
27	137.7693	—0.0281	—4.6747	85.2872	—0.7672	—3.5959
28	148.7864	—1.6596	—4.3710	66.2307	—0.6166	—3.1683

This leads to curves known as cycloids which in parametric form can be written as

$$\left. \begin{array}{l} y_{(t)} = a - b \cos t, \\ z_{(t)} = at - b \sin t. \end{array} \right\} \quad (3)$$

The new vertical scale $z_{(t)}$ will have to be multiplied by a constant to fit the actual observed vertical scale. The determination of the constants a and b could be done by trial and error to provide the best fit. Excellent results were also obtained by choosing a such that the area of the cycloid corresponded to the known limestone percentage of a complete oscillation, and b such that the cycloid reached a maximum of 100%. This method, which is simple to apply, can sometimes lead to negative limestone percentages, which are of course unrealistic. They have been shown as zero limestone (Fig. 5).

It is believed that the main weakness of the method lies in the assumption that the lithologic response to an environmental variable is continuous, whereas, in fact, it is more likely that lithology can change abruptly if the environmental variation has reached certain critical values. In the case of the limestone—shale alternations, this could account for the truncated appearance of the variation curves.

The Lateral Variation

As in the study of vertical variation, we can again introduce smoothing as a device to bring out the essential part of lateral variation in lithology. The method employed is the well-known technique of trend surface fitting, whereby polynomials of a specified degree are fitted to the data by the least squares method [Miller, 1956]. In this study the lack of exposures (only 14 reliable localities) made it undesirable to attempt any high-order curve fitting. Consequently, only a second-order trend surface was fitted to the basic data, which were averaged for the whole middle limestone (group 20-28) (Fig. 2). The shape of this trend surface makes it clear that the lithologic variation (expressed in limestone percentages) is definitely related to the basin geometry. We know from the paleogeography of the area that the Sligo basin was bordered on the northwest by the mountains of Donegal and on the southeast by the Ox mountains; it probably was open to the southwest. The investigation of individual oscillations is even more informative. If we accept the field correlations as correct, then each of these stratigraphical units represents a time interval which is constant throughout the area, and thickness measurements should be directly proportional to sedimentation velocities. Since we have differentiated two lithologies, we can separate the thickness of shale and the thickness of limestone laid down in a single oscillation. Linear trend surfaces have been calculated on both limestone and shale thickness for 22 consecutive oscillations.

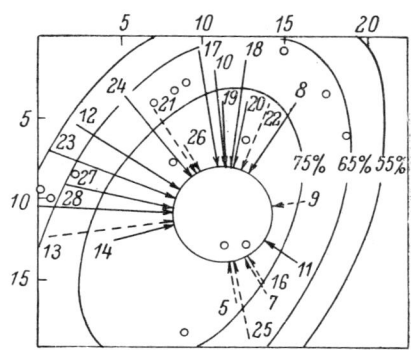

Fig. 2. Second-order trend surface of limestone percentages and gradients for linear shale thickness trends for oscillations 7-28. Solid arrows, distance of a 25-cm decrease in thickness; broken arrows, a 5-cm gradient. Localities of the sections are shown by small open circles. For scale see text.

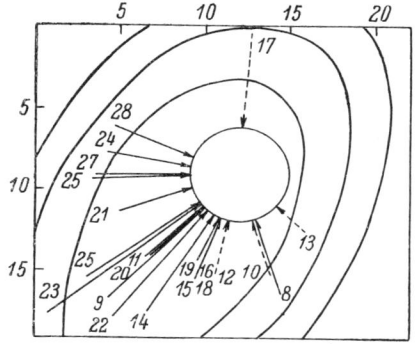

Fig. 3. Linear limestone gradients for oscillations 7-28.

The equation for the lithological variable $y(\varphi, \lambda)$ in this case is very simple:

$$y(\varphi, \lambda) = a_0 + a_1\varphi + a_2\lambda, \qquad (4)$$

where φ and λ are geographic coordinates measured from an arbitrary zero to the south and east, respectively in units of 0.595 km, and a_0, a_1, a_2 are constants which are given in Table 2.

The trend surfaces can be represented by a vector which gives the magnitude and direction of the gradient, the arrow pointing toward decreasing thickness and thus suggesting the direction of sediment transport. Dealing with linear surfaces we cannot exclude the possibility that the trend surface inclination is the resultant of two or more sources supplying sediment to one particular oscillation. However, the strong concentration of north and northwest directions of shale thickening (Fig. 2) strongly suggests that the source area of the clastic component lies in these directions. Geologically this is reasonable, since we know that the Donegal mountains supplied large amounts of deltaic sands from this direction in the lower parts of the succession. It is also possible to identify the southeast direction as the Ox mountain source. The gradients from this direction, if they occur, are as a rule much lower. It is very likely that both sources were continuously active and that the southeast source was overwhelmed for most of the time by the more massive influx from the northwest. There does not seem to be any regular alternation between the two sources. The only progressive development which can be made out is an anticlockwise shift from north-northwest to the northwest when the section enters the middle limestone (group 20-28).

The directions of the limestone gradients (Fig. 3) appear to occupy only one maximum in the southwestern quadrant. This maximum corresponds roughly to the longitudinal axis of the basin and suggests that the main limestone source was in the deeper southwestern parts of the basin. Paleoecological and paleocurrent studies will be carried out eventually, and they should clarify this problem. It is already known that lithostrotion and *caniniide* corals play an important role in the limestone distribution.

The limestone gradients also rotate when they enter the middle limestone but in an opposite sense to the shale gradients; this could be caused by a change in basin shape or possibly by faunal migration.

An important result of the limestone—shale trend-surface analysis is that the depositions of the two lithologic types were apparently quite independent of each other; there exists, therefore, no simple relationship between shale percentage of a stratigraphic unit and thickness of the unit. The previously postulated difference between shale and limestone sedimentation velocities is not contradicted by this, but there are too many variables to obtain quantitative data on this relationship.

Geological Considerations

We found that the vertical record of lithologic variation, at least for the purpose of analysis, can be split into three components. The purely random component, which accounts for roughly 50% of the variation, needs no further explanation, since we simply regard this as a kind of experimental error which includes our own error in measuring the section plus the instrumental error of geologic processes in recording environmental conditions. The two components $u_{(z)}$ and $T_{(z)}$ must be explained in geologic terms. The correlogram analysis of $u_{(z)}$ shows that this may be described by a random process that is persistent (ergodic) but not stationary, since the autocorrelation function changes for different parts of the section. One immediately suspects that the latter is caused by the vertical-scale—time relationship. In other words, we are almost certainly dealing with a time series in which the time scale itself undergoes fluctuations. This leads us to a situation in which mathematical analysis of the data has to stop, since we could produce a wide range of functions $u_{(t)}$ that could be transformed into the observed $u_{(z)}$. Also unsatisfactory is the separated component $T_{(z)}$, which, although it has been expressed in simple terms, does not suggest any obvious geological explanation. $T_{(z)}$ as a harmonic trend is only defined for the extent of the section and, of course, we must not extrapolate with this function.

The analysis can only proceed by making definite assumptions about $u_{(t)}$ and $T_{(t)}$. One possible assumption which we will investigate is that $u_{(t)}$, which is responsible for $u_{(z)}$, is a periodic function, that is,

$$u(t, p) = u(t + rp, p), \quad r = 0, 1, 2, 3 \ldots$$

We believe that there is some justification for this assumption. Most work on geological "cycles" has been carried out on rock sequences in which clastic sediments are very important, and it is usual to account for such "cycles" as being produced by changes in water depth, this being the easiest way out to explain the sequence. Any kind of tectonic movement is unlikely to be time periodic in the strict sense, and it would certainly be more appropriate to chose some autocorrelated oscillating model for such a process. In contrast to this, many cycles which have been observed in carbonate rock sequences are not easily explained by fluctuations in water level alone, as they are extremely constant over wide areas and frequently quite persistent through different facies developments. The typical grouping of 5-7 limestone bands [Schwarzacher, 1954] also seems to be a feature which occurs in many different formations and in all parts of the world. We believe, therefore, that the cause for this type of oscillating sedimentation must be external, i.e., not determined by the immediate environment of the basin. The most likely factors are climatic changes with all their related effects. There is still very little information available about long climatic cycles, but the effects of astronomical variations cannot be excluded, and these, at least over considerably long times, can be regarded as periodic functions.

The Periodic Model

If we accept the hypothesis that oscillations represent equal time intervals, $u_{(t)}$ is a periodic function and any unpredictable part of the variation must be taken up by the trend T. If we take one oscillation as a time unit, the trend can be accurately expressed as a function of time. We can now make use of the trend-surface analysis and obtain the average data of limestone percentages for all 14 localities. Naturally this average trend will vary for each locality, but it can be easily obtained for any chosen locality from the equations of the trend surfaces. Again it is more profitable to investigate the absolute amount of limestone and shale laid down in one cycle rather than the limestone percentage.

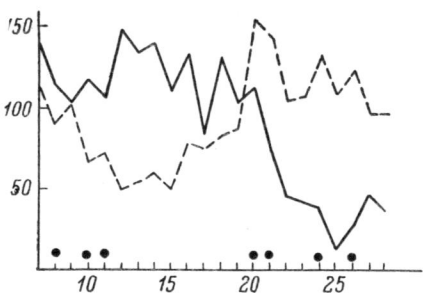

Fig. 4. Calculated time trend for locality 7.00, 9.00. Solid line, shale; broken line, limestone. Horizontal scale, oscillation numbers; vertical scale, thickness of sediment in cm. Black dots mark coral beds.

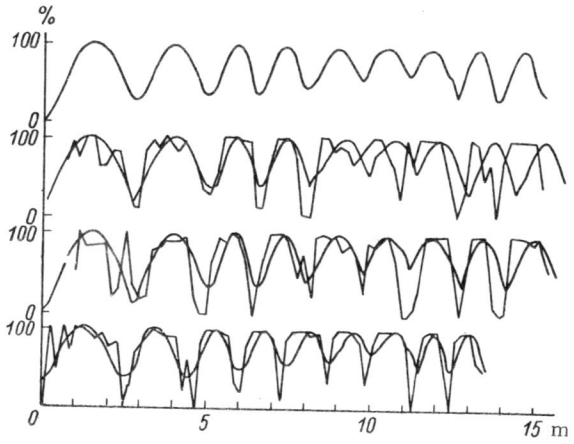

Fig. 5. Calculated and observed (basic data) limestone percentages for oscillations 20-28. From top to bottom: locality 7.00, 9.00; section 14; section 4; section 0. Horizontal scale in meters and vertical scale in percentage of limestone.

The trend was calculated for the imaginary locality $\varphi = 7.00$, $\lambda = 9.00$, which is close to the mean coordinates of all sample localities (see Fig. 4). The series is too short for any detailed analysis, but it seems obvious that both the shale and limestone trends are not pure random fluctuations. Geologic considerations make it likely that the shale trend follows a random process produced by a moving average; a longer but less reliable series, because it was derived from only one section, did in fact give the appropriate correlogram [Schwarzacher, 1964]. The moving average model was first introduced by Vistelius [1949] in context with sedimentation and it may be interpreted in geologic terms as follows. The source of any clastic sediment will not change in a purely random fashion as the amount of erosion is always, to a certain extent, determined by the previous history of the area. It is possible that more or less random tectonic movements can keep such a system active, but it is also possible that the random events that occur during the geomorphological development of the source and deposition areas are sufficient, without invoking major crustal movements.

The absence of any regularity in the supply from either the Donegal or the Ox mountains can be used as an argument against regional tectonic control of the source areas.

The limestone trend is probably very much related to the development of organisms in the basin. This trend shows several sharp peaks, which roughly correspond to beds in which numerous corals occur, and, although one cannot yet decide whether the corals are responsible for limestone deposition or vice versa, there is definitely a relationship between these two. It seems likely that the limestone trend can eventually be understood as a random process generated by living populations.

To investigate the lithologic variation within a cycle we make use of the vertical-scale—time relationship (3) under the assumption that the time change of environment follows in fact a sine wave. We have calculated the vertical variation of lithology under this assumption for all sections by fitting for each cycle the appropriate cycloid. In order to give the section continuity, the first value of a cycloid ($t = 0$) has been taken as the mean of its own initial value and the last value ($t = 2\pi$) of the preceding cycloid. The whole procedure has been computer programmed and it is thus possible to calculate vertical-variation curves for any locality needing as data only the constants given in Table 2 and, of course, the coordinates of the locality. Calculated results for sections 0, 4, and 14 (oscillations 20-28) are shown in Fig. 5, in which also the basic data for the sections are plotted. Apart from some slight phase shifts of the two curves the representation of the lithological variation is surprisingly good, so that one may hope that the method can be used for predicting the lithologic variation for localities which are not exposed. For example, the section for the locality $\varphi = 7.00$, $\lambda = 9.00$ has been calculated and is shown on the same figure. Calculated longer sections (oscillation 7-28) did not give such good correspondence to the actual sections because the phase shift, which is continuously added, becomes more noticeable. However, this is entirely due to the inadequate representation of the lateral variation by a linear trend surface and, if more localities were available, this could be corrected.

It is not possible to prove the correctness of the periodic model, but we have shown that it can be used to represent the lithologic variation with reasonable accuracy and that it may be even of practical use. To the author at least, it seems in this particular case more satisfactory to accept $u_{(t)}$ as a periodic function and replace $T_{(t)}$ by a random process which operates throughout the whole area.

REFERENCES

Kheiskanen, K. I., "Some features of sedimentational dynamics of the Middle and Upper Jatulian basin in Central Karelia," Sov. Geol., No. 12, p. 58 (1964).

Miller, R. L., "Trend surfaces: Their application to analysis and description of environments of sedimentation," J. Geol., Vol. 64, p. 425 (1956).

Schwarzacher, W., "Die Grossrhytmik des Dachstein Kalkes von Lofer," Tschermaks Min. Pet. Mitt., Vol. 4, p. 44 (1954).

Schwarzacher, W., "An application of statistical time-series analysis of a limestone—shale sequence," J. Geol., Vol. 72, p. 195 (1964).

Vistelius, A. B., "The mechanism of correlation of stratification," Dokl. Akad. Nauk SSSR, Vol. 65, No. 4, p. 535 (1949).

Vistelius, A. B., "Sedimentation time trend functions and their applications for correlations of sedimentary deposits," J. Geol., Vol. 69, p. 703 (1961).

USE OF A HARMONIC MODEL FOR ANALYSIS OF THE DYNAMIC SYSTEM OF SEDIMENTATION IN THE JATULIAN OF CENTRAL KARELIA

K. I. Kheiskanen

Institute of Geology
Ministry of Geology of the USSR
Petrozavodsk, USSR

Operation of the dynamic system of sedimentation, represented by the actual distribution of the mean grain size of sediment in the Lower Jatulian (Middle Proterozoic) of central Karelia, is satisfactorily described by means of a harmonic process adopted as a mathematical model. The conclusions are confirmed by data on cross-bedding of the corresponding deposits, by the areal distribution of weathering zones and facies types of basal deposits.

It was shown earlier [Kheiskanen, 1964] that of four possible types of models, the harmonic process best describes the operation of a natural dynamic system (in terms of the theory of automatic control), generating "at the output" the observed distribution of an investigated parameter of sediments in the section. The purpose of the present paper is to show some results obtained by means of the chosen model during a study of the distribution of mean grain size [Kheiskanen, 1964] of sediments in the clastic deposits of the Lower Jatulian (Middle Proterozoic) of central Karelia.

Material for the study comes from field descriptions of 22 sections of the Lower Jatulian rocks, made in 1961-64 by members of the lithologic party of the Geological Institute (Petrozavodsk): V. A. Sokolov, L. P. Galdobina, A. V. Ryleev, Yu. I. Satsuk, A. P. Svetov, and K. I. Kheiskanen. The description of section K was taken from data of the Porosozero party of Northwestern State University (SZGU), which worked in this area in 1960-61 under the direction of A. I. Kairyak.

The data were processed at the computer center of Petrozavodsk State University.

Parameters of the harmonic process adopted as a model were determined by approximation of the empirical distribution of mean grain size of the sediment (y) in each section of a mesorhythm curve [Vistelius, 1962]:

$$y = A \cos(\omega x - \varphi),$$

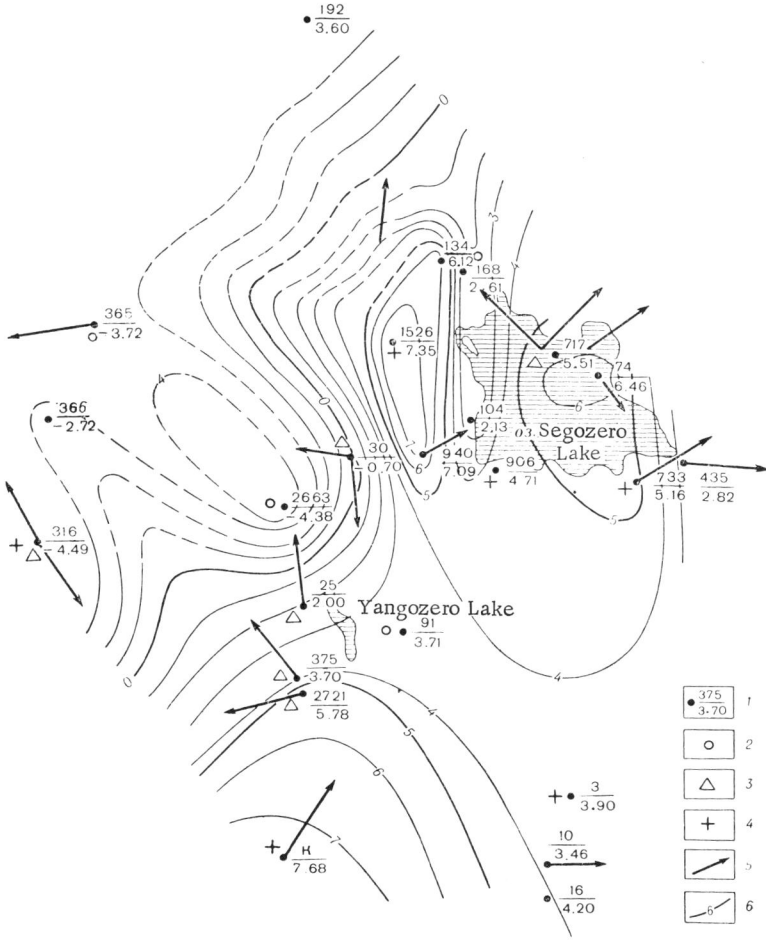

Fig. 1. Structure-contour map of the initial phase of the mesorhythm at the base of the Jatulian. 1) Section locality (numerator of fraction represents section number; denominator represents value of the initial phase of the mesorhythm at the base of the Jatulian); 2) appreciable amounts of conglomeratic rock in the section; 3) appreciable amounts of coarse arkosic rock in the section; 4) weathering zones on basement rock; 5) summed vector of dip direction of current cross-bedding in the basal Jatulian deposits (length of arrow is proportional to mean absolute value of the vector); 6) structure contours on basal phase of the Jatulian.

where A is the amplitude of the curve; $\omega = 2\pi/l$ is the frequency, inversely proportional to l; φ is the phase, and x is the thickness.

The values of the parameters were plotted on a map at points corresponding to the location of the sections, and between these structure contours were drawn. After this, the map was compared with different paleogeographic and lithologic data for the investigated region in order to study the geologic significance of the mesorhythm parameters, in particular, the phase of the beginning of the mesorhythm and the phases of different stratigraphic boundaries. Examples of such comparison are given below.

Let us examine the structure-contour map of the basal base of the Jatulian (Fig. 1), which is a rather complex surface with highs (in the vicinity of the K, 1526-940, and 717-74 sections), valley-like lows (the 365-2663-91, and 168-104 sections), and saddles (sections 91-3 and 940-906).

Since correlation of sections for extreme points of the mesorhythm curves is confirmed geologically, it may be stated that sedimentation began earliest in zones with the fewest phases of basal surfaces and latest in those with the greatest number of phases, i.e., the value of the beginning phase of a mesorhythm at the base of a section probably reflects in some measure the relief of the zone of sedimentation at the time the basal sediments accumulated.

The facts actually support this view.

1. Chemical weathering of the Pre-Jatulian basement has been established at a number of places (in the vicinity of sections 1526, 316, K, 906, 733, 3). Such places coincide with zones where the gradient of change in the initial phase of the mesorhythm has a low absolute value. At the same time, signs of weathering in the formation underlying the basal horizon is either absent (sections 104, 940, 30) or relatively weak (sections 717, 74, 375) where the absolute value of the gradient of the initial phase is large.

2. The thickness of quartzose conglomerates, widespread at the base of the Jatulian, is no more than a few meters over most of the region, but in sections coincident with the valley-like lows of the initial phases, it reaches hundreds of meters (180 m in section 91). In this section the conglomerates contain the coarsest fractions. In section 2663, the corresponding part, which is over 500 m thick, consists of sandstones and thin interbeds of conglomerate. In section 365 the thickness of synchronous deposits is still greater, but coarse-clastic rocks occupy but 50 m of the lower part of the section. It is clear that thick quartz conglomerates here occupy a greatly elongated zone, areally, along the axial line of which the grain size declines toward the northwest, and conglomerate gives way to sandstone.

On both sides of this zone the basal Jatulian deposits are coarse arkosic rocks, increasing in thickness where the gradient of change in the initial phase of the basal Jatulian is greatest (Fig. 1). In section 30, the thickness of coarse rocks is 350 m; between sections 940 and 91, according to V. Z. Negruts, it is about 200; in section 2721, 30 m; section 375, 40 m; section 25, 120 m; and, lastly, in section 316, individual beds of coarse rocks are found in a section of sandstone 480 m thick.

Thus, the axial zone of the valley-like lows of the basal Jatulian coincides with the zone of accumulation of monominerallic quartzose conglomerate, with definite facies transitions along the "valley," whereas on the "slopes," according to their steepness, more or less coarse arkosic rocks accumulated.

On a smaller scale, the same patterns are observed in the "valley" formed by structural contours on the initial phase between points in sections 940, 134, and 74.

3. The summed vectors of dip directions of cross-bedding (Fig. 1) observed in the basal deposits generally intersect the structural contours of the initial phase of the basal Yatulian nearly at right angles, directed downslope, i.e., in the direction of streams that deposited the cross-bedded sediments, and they thus agree with the assumptions we have made.

The list of such coincidences (of lithologic, structural-geologic, and other features of the Jatulian deposits in the investigated region) with the specific aspects of structural contours drawn on the basal phase could be enlarged, but the facts given above are sufficient for maintaining firmly that the structure contours on the initial phase of the mesorhythm at the base of the Jatulian reflect the principal relief features of the zone of sedimentation and that they may

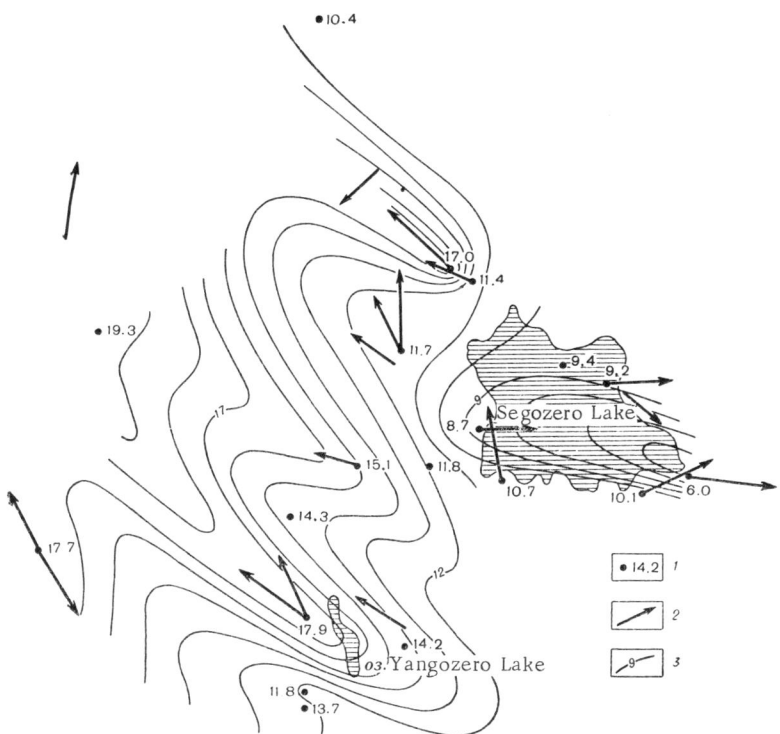

Fig. 2. Structure contours on the basal phase of the quartzites. 1) Location of section and value of basal phase of the quartzites (section numbers as in Fig. 1); 2) the same as (5) in Fig. 1, but for rocks of the quartzite unit; 3) structure contours on the basal phase of the quartzites.

serve as a basis for paleogeographic constructions and also for predicting occurrences of placer and other mineral deposits.

Let us examine still another stratigraphic boundary, the base of the quartzites, which form an exceptionally persistent lithic–stratigraphic unit in the region [Kheiskanen, 1964]. Structure contours on the base of the quartzites are shown in Fig. 2, where they form a series of embayments convex to the northwest, passing through sections 375-316, 2663-365, and 1526.

On the basis of the general geologic significance of the initial phase of the mesorhythm, mentioned above, it must be assumed that the indicated structure contours reflect a "front" of sediments, formed from quartzites, advancing from lower to higher phases. This assumption agrees beautifully with data on cross-bedding in the quartzites throughout the entire region except for the vicinity of Segozero Lake. In this latter locality, deviations are explained by the absence here of a complete unit of the sandstones that elsewhere occur above the quartzites. Either they were not formed in this locality or they were eroded before deposition of the sediments derived from the quartzites. In either case, the basal phase of the quartzites must have been reduced.

Another proof of the validity of the assumption that this "front" reflects actual conditions may be found in the general decrease in grain size of the quartzitic rocks toward the northwest, a fact established by lithologic studies.

An interesting detail appears when we compare the two structure-contour maps (Figs. 1 and 2). Advancement of the sands that formed after the quartzites were deposited took place more rapidly along valleys of tectonic relief, as revealed in maps of the basal Jatulian phases. Some of the valleys have been preserved since the beginning of Jatulian time (valleys passing through points in the sections 2663-365 and to the north of section 168); others (375-316, to the northwest of 1526) were reestablished and may not be reflected in the surface of the basal Jatulian plotted for few scattered points.

The minimum in the vicinity of Segozero Lake on Fig. 2 may be explained by a tendency of this zone to rise, as already noted for the beginning of Jatulian time.

Representing the operation of the dynamic system of sedimentation as a harmonic process thus gives results that agree well with lithologic data on the Lower Jatulian deposits of central Karelia. Consequently, in adopting a mathematical model constructed on objective data (distribution of mean grain size in sediments of the sections), we may satisfactorily describe the natural systematic patterns of Jatulian sedimentation.

This method, first used in studying the porosity distribution in carbonate rocks [Vistelius, 1945], may be recommended for studying different sequences of sedimentary rocks. It may prove to be especially useful when employing borehole data.

Completion of this work was aided by suggestions of A. B. Vistelius, advice and remarks of V. A. Sokolov, consultations with L. P. Galdobina, and the assistance of D. I. Dubrovskaya. The author takes this opportunity to express his sincere thanks to these people.

REFERENCES

Kheiskanen, K. I., "Some features of dynamic sedimentation in the Middle and Upper Jatulian basin of central Karelia," Sov. Geol., No. 12 (1964).

Vistelius, A. B., "Expression of the results of fossilization of fluctuating movements of the earth's crust by means of the series $\sum_{i=0}^{k} e^{a_i x + b_i} \cos(\omega_i x + y_i)$." Dokl. Akad. Nauk SSSR, Vol. 49, No. 7, pp. 531-535 (1945).

Vistelius, A. B., Phase Differentiation of Paleozoic Deposits of the Middle Volga and Trans-Volga Regions, Izd. Akad. Nauk SSSR, Moscow (1962).

ANALYSIS OF SEQUENCES OF MINERAL GRAINS IN GRANITES OF THE KYZYLTAS MASSIF (CENTRAL KAZAKHSTAN) AS A MANIFESTATION OF THE MARKOV PROCESS

D. N. Ivanov

Laboratory of Mathematical Geology
V. A. Steklov Mathematical Institute
Academy of Sciences of the USSR
Leningrad, USSR

This is a discussion of the results of studying sequences of rock-forming minerals (potassium feldspar, quartz, and plagioclase) in leucocratic granites of the Kyzyltas massif (central Kazakhstan). It has been established that these sequences in many cases correspond to simple homogeneous Markov chains. Analysis of the laws of these chains, obtained empirically, permits some conclusions to be drawn concerning aspects of crystallization of the investigated granites.

Early in 1964, A. B. Vistelius advanced the idea that, when studying sequences of the chief rock-forming minerals in magmatic rocks, information of a genetic character may be obtained. In this study it was thought that the sequences could be considered Markov chains. This view was confirmed in a study of grain sequences of the principal minerals in a thin section of granodiorite from Belaya (White) Mountain in the Central Range of Kamchatka [Vistelius, 1966a]. It was found that the investigated sequence may be transformed into a simple homogeneous Markov chain, which corresponds to a simple stochastic model of crystallization of a granodiorite magma.

A further step in this direction was development of a stochastic model for the crystallization of alaskitic granites. The model was based on the well-known diagram of Tuttle and Bowen [1960], characterized by a ternary eutectic in the center of the diagram. All possible crystallization paths were considered, depending on the initial position on the diagram and, for each of these, formulas were derived for the transitional possibilities characterizing theoretical sequences of mineral components on the assumption that they are simple Markov chains [Vistelius, 1966b]. Somewhat later, on the basis of the derived formulas, criteria were selected for determining correspondence between actual sequences and the theoretical model. If the sequence of mineral grains in granites corresponds statistically to a simple Markov chain and satisfies conditions that indicate no contradiction between the investigated rock and the scheme of Tuttle and Bowen, then these criteria make it possible to establish the most probable crystallization path and to compute the percentage contents of individual mineral grains at any particular stage of the crystallization process. Such a study was specifically carried out on an alaskite along the Karakul'dzhur River in the central Tien Shan [Vistelius, 1967].

In the present paper we shall examine data obtained in a study of mineral sequences in leucocratic granites and shall attempt to treat these data genetically.

Characteristics of the Investigated Granites

The Kyzyltas granite massif belongs to the so-called Permian plutons that have been rather thoroughly studied by several investigators in central Kazakhstan [Shcherba, 1948; Gokoev, 1949; Shevchenko, 1951; Koptev-Dvornikov, 1952; Monich, 1957; et al.]. The mass occupies an area of about 70 km^2 and lies within the free part of a horst-anticlinal uplift on the western limb of the Tokrau anticlinorium. The country rocks are Upper Devonian—Lower Carboniferous volcanic rocks, partly silicified. Structurally the mass is a central intrusive, with the center displaced somewhat toward the west-southwestern edge of the mass, and is considered one of those plutons of volcanic—plutonic associations that formed at the culmination of volcanic activity of appropriate volcanic centers [Zeilik and Vin'kovetskii, 1965].

Three varieties of granite are widespread within the Kyzyltas massif: coarse-, medium-, and fine-grained. These varieties are present in all the Permian granite intrusions of central Kazakhstan belonging to the Akchatau complex, and they have been the principal objects of past and present studies on these intrusions. Associated dikes are chiefly of intermediate and silicic composition, but pegmatites, aplites, and quartz veins are also present. Greisenization and potassium feldspathization are locally observed in the granites. We are interested in the three principal varieties of granites unaffected by secondary processes, and in the remainder of this paper we shall consider only these.

The coarse-grained granites form individual isolated outcrops up to 1.5 km^2 in area, confined chiefly to the marginal parts of the massif. Megascopically these are slightly porphyritic coarse-grained rocks with average grain size of 0.3-0.6 cm, but some grains are as much as 1.5 cm across. Besides the three principal minerals (potassium feldspar, quartz, and plagioclase), biotite and muscovite are sparsely present, and, even more sparsely, rare grains of epidote, sphene, apatite, fluorite, and ore mineral. Together these accessories constitute no more than 1-2% of the rock.

The potassium feldspar makes up 35-45% of the rock and occurs both as individual large grains and as aggregates of 2-3 grains. It is characterized by simple twinning, development of perthite, and alteration to clay. The potassium feldspar is the most xenomorphic mineral in the coarse-grained granites. At contacts with plagioclase it has irregular outlines and appears to corrode it. Against quartz the potassium feldspar forms comparatively straight and clean boundaries.

Quartz constitutes 35-40% of the rock and is found both in individual rounded grains and in aggregates. It is a later mineral than the plagioclase.

Plagioclase constitutes up to 30% (generally 25-27%) of the rock. It is chiefly oligoclase. The grains are generally sericitized in the central part, and they occur in individual grains and in small aggregates. Relative to quartz and potassium feldspar, the plagioclase is idiomorphic.

The coarse-grained granites are typically hypidiomorphic granular, with plagioclase relatively idiomorphic in comparison to quartz and to potassium feldspar.

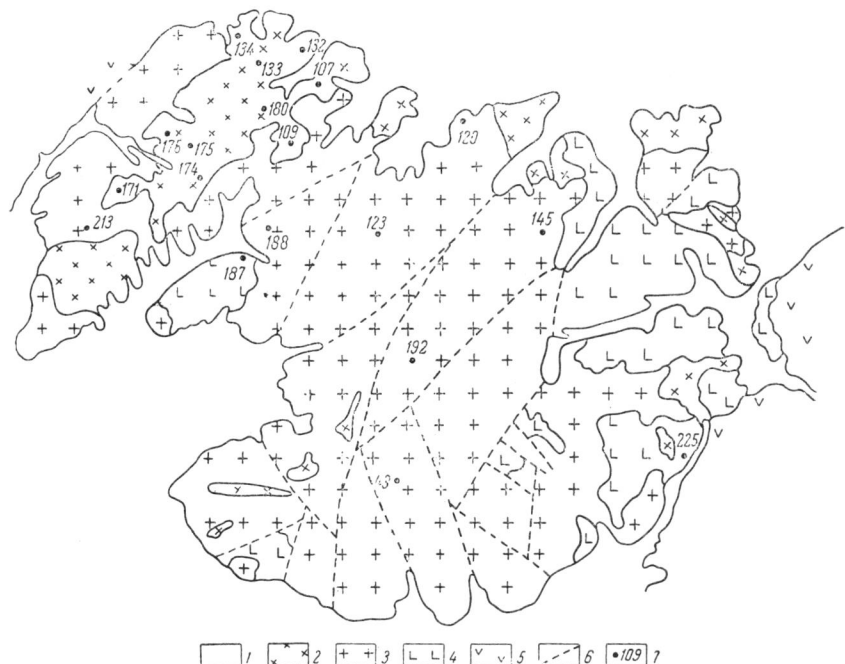

Fig. 1. Geologic sketch map of the Kyzyltas granitic massif, central Kazakhstan. 1) Recent deposits; 2) fine-grained granite; 3) medium-grained granite; 4) coarse-grained granite; 5) country rock; 6) fault; 7) sample locality.

The medium-grained granites form the great bulk of the Kyzyltas mass. They are porphyritic varigrained rocks, and they are thought to be younger than the coarse-grained granites. They are similar to the coarse-grained granites in their mineral relations. The principal minerals are again potassium feldspar, quartz, and plagioclase. The amount of mica and other accessory minerals is no greater than 1-2%.

Potassium feldspar makes up 35-40% of the rock and commonly occurs in aggregates of 2-4 grains, although individual grains are also present. Simple twins are present, and all grains exhibit perthitic intergrowths and kaolinitic alteration. The character of the contacts with plagioclase and quartz is similar to that found in the coarse-grained granites, i.e., potassium feldspar is later than the plagioclase, but it was impossible to determine definitely the relation with quartz.

Quartz constitutes up to 35% of the rock and is found chiefly in aggregates. Occasionally one may find small rounded inclusions of quartz in the marginal parts of large grains of potassium feldspar.

Plagioclase (oligoclase) forms 20-25% of the rock. The grains are sericitized and are found in aggregates; rarely one may find isolated individuals. At the contact with potassium feldspar, the plagioclase appears to be corroded. The contacts with quartz are generally smoother, but locally the penetration of quartz into plagioclase may be observed.

The medium-grained granites are hypidiomorphic granular. The potassium feldspar is the most xenomorphic of the principal minerals, and plagioclase exhibits the highest degree of idiomorphism.

The fine-grained granites are found chiefly in the marginal parts of the massif. Tabular (horizontal) and dike-like bodies of fine-grained granite have been observed, the latter with no porphyritic structure. We shall examine the fine-grained granites of a rather large tabular body in the western part of the massif. It consists of dense, distinctly porphyritic rock, with phenocrysts of potassium feldspar and quartz, rarely of plagioclase. This granite is younger than the coarse- and medium-grained granites, as indicated by apophyses and dikes of fine-grained granite in the coarse- and medium-grained varieties and also by the presence of xenoliths of the coarser varieties in the fine-grained granite. Some fine-grained varieties are but slightly porphyritic (less than 5% phenocrysts); others have up to 20% phenocrysts, and a few consist of up to 50% phenocrysts. The slightly porphyritic varieties were studied, and thin sections were prepared to avoid phenocrysts in the computation.

The mineral content of the fine-grained granites differs little from that of the other two varieties. The chief minerals are again potassium feldspar, quartz, and plagioclase, and biotite and accessory minerals are present in small quantities.

Potassium feldspar constitutes up to 45% of the rock and is found in phenocrysts and in the groundmass in individual grains, rarely as aggregates. Simple twins and fine perthitic intergrowths are present. Relative to quartz and plagioclase, this mineral is xenomorphic, but distinct corrosion of plagioclase, as observed in the medium- and coarse-grained granites, was not observed. Locally, rounded and vermicular intergrowths of quartz and potassium feldspar may be found, but these were not considered in the computation.

Quartz makes up 35-40% of the rock and, like the potassium feldspar, is present both as phenocrysts and as part of the groundmass. The grains are generally rounded and rarely form aggregates.

Plagioclase constitutes up to 20-25% of the rock, is oligoclase, and is generally sericitized in the central part of the grains. Idiomorphism of plagioclase relative to quartz and potassium feldspar is not always clearly demonstrated in the fine-grained granites. The plagioclase grains more commonly form aggregates than either the quartz or potassium feldspar.

The texture of the fine-grained granites varies in different parts of the body, ranging from hypidiomorphic granular in the central parts to aplitic and granophyric in the marginal parts. Chiefly samples with hypidiomorphic texture were studied, rarely ones with aplitic texture.

Collecting and Processing the Data

To study the nature of the sequences of principal minerals in the investigated varieties of leucocratic granites, samples were collected from various parts of the Kyzyltas mass. Samples of coarse-grained granite were taken from two small bodies, one on the eastern end, one on the western end of the mass; samples of medium-grained granite were collected throughout the entire mass, forming north-south and east-west profiles, so to speak. All samples of fine-grained granite, as already noted, were collected from a comparatively large tabular body at the western edge of the massif (see Fig. 1). In collecting samples, special attention was given to distance of locality from fracture zones, secondary alteration, quartz veins, and dikes. Thin sections were prepared from each sample, in an arbitrary direction, and continuous sequences of the principal minerals were examined on an integrating stage. The type of mineral (potassium feldspar, quartz, or plagioclase) under the intersection of the cross hairs was identified at intervals as the thin section was moved along its traverse. At the end of each traverse, at a distance of 1-2 grains from the edge of the thin section, the thin section was shifted perpendicular to this direction, a distance generally greater than twice the average grain size of the rock. In this shift the determination of minerals was continued, keeping continuity of the sequence, and the identification proceeded along the next traverse in a direction opposite to the previous one. In this way the entire thin section was counted. The number of counted grains depended both on grain size and on size of thin section, and it ranged from 120 in a slide of coarse-grained granite to 700-800 in a slide of fine-grained granite.

Only the principal minerals of the investigated rocks were counted, specifically the potassium feldspar, quartz, and plagioclase. Perthitic intergrowths of albite in the potassium feldspar, small rounded inclusions of quartz, and rare grains of dark and accessory minerals were not counted. During the counting, the minerals were considered states of continuous sequences and were indexed by numbers (1, 2, and 3). The successions of states thus obtained were tested for correspondence with a simple homogeneous Markov chain. The test was made on a BESM-2 electronic computer according to the program of the Laboratory of Mathematical Geology of the Order of Lenin Mathematical Institute of the Academy of Sciences, USSR. Among the operations provided by the program, the following are of interest to us.

1. Test of a sequence for homogeneity.

For this test, the sequence of states was broken down into two or three equal parts, depending on the length, and a matrix of transitional frequencies was constructed for each part. Each matrix was then compared with the matrices of all sequences by means of the χ^2 criterion for six degrees of freedom. When the compared matrices were similar, the sequence was considered homogeneous.

2. Test of a sequence for simplicity (simple hypothesis).

This test consisted of calculating the correlation between two indices separated by an index of a specific type. Thus, a simple hypothesis was established when potassium feldspar, quartz, and plagioclase were identified. From the sequence those three indices were selected for which the mean was a determined index (mineral), and matrices of transitional frequencies were constructed for the extreme indices of the selected three. The χ^2 criterion with four degrees of freedom was then used in these matrices to test for the absence of contingency, i.e., the constructed matrices were compared with theoretical matrices in which independence of frequency was specified. If no appreciable differences appeared between constructed and theoretical matrices, the sequence of states was considered to be a simple chain [Romanovskii, 1940].

3. Test of the coincidence of the second and third degree of the initial matrix of transitional frequencies with the matrices of transitional frequencies after one and two intervals, respectively (the test of Markov character).

In case of a simple Markov chain, this coincidence should take place according to the formula for total probability [Gnedenko, 1965]. Similarity of the obtained matrices, as for the test for homogeneity, is determined by the χ^2 test for six degrees of freedom.

These three tests are sufficient to discover if the constructed sequence of states is contradicted by a simple homogeneous Markov chain.

The discussed scheme was used to test sequences of mineral grains in thin sections from two samples of coarse-grained granite (225 and 187), seven samples of medium-grained granite (43, 120, 123, 145, 188, 192, 213), and 10 samples of fine-grained granite (107, 109, 132-134, 171, 174-176, 180). A matrix of transitional frequencies and the vector of *a priori* frequencies were constructed for each of the 19 sequences. The tests revealed correspondence of the discriminated sequences with a simple homogeneous Markov chain in the two samples of coarse-grained granites, in four of the medium-grained granites, and in five of the fine-grained granites. Homogeneity was lacking in sample 180, and the simple hypothesis for minerals 1 or 2 was disturbed in samples 109, 123, 132, 134, 145, 176, 180, and 213. Results of the tests are shown in Table 1.

Table 1. Results of Testing Sequences of Mineral Grains in Thin Sections of Granites in the Kyzyltas Massif for Correspondence with a Simple Homogeneous Markov Chain

Sample No.	Matrices of trans. freq. (1, potassium feldspar; 2, quartz; 3, plagioclase)	Vectors of a priori freq. and No. of transitions	Homogeneity test, $\chi^2_{0.05}$ = 12.6 (1, 2, 3- indices of parts of sequence)	Simple hypoth. test, $\chi^2_{5,.05}$ = 9.6 (1, 2, 3- indices of minerals)	Markov-charac. test, $\chi^2_{0.05}$ = 12.6 (2, 3- indices of square and cube of the matrix)
	Fine-grained granite				
107	$\begin{pmatrix} & 1 & 2 & 3 \\ 1 & 0.09 & 0.50 & 0.41 \\ 2 & 0.42 & 0.11 & 0.47 \\ 3 & 0.35 & 0.49 & 0.16 \end{pmatrix}$	$\begin{pmatrix} 0.30 \\ 0.36 \\ 0.34 \end{pmatrix}$ 516	$\chi^2_1 = 3.18$ $\chi^2_2 = 3.13$	$\chi^2_1 = 6.28$ $\chi^2_2 = 3.79$ $\chi^2_3 = 2.74$	$\chi^2_2 = 2.70$ $\chi^2_3 = 3.50$
109	$\begin{pmatrix} & 1 & 2 & 3 \\ 1 & 0.11 & 0.50 & 0.39 \\ 2 & 0.55 & 0.12 & 1.36 \\ 3 & 0.52 & 0.37 & 0.11 \end{pmatrix}$	$\begin{pmatrix} 0.38 \\ 0.33 \\ 0.29 \end{pmatrix}$ 708	$\chi^2_1 = 5.04$ $\chi^2_2 = 1.36$ $\chi^2_3 = 7.18$	$\chi^2_1 = 2.78$ $\chi^2_2 = 3.64$ $\chi^2_3 = 12.0$	$\chi^2_2 = 4.32$ $\chi^2_3 = 5.98$
132	$\begin{pmatrix} & 1 & 2 & 3 \\ 1 & 0.01 & 0.49 & 0.50 \\ 2 & 0.39 & 0.14 & 0.47 \\ 3 & 0.36 & 0.44 & 0.20 \end{pmatrix}$	$\begin{pmatrix} 0.27 \\ 0.35 \\ 0.38 \end{pmatrix}$ 815	$\chi^2_1 = 8.18$ $\chi^2_2 = 3.10$ $\chi^2_3 = 8.65$	$\chi^2_1 = 2.57$ $\chi^2_2 = 8.56$ $\chi^2_3 = 9.93$	$\chi^2_2 = 12.2$ $\chi^2_3 = 7.09$
133	$\begin{pmatrix} & 1 & 2 & 3 \\ 1 & 0.12 & 0.44 & 0.44 \\ 2 & 0.47 & 0.21 & 0.32 \\ 3 & 0.45 & 0.31 & 0.24 \end{pmatrix}$	$\begin{pmatrix} 0.34 \\ 0.33 \\ 0.33 \end{pmatrix}$ 528	$\chi^2_1 = 3.31$ $\chi^2_2 = 3.61$ $\chi^2_3 = 5.24$	$\chi^2_1 = 1.53$ $\chi^2_2 = 2.37$ $\chi^2_3 = 6.41$	$\chi^2_2 = 1.21$ $\chi^2_3 = 2.59$
134	$\begin{pmatrix} & 1 & 2 & 3 \\ 1 & 0.07 & 0.52 & 0.41 \\ 2 & 0.43 & 0.13 & 0.44 \\ 3 & 0.40 & 0.50 & 0.10 \end{pmatrix}$	$\begin{pmatrix} 0.31 \\ 0.37 \\ 0.32 \end{pmatrix}$ 624	$\chi^2_1 = 4.17$ $\chi^2_2 = 8.78$ $\chi^2_3 = 7.57$	$\chi^2_1 = 7.90$ $\chi^2_2 = 17.1$ $\chi^2_3 = 3.20$	$\chi^2_2 = 11.0$ $\chi^2_3 = 12.4$
171	$\begin{pmatrix} & 1 & 2 & 3 \\ 1 & 0.18 & 0.43 & 0.39 \\ 2 & 0.47 & 0.29 & 0.24 \\ 3 & 0.54 & 0.28 & 0.18 \end{pmatrix}$	$\begin{pmatrix} 0.38 \\ 0.34 \\ 0.28 \end{pmatrix}$ 504	$\chi^2_1 = 6.38$ $\chi^2_2 = 5.15$ $\chi^2_3 = 6.59$	$\chi^2_1 = 2.17$ $\chi^2_2 = 5.20$ $\chi^2_3 = 1.06$	$\chi^2_2 = 1.84$ $\chi^2_3 = 7.17$
174	$\begin{pmatrix} & 1 & 2 & 3 \\ 1 & 0.10 & 0.42 & 0.48 \\ 2 & 0.48 & 0.23 & 0.28 \\ 3 & 0.50 & 0.34 & 0.16 \end{pmatrix}$	$\begin{pmatrix} 0.35 \\ 0.34 \\ 0.31 \end{pmatrix}$ 696	$\chi^2_1 = 4.41$ $\chi^2_2 = 2.62$ $\chi^2_3 = 1.15$	$\chi^2_1 = 8.57$ $\chi^2_2 = 1.90$ $\chi^2_3 = 3.62$	$\chi^2_2 = 1.97$ $\chi^2_3 = 6.93$
175	$\begin{pmatrix} & 1 & 2 & 3 \\ 1 & 0.07 & 0.41 & 0.52 \\ 2 & 0.50 & 0.10 & 0.40 \\ 3 & 0.50 & 0.33 & 0.17 \end{pmatrix}$	$\begin{pmatrix} 0.35 \\ 0.29 \\ 0.36 \end{pmatrix}$ 585	$\chi^2_1 = 4.55$ $\chi^2_2 = 1.93$ $\chi^2_3 = 2.03$	$\chi^2_1 = 1.03$ $\chi^2_2 = 5.44$ $\chi^2_3 = 7.82$	$\chi^2_2 = 1.60$ $\chi^2_3 = 5.68$
176	$\begin{pmatrix} & 1 & 2 & 3 \\ 1 & 0.06 & 0.48 & 0.46 \\ 2 & 0.46 & 0.14 & 0.40 \\ 3 & 0.43 & 0.47 & 0.10 \end{pmatrix}$	$\begin{pmatrix} 0.32 \\ 0.35 \\ 0.33 \end{pmatrix}$ 444	$\chi^2_1 = 3.89$ $\chi^2_2 = 3.89$	$\chi^2_1 = 3.84$ $\chi^2_2 = 2.65$ $\chi^2_3 = 12.15$	$\chi^2_2 = 1.51$ $\chi^2_3 = 3.45$
180	$\begin{pmatrix} & 1 & 2 & 3 \\ 1 & 0.07 & 0.42 & 0.51 \\ 2 & 0.33 & 0.23 & 0.44 \\ 3 & 0.44 & 0.32 & 0.24 \end{pmatrix}$	$\begin{pmatrix} 0.30 \\ 0.32 \\ 0.38 \end{pmatrix}$ 528	$\chi^2_1 = 15.4$ $\chi^2_2 = 7.07$ $\chi^2_3 = 8.78$	$\chi^2_1 = 6.79$ $\chi^2_2 = 8.51$ $\chi^2_3 = 13.2$	$\chi^2_2 = 5.74$ $\chi^2_3 = 3.95$
	Medium-grained granite				
43	$\begin{pmatrix} & 1 & 2 & 3 \\ 1 & 0.31 & 0.41 & 0.28 \\ 2 & 0.54 & 0.27 & 0.19 \\ 3 & 0.42 & 0.33 & 0.25 \end{pmatrix}$	$\begin{pmatrix} 0.42 \\ 0.34 \\ 0.24 \end{pmatrix}$ 165	$\chi^2_1 = 1.76$ $\chi^2_2 = 1.63$	$\chi^2_1 = 3.23$ $\chi^2_2 = 2.10$ $\chi^2_3 = 4.64$	$\chi^2_2 = 3.47$ $\chi^2_3 = 3.14$
120	$\begin{pmatrix} & 1 & 2 & 3 \\ 1 & 0.14 & 0.51 & 0.35 \\ 2 & 0.43 & 0.32 & 0.25 \\ 3 & 0.39 & 0.32 & 0.29 \end{pmatrix}$	$\begin{pmatrix} 0.33 \\ 0.38 \\ 0.29 \end{pmatrix}$ 288	$\chi^2_1 = 5.67$ $\chi^2_2 = 4.66$	$\chi^2_1 = 3.23$ $\chi^2_2 = 2.24$ $\chi^2_3 = 5.67$	$\chi^2_2 = 4.31$ $\chi^2_3 = 7.78$
123	$\begin{pmatrix} & 1 & 2 & 3 \\ 1 & 0.25 & 0.32 & 0.43 \\ 2 & 0.32 & 0.42 & 0.26 \\ 3 & 0.45 & 0.26 & 0.29 \end{pmatrix}$	$\begin{pmatrix} 0.34 \\ 0.34 \\ 0.32 \end{pmatrix}$ 238	$\chi^2_1 = 5.53$ $\chi^2_2 = 5.53$	$\chi^2_1 = 10.7$ $\chi^2_2 = 3.11$ $\chi^2_3 = 4.00$	$\chi^2_2 = 8.28$ $\chi^2_3 = 1.32$
145	$\begin{pmatrix} & 1 & 2 & 3 \\ 1 & 0.25 & 0.32 & 0.43 \\ 2 & 0.30 & 0.47 & 0.23 \\ 3 & 0.46 & 0.23 & 0.31 \end{pmatrix}$	$\begin{pmatrix} 0.34 \\ 0.34 \\ 0.32 \end{pmatrix}$ 288	$\chi^2_1 = 4.66$ $\chi^2_2 = 4.53$	$\chi^2_1 = 6.08$ $\chi^2_2 = 1.15$ $\chi^2_3 = 6.26$	$\chi^2_2 = 5.47$ $\chi^2_3 = 3.63$

Table 1 (continued)

Sample No.	Matrices of trans. freq. (1, potassium feldspar; 2, quartz; 3, plagioclase)	Vectors of *a priori* freq. and No. of transitions	Homogeneity test, $\chi^2_{0.05} = 12.6$ (1, 2, 3- indices of parts of sequence)	Simple hypoth. test, $\chi^2_{0.05} = 9.6$ (1, 2, 3- indices of minerals)	Markov-character test, $\chi^2_{0.05} = 12.6$ (2, 3- indices of square and cube of the matrix)
188	$\begin{array}{c} \\1\\2\\3\end{array}\begin{pmatrix}0.31 & 0.35 & 0.34\\0.47 & 0.36 & 0.17\\0.54 & 0.18 & 0.28\end{pmatrix}$	$\begin{pmatrix}0.42\\0.31\\0.27\end{pmatrix}$ 360	$\chi^2_1 = 3.14$ $\chi^2_2 = 3.29$	$\chi^2_1 = 2.92$ $\chi^2_2 = 3.14$ $\chi^2_3 = 4.13$	$\chi^2_2 = 5.52$ $\chi^2_3 = 0.33$
192	$\begin{array}{c}1\\2\\3\end{array}\begin{pmatrix}0.21 & 0.48 & 0.31\\0.49 & 0.34 & 0.17\\0.46 & 0.20 & 0.34\end{pmatrix}$	$\begin{pmatrix}0.38\\0.35\\0.27\end{pmatrix}$ 252	$\chi^2_1 = 4.25$ $\chi^2_2 = 4.38$	$\chi^2_1 = 1.46$ $\chi^2_2 = 7.43$ $\chi^2_3 = 2.61$	$\chi^2_2 = 3.32$ $\chi^2_3 = 3.29$
213	$\begin{array}{c}1\\2\\3\end{array}\begin{pmatrix}0.22 & 0.47 & 0.31\\0.43 & 0.37 & 0.20\\0.47 & 0.31 & 0.22\end{pmatrix}$	$\begin{pmatrix}0.36\\0.39\\0.24\end{pmatrix}$ 300	$\chi^2_1 = 2.05$ $\chi^2_2 = 2.05$	$\chi^2_1 = 1.21$ $\chi^2_2 = 7.03$ $\chi^2_3 = 12.0$	$\chi^2_2 = 8.82$ $\chi^2_3 = 2.03$
Coarse-grained granite					
187	$\begin{array}{c}1\\2\\3\end{array}\begin{pmatrix}0.25 & 0.54 & 0.21\\0.58 & 0.30 & 0.12\\0.60 & 0.24 & 0.16\end{pmatrix}$	$\begin{pmatrix}0.44\\0.39\\0.17\end{pmatrix}$ 324	$\chi^2_1 = 3.55$ $\chi^2_2 = 3.10$	$\chi^2_1 = 2.03$ $\chi^2_2 = 6.05$ $\chi^2_3 = 5.48$	$\chi^2_2 = 3.73$ $\chi^2_3 = 0.88$
225	$\begin{array}{c}1\\2\\3\end{array}\begin{pmatrix}0.30 & 0.44 & 0.26\\0.51 & 0.28 & 0.21\\0.23 & 0.30 & 0.47\end{pmatrix}$	$\begin{pmatrix}0.36\\0.34\\0.30\end{pmatrix}$ 120	$\chi^2_1 = 2.06$ $\chi^2_2 = 2.42$	$\chi^2_1 = 2.73$ $\chi^2_2 = 5.02$ $\chi^2_3 = 1.05$	$\chi^2_2 = 0.68$ $\chi^2_3 = 2.72$

Analysis of the Results Obtained

The mineral sequences in the leucocratic granites of the Kyzyltas massif were studied in the following way. By means of criteria already worked out [Vistelius, 1967], the sequences were first compared (those that did not contradict the simple homogeneous Markov chain) with the probability model of alaskite crystallization, and all sequences were then analyzed without regard to whether they corresponded to the model.

As a result of comparing sequences not contradicting the simple homogeneous Markov chain with the theoretical model of alaskite crystallization, it was found that this model corresponds best of all to the sequence in the fine-grained granites. When the criteria indicate good agreement with the model (such correspondence was observed for samples 107, 133, and 175), the following scheme of crystallization appears most probable: $ab^{(1)}$, $ab^{(2)}$, $Q^{(2)}$, $ab^{(3)}$, $Q^{(3)}$, $Or^{(3)}$.*

In other words, crystallization of the fine-grained granites (or, more precisely, those parts of the fine-grained granites for which correspondence with the model was observed) began with the formation of plagioclase and continued along the plagioclase—quartz line. Finally, at the ternary eutectic point, all three minerals were precipitated (plagioclase, quartz, and potassium feldspar). The tendency for potassium feldspar to form in the final stage of crystallization is reflected in the low transitional frequencies of potassium feldspar—potassium feldspar in all matrices for the fine-grained granites, regardless of whether they corresponded with the model. At the same time, comparatively low frequency values of the quartz—quartz and plagioclase—plagioclase transitions were also observed. These latter are due chiefly to the nearness of points of initial crystallization to the eutectic point, for which reason the eutectic stage of crystallization was dominant in such rocks.

---

*Here and elsewhere in the article the following symbols apply: Or, potassium feldspar; Q, quartz; ab, plagioclase. The figures in parentheses indicate stage of crystallization.

The following picture holds for the medium- and coarse-grained granites: the matrices of transitional frequencies in all samples are characterized by relatively high values along the principal diagonal for all minerals. In this case, therefore, comparison with the theoretical model of alaskite crystallization was not made, since one of the fundamental premises on which the model is based — that grains of a single kind of a single generation cannot crystallize in a series — is not fulfilled.

It is thus shown that the theoretical model of alaskite crystallization "operates" only in individual cases for one variety of the investigated granites. The other two varieties do not generally fit the model. It does not follow from this, of course, that the model is fundamentally untrue, but it follows simply that it does not consider all possible deviations from comparatively ideal conditions provided for in the diagram of Tuttle and Bowen. The model should be considered primarily a kind of standard. Lack of correspondence with it for any mass means merely that there were deviations from conditions for which the diagram was constructed. The explanation and statistical evaluation of these deviations would clearly permit us to apply the model to a wider group of rocks.

We shall now consider the mineral sequences in the investigated granites as characteristics of rock properties without regard to whether the sequences correspond to the crystallization model. For this we shall examine the matrices of transitional frequencies and the vectors of *a priori* frequencies and, if this proves possible, we shall discriminate features in the matrices common to the individual varieties of granites. In analyzing the matrices we shall compare the transitional frequencies within columns in order to eliminate the effect of relative contents of any particular mineral on the transitional frequencies.

Let us turn to the data in the table. For the individual granite varieties, the following features are characteristic.

Fine-Grained Granite

1. The *a priori* frequencies for all three minerals are similar.

2. $P(Or, Q) > P(ab, Q)$, i.e., the probability of finding quartz side by side with potassium feldspar is greater than the probability of finding quartz with plagioclase. This means that quartz and potassium feldspar crystallized together in the fine-grained granites more readily than quartz and plagioclase. This aspect of the fine-grained granites is associated with the following.

3. $P(ab, ab) > P(Or, Or)$. It is this relation, due to the fact that potassium feldspar in the fine-grained granites crystallized last, plagioclase first, that caused plagioclase to form aggregates more readily than potassium feldspar; and this means that the amount of plagioclase at contacts with other minerals (including quartz) will be less (other conditions remaining the same) than the amount of potassium feldspar, which rarely forms aggregates.

4. $P(i, i) < P(i, j)$. In words, the probability of finding two grains of the same mineral side by side is less than the probability of finding two different minerals side by side. As pointed out above, this feature primarily reflects nearness of the initial point of crystallization of the fine-grained granites to the eutectic point.

Medium-Grained Granite

1. The *a priori* frequencies for potassium feldspar are near those for quartz, and both are greater than the *a priori* frequencies for plagioclase.

2. $P(Or, Q) > P(ab, Q)$, i.e., here is an aspect already noted for the fine-grained granites. But, clearly, the explanation must be different because observations establish no rule that

$P(ab, ab)$ is greater than $P(Or, Or)$ in all samples. In this case the feature is most likely due to the fact that a considerable part of the plagioclase occurs at contacts with potassium feldspar.

This conclusion is supported by the following relation.

3. $P(Or, ab) > P(Q, ab)$. In words, the amount of potassium feldspar coming in contact with plagioclase is greater than the amount of quartz in contact with plagioclase on the condition that the *a priori* frequencies of quartz and potassium feldspar are equal. All this directs one's mind toward the probability that potassium feldspar formed after crystallization of the rock (for example, by autometasomatism). This process naturally took place by replacement of previously formed minerals, and the most accessible mineral for replacement was plagioclase. This view is confirmed by observations in thin sections of the medium-grained granites, where corrosion of plagioclase by potassium feldspar is noted.

Coarse-Grained Granite

Not much in the way of conclusions may be drawn concerning this variety because of the few samples studied. The following features appear to be most characteristic, however:

1. $P(Or) > P(Q) > P(ab)$.
2. $P(Or, Q) > P(ab, Q)$.
3. $P(Or, ab) > P(Q, ab)$.

The second and third relations are found in the varieties investigated above, and nothing new may be added to the already discussed matrix of the coarse-grained granites.

Conclusion

1. All three varieties of granites differ from each other in *a priori* frequencies.

2. A comparison of mineral sequences in the granites of the Kyzyltas massif with the probability model of alaskite crystallization has shown that the most probable order of crystallization in the fine-grained granites is the following: $ab^{(1)}, ab^{(2)}, Q^{(2)}, ab^{(3)}, Q^{(3)}, Or^{(3)}$.

3. From an analysis of the matrices of transitional frequencies, it was possible to ascertain that crystallization of the fine-grained granites took place near the eutectic point and that potassium metasomatism took place in the medium- and coarse-grained granites, leading to partial replacement of the plagioclase by potassium feldspar. This process was clearly responsible for the lack of correspondence between empirically established sequences and the theoretical model.

REFERENCES

Gnedenko, G. V., A Course in the Theory of Probability, 4th edition, Gos. Izd. Fiz.-Mat. Lit., Moscow (1965).

Gokoev, A. G., "Comparative characteristics of some Permian intrusions of Sary-Ark (central Kazakhstan)," Izv. Akad. Nauk KazSSR, Ser. Geol., No. 11, pp. 108-117 (1949).

Koptev-Dvornikov, V. S., "Some systematic patterns in the formation of intrusions of granitoidal complexes," Izv. Akad. Nauk SSSR, Ser. Geol., No. 4, pp. 63-80 (1952).

Monich, V. K., Petrology of the Granitic Intrusions of the Bayan-Aul Region in Central Kazakhstan, Izd. Akad. Nauk KazSSR, Alma-Ata (1957).

Romanovskii, V. I., Statistical Problems Associated with Markov Chains, Izd. Uzb. Fil. Akad. Nauk SSSR, Tashkent (1940).

Shcherba, G. N., "Granitic intrusions of Akchatau," in: Granitic Intrusions of Kazakhstan, Izd. Akad. Nauk KazSSR, Alma-Ata (1948).

Shevchenko, E. V., Structural and Petrographic Features of Some Permian Intrusions of Central Kazakhstan, Izd. L'vovsk. Univ. (1951).

Tuttle, O. F., and Bowen, N. L., "Origin of granite in the light of experimental studies in the system $NaAlSi_3O_8-KAlSi_3O_8-SiO_2-H_2O$," Geol. Soc. Am. Mem., Vol. 74 (1960).

Vistelius, A. B., "Formation of the Belaya granodiorites on Kamchatka," Dokl. Akad. Nauk SSSR, Vol. 167, No. 4, pp. 1115-1118 (1966a).

Vistelius, A. B., "A stochastic model of crystallization of alaskites and the corresponding transitional probabilities," Dokl. Akad. Nauk SSSR, Vol. 170, No. 3, pp. 653-656 (1966b).

Vistelius, A. B., "Crystallization of alaskites from the Karakul'dzhur River," Dokl. Akad. Nauk SSSR, Vol. 172, No. 1, pp. 165-167 (1967).

Zeilik, V. A., and Vin'kovetskii, Ya. A., "The trend in prospecting for 'blind' ore deposits in connection with intrusive magmatism and vulcanism in central Kazakhstan," Izv. Akad. Nauk KazSSR, Ser. Geol., No. 2, pp. 15-18 (1965).

USE OF THE COMPUTER FOR QUANTITATIVE ANALYSIS OF FOSSIL DISTRIBUTION

W. T. Fox

Department of Geology
Williams College
Williamstown, Massachusettes

Time-trend analysis is a mathematical technique used to smooth out irregularities in geologic data based on a sequence of observations taken along a traverse or on a vertical measured section. A high-speed digital computer is used to calculate smoothed values and plot the time-trend curves, which in turn are used to study the underlying trends of sedimentation. Nine smoothing equations containing from 5 to 21 terms are available to compute curves with varying degrees of smoothing. The smoothing sum of squares is a measure of the total variability accounted for by each of the smoothing equations and can be used to detect a rhythm or cycle in the data. Several closely spaced, measured sections are being investigated in the Upper Ordovician of southeastern Indiana to reconstruct the sequence of environments which existed during their deposition. Time-trend curves are being used to study the relationship between physical properties of the environment and faunal distribution. The percentage of limestone in a sequence of thin limestone and shale layers is used as an index of environmental conditions. The corresponding time-trend curves are plotted for ten genera of brachiopods to examine their vertical distribution. There appears to be a definite relationship between the amount of limestone in the section and the occurrence of certain brachiopod genera. The maximum on the limestone percentage curve corresponds closely to the disappearance of one brachiopod genus and the introduction of two new genera, indicating a significant shift in environmental conditions.

Introduction

The shifting patterns of ancient environments are reflected in the physical, chemical, and biological properties of a sedimentary rock. As the sea transgresses over an area, the depth of water changes and affects the distribution of waves and currents and the amount of sunlight that filters through the water and reaches the bottom. With a transgression, daily movements of the tides are also altered, which accounts for local differences in salinity. Small raised mounds in the bottom topography form shoal areas which may break the surface to become small islands. Benthonic organisms which live in rather restricted environments are also profoundly affected by subtle changes in the environment. When the environment changes beyond the lethal limits of an organism, the organism must either emigrate to another area, adapt to the new environment, or be killed.

There are several types of evidence preserved in the sediments which allow the geologist to make a reasonable interpretation of the environmental conditions that existed when the rocks were deposited. The amount and type of matrix material or secondary cement which fills the interstitial space of a sandstone or limestone can be used as an indication of the amount of turbulence in the depositional environment. The presence of dolomite in dominantly calcitic limestone may be indicative of increased salinity. Primary sedimentary structures such as ripple marks, cross-bedding, and mud cracks, can be used directly for environmental interpretation. By studying the sequence of changes in the physical attributes of a sediment through a vertical stratigraphic section, it is possible to plot the changes in environment which were responsible for the present configuration of the strata.

A majority of the fossils that are preserved on a single bedding plane or within a thin sedimentary unit represent the bottom-dwelling organisms which existed at the instant in geologic time when the rocks were deposited. Many members of the original ecologic community were not fossilized for various reasons, and some of the fossils on the bedding plane may have been transported into the area after their death. If the sediment is relatively undisturbed and most of the fossils are unbroken with some in life positions, it is reasonably safe to assume that the fossil assemblage constitutes a fair representation of the benthonic community.

In the final paleoecologic analysis, the interpretations based on the lithologic aspects of the sediment are combined with the paleontologic interpretations to reconstruct the succession of environments and to show the effect they had on the distribution of organisms. Where possible, it would also be interesting to study the effect which rate of change in environment had on the rate of evolution within a species or a larger taxonomic unit. It also might be possible to reverse the procedure and interpret changes in environment from changes in the faunal assemblage where the lithology is relatively homogeneous. This would give the geologist another delicate tool to be used in interpreting minor transgressions and regressions of the sea which are useful in locating the position of the shoreline.

One of the major problems encountered in a study of faunal assemblages within a vertical sequence of beds is the portrayal of the distribution of the various fossils. If a bed-by-bed record is kept of faunal abundance, a simple plot such as a bar graph will be cluttered by the widespread fluctuations in abundance. It is important to plot the relative abundances in just enough detail to emphasize the significant trends without burying the trends in a mass of local variations. Where an attempt is made to relate the occurrence of fossils to the conditions of the depositional environment, it is necessary to correlate the major trends in fossil abundance with physical and chemical factors of the sediment. The most satisfactory method found so far in plotting the fossil data is by means of smoothing the data by using time-trend functions.

Time-Trend Analysis

The major objective of time-trend analysis is to smooth out small-scale fluctuations in the raw data which tend to conceal the major trends of sedimentation. Several different methods are available for smoothing geologic data and have been tried by different workers in the field. The simplest method of smoothing is to use a moving average which covers a wide enough interval to iron out minor irregularities in the raw data. This technique has been used successfully by several workers in different phases of stratigraphy. In a recent paper on the Upper Ordovician limestones and shales of southern Ohio, Weiss, Edwards, Norman, and Sharp [1965] used a three-foot interval to construct moving average curves for clastic ratio, bedding indices, and relative abundance of certain limestone types. Because the moving average technique places equal weight on all the beds within the interval, it loses its value as a sensitive indicator of environmental changes. In using moving averages, the shape of the smoothed curve is somewhat distorted by high values at either end of the smoothing interval.

Time-trend analysis, which is in essence the use of a weighted moving average, uses a smoothing equation to give a more accurate representation of the underlying trend of sedimentation. Two distinct methods are available for determining the constants which are used in the smoothing equations. A brief discussion of these methods follows with a more complete examination of their derivation and application given by Whittaker and Robinson [1929, pp. 285-296] and W. E. Milne [1949, p. 279].

The first smoothing formula to be widely accepted was derived by Woolhouse in 1870. He attempted to smooth a series of observations by plotting the values on a graph and drawing a series of equally spaced parabolas through the points. To obtain a single smoothed value, he used a sequence of 15 observations, 7 on each side of the value being smoothed. Five parabolas were drawn, each passing through three points in the sequence. The arithmetic mean of the points of the parabolas passing through the midpoint was used as the smoothed or graduated value. The graphical procedure carried out by Woolhouse was transformed into a summation formula which is

$$u_0' = \frac{1}{125} [25u_0 + 24(u_1 + u_{-1}) + 21(u_2 + u_{-2}) + 7(u_3 + u_{-3}) + 3(u_4 + u_{-4}) - 2(u_6 + u_{-6}) - 3(u_7 + u_{-7})], \quad (1)$$

In the Woolhouse formula and the formulas that follow, u_0 is the value being smoothed, and u_n is the n-th value above or below the value being smoothed.

The Woolhouse formula is one of a class of smoothing formulas which are called "summation formulas." A more widely accepted formula in the same group was developed by Spencer in 1893. Spencer's formula, which uses 10 terms on either side of the value being smoothed, has a total of 21 terms:

$$u_0' = \frac{1}{350} [60u_0 + 57(u_1 + u_{-1}) + 47(u_2 + u_{-2}) + 33(u_3 + u_{-3}) + 18(u_4 + u_{-4}) +$$
$$+ 6(u_5 + u_{-5}) - 2(u_6 + u_{-6}) - 5(u_7 + u_{-7}) - 5(u_8 + u_{-8}) - 3(u_9 + u_{-9}) - (u_{10} + u_{-10})]. \quad (2)$$

The working process of Spencer's 21-term formula is given by Whittaker and Robinson [1929, p. 291] and a geologic example is given by Vistelius [1961, p. 708].

The Spencer 21-term formula has been used extensively by Vistelius [1961] for smoothing lithologic observations from the Tertiary red-bed sequence of Cheleken Peninsula (the Caspian Sea) and several other areas in the USSR. Using the special computing scheme outlined in Vistelius [1961] along with simple computation equipment, auxiliary tables, and stencils, a single operator could smooth approximately 600 points in one day. For ease in computation, range numbers are assigned to the distinct lithologies in multiples of seven. Vistelius was quite successful in using smoothed curves for correlating between closely spaced measured sections and boreholes. The same techniques were also used by Romanova [1957] in a study of the Cheleken red-bed deposits.

A second class of smoothed curves derived by Sheppard in 1912 is based on a completely different principle. Sheppard developed a series of smoothing equations by fitting a parabolic curve to the points within a smoothing interval and using the ordinate of the parabolic curve as the smoothed value. The constants of the parabolic curve are determined by the method of least squares. A complete table listing the smoothing equations with the number of terms varying from 1 to 21 is given in Whittaker and Robinson [1929, p. 279]. As examples of this method of smoothing, the 5-term and 7-term formulas of Sheppard are given below.

$$u_0' = \frac{1}{35} [17u_0 + 12(u_1 + u_{-1}) - 3(u_2 + u_{-2})], \quad (3)$$

$$u_0' = \frac{1}{21} [7u_0 + 6(u_1 + u_{-1}) + 3(u_2 + u_{-2}) - 2(u_3 + u_{-3})]. \quad (4)$$

The smoothing equations with the same number of terms derived by Spencer and Sheppard have different constants and produce quite distinct smoothed curves. Sheppard's 5-term equation was used extensively by Vistelius [1944] in a study of the porosity distribution in the Upper Permian beds of the Buguruslan region. The 5-term equation by Sheppard was used by Fox and Brown [1965] to study small-scale fluctuations in the environmental interpretation of Upper Ordovician limestones in southeastern Indiana. In the same paper, Spencer's 21-term equation was used for the major underlying trend of sedimentation. The author selected these two curves because of their ease in computation. The 5-term equation is relatively easy to calculate, and the computational scheme is available for Spencer's 21-term equation.

With the availability of high-speed digital computers, it is now possible to experiment with different smoothing curves and to select the equation which fits a particular problem, instead of selecting an equation because of its computational ease.

With the use of the computer, it is now possible to calculate and plot the curves with greater speed and accuracy, and this allows the geologist the opportunity to experiment with different smoothing intervals and equations in order to determine which best fits the data.

Computer Program for Smoothing and Plotting Curves

With the widespread application of high-speed digital computers to the field of geology, a computer program was compiled to calculate and plot smoothed curves based on a series of observations taken from a traverse or stratigraphic section. The program was written in FORTRAN (FORmula TRANslator) computer language for use on the IBM (International Business Machines Corporation) Model 7094 computer system. The program requires 24,757 storage locations in the core and takes about 1 min to calculate and plot 600 points on a smoothed curve. A complete listing of the computer program [Fox, 1964] and instructions for handling the data are available from the Kansas Geological Survey, Lawrence, Kansas.

When observations have been made on several variables, including the abundances of different fossil species and mineralogic composition of the rock, the data are punched onto a single card or a group of cards for each sedimentation unit or stratigraphic interval. The computer will calculate and plot the smoothed curves for each of ten variables with up to 500 observations per variable. With some minor modifications of the program since it was published in 1964, it is now possible to plot the curves for 30 variables with a single pass through the computer. This is quite useful when observations have been made on several variables within each sedimentation unit.

The computer program is subdivided into a main program and several subroutines for processing the data. The main program is used to read in the data and the program control cards which are used to select or reject various program options. The subroutines are used to establish the smoothing interval, to compute the average value for each variable within consecutive increments of the smoothing interval, and to calculate and plot the smoothed curves.

The first subroutine, FIXINT (FIX INTerval), is used to determine the constant thickness increment or "smoothing interval" which is used in the subsequent subroutines. The thickness of sedimentation units may vary considerably within a stratigraphic section and must be converted to equal increments for smoothing. Two options are available with the program for determining the smoothing interval. The first method of obtaining the smoothing interval is to use a multiple of the average thickness for the sedimentation units. In dealing with a section where there are only slight lithologic differences in adjacent sedimentation units, the average thickness can be used directly as the smoothing interval. Where two distinct lithologic types, such as limestone and shale, alternate with a fair degree of regularity, they may be considered a sedimentation pair. In smoothing a sequence of sedimentation pairs, each pair is considered as a sedimentation unit. Two times the average rock unit thickness is therefore used for the smoothing interval.

The second subroutine, SETVAL (SET VALue) is used to calculate the average for each variable within each consecutive smoothing interval. To compute the average within the smoothing interval, the sum of the thickness times the observed value for all the layers in the interval is divided by the thickness of the smoothing interval. Since the thickness and observed values for each sedimentation unit are punched onto data cards for processing the program, all the calculations are handled by the computer. By changing commands on the control card, it is possible to use a different smoothing interval and have the computer calculate a new set of average values for use by the smoothing equations. In using the average value for an interval in contrast with a single observation, all the data collected in the field are represented on the final curve. In using the smoothed curves to correlate a series of measured sections, it usually is not essential to include each small unit in order to trace the major trends. In making environmental interpretation, on the other hand, it is necessary that the average value for the interval include all the fossil species present in thin units. If single observations are made at constant intervals throughout the section, it is possible to pass over relatively thin sedimentation units which contain ecologically significant species.

Nine distinct smoothing equations, which operate with varying degrees of smoothing, are available with the program. Subroutine SMOOTH is used to select one or more of the smoothing equations, as indicated by the control card read with the data in the main program. The nine equations which are used with the program published by Fox [1964] are those derived by Sheppard and listed by Whittaker and Robinson [1929, p. 279]. The equations range from a 5-term equation which uses two terms on each side of the value being smoothed to a 21-term equation which includes ten values on each side of the central value.

As an example of how the program works, a series of curves which were plotted by the computer using each of the nine smoothing equations on the same set of observations is included as Fig. 1. The curves are based on the percentage of limestone within an alternating sequence of thin-bedded limestone and shale layers. Two times the average bedding thickness (11.21 cm), or the average thickness of a limestone—shale pair, was used for the smoothing interval. The curve in the first column under raw data is based on the percentage of limestone present within consecutive smoothing increments. These values are used as the raw data for computing the smoothed curves in the remaining columns which result from applying smoothing equations with an increasing number of terms. With an

Table 1. Statistical Data for Curves in Fig. 1

Terms in smoothing equation	Smoothing sum of squares	Percentage of total sum of squares	Percentage of total sum of squares	Percentage increase
Raw data	1014.4	100.0	—	—
5	322.5	32.7	67.3	23.9
7	548.3	56.6	43.4	2.4
9	564.4	59.0	41.0	8.1
11	630.8	67.1	32.9	5.3
13	685.2	72.4	27.6	—0.7
15	673.7	71.7	28.3	4.9
17	719.7	76.6	23.4	9.1
19	786.3	84.7	15.3	—1.4
21	763.4	83.3	16.7	—

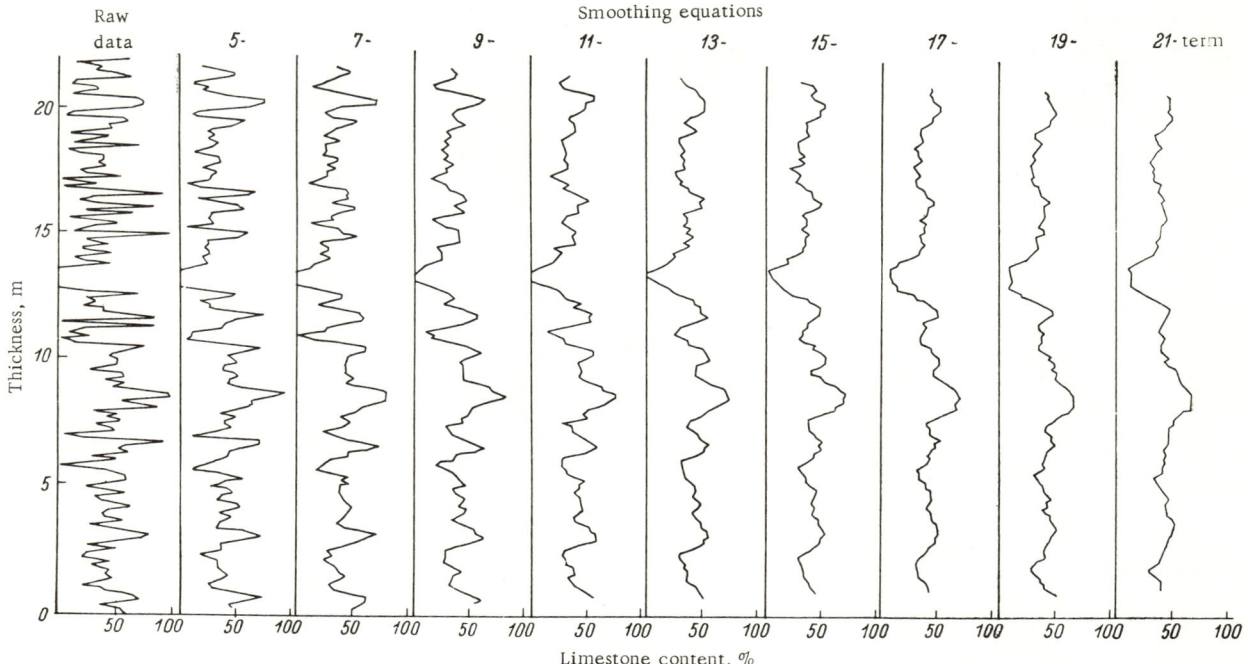

Fig. 1. Limestone percentage curves based on the raw data and nine smoothing equations with from 5 to 21 terms.

increasing number of terms in the smoothing equation, the fluctuations in the original data are subdued and the underlying trends of sedimentation are accentuated. In using the smoothing equations derived by Sheppard, there are still minor fluctuations in the smoothed curves even when the 21-term equation is used. This is not dependent on the data but is a characteristic of the smoothing equations. When the Woolhouse 15-term equation, (1), or the Spencer 21-term equation, (2), is used, the minor fluctuations drop out and the resulting curve is completely smoothed.

In order to compare the smoothing effect of the different smoothing equations, the program computes several statistical parameters. The mean, variance, total sum of squares, and standard deviation are calculated for the raw data on each of the variables. For each smoothed curve, the total sum of squares of deviations from the curve is computed and referred to as the "smoothing sum of squares." By comparing the smoothing sum of squares for a particular set of curves with the total sum of squares of deviations from the mean, it is possible to measure the amount of smoothing which can be attributed to each of the smoothing equations. By using the smoothing sum of squares, it is possible to select smoothed curves which most closely correspond to natural cycles in the observed data.

The smoothing sum of squares and percentage data for the curves presented in Fig. 1 are given in Table 1. The first column lists the number of terms in each smoothing equation. The second column gives the smoothing sum of squares computed for each smoothed curve, with the sum of squares of deviations from the mean given for the raw data. The percentage of the total sum of squares which remains after each curve is smoothed is given in column 3. The actual percentage of the total sum of squares accounted for by each curve is listed in the fourth column. The fifth column shows the additional percentage of the total sum of squares which can be accounted for by moving to a smoothing equation with fewer terms. As can be seen from column 5 in Table 1, the 5-, 9-, and 17-term equations account for the greatest increases in the total sum of squares. This would indicate that the smoothed curves resulting from these equations most closely match the natural cycles or rhythms of sedimentation in that particular section. In computing a series of curves for a section which includes data on fossil abundance and various parameters, it is advisable to compute the complete set of nine smoothed curves for the major lithologic attribute which, in this case, is the percentage of limestone. The curves which give the highest values for the percentage increase in smoothing sum of squares for lithology are used with the remaining variables.

The Use of Time-Trend Curves for Paleontologic Analysis

The Richmond Group of Upper Ordovician age, which crops out in the steep cliffs and narrow valleys along the banks of the Ohio River in southeastern Indiana, was selected for detailed paleoecologic analysis. The paleontology of the unit had been studied in detailed by Cumings [1908, 1922], Foerste [1909, 1910], and Nickles [1903], and the stratigraphy had been worked out by Cumings and Galloway [1912], Patton, Perry, and Wayne [1953], and Fox [1952]. Little previous attention had been given to working out the relationships between the lithology and the occurrence of specific fossil groups. With this in mind, the author undertook a three-year research project under the auspices of a National Science Foundation research grant, G-23333, to use this unit as a laboratory in the field for experimenting with various techniques of paleoecologic analysis.

Particular attention was focused on the Tanners Creek Formation, which occupies the lower half of the Richmond group. The Tanners Creek Formation was defined by Fox [1962] to include the alternating sequence of fossiliferous limestones and calcareous shales between the top of the rubbly limestone of the Mount Auburn Formation, containing abundant specimens of the brachiopod *Platystrophia ponderosa,* and the base of the massive earthy dolomitic limestone of the Whitewater Formation. In the area studied, about 12 km north of Madison, Indiana, the Tanners Creek Formation is about 70 m thick. The thin limestone beds with an average thickness of about 5 cm crop out as a series of broad, flat ledges in the stream beds. The intervening, medium-gray shale layers, which average about 9 cm in thickness, are exposed beneath the limestone ledges. Well-preserved specimens of brachiopods, bryozoans, corals, and other taxonomic groups are etched out in relief on the limestone bedding-plane surfaces. Several well-exposed limestone slabs were collected for more detailed study in the laboratory. Many beautifully preserved fossil specimens that have weathered out of the clay shale have found their way into Upper Ordovician fossil collections throughout the world.

The field observations were recorded in numerical notation on a chart so that they could be rapidly transferred to punched cards for processing on the digital computer. The format of the chart has gone through several stages of evolution, and each stage was field-tested to determine its strong points and weaknesses. The left portion of the chart is used to record lithologic data, including the unit number, thickness, cumulative thickness, lithology, color, and type of matrix. A numerical scale has been worked out for each variable similar to the range numbers proposed by Vistelius [1961, p. 777]. For most of the variables, the scale ranges from 1 to 10, so a single number is recorded in the field and later punched on the card.

The right portion of the field chart is used to record relative fossil abundance. Ten major genera of brachiopods, including *Hebertella, Plaesiomys, Platystrophia, Leptaena, Rafinesquina, Resserella, Rhynchotrema, Sowerbyella, Strophomena,* and *Zygospira,* were placed on the chart. In order to get a fairly accurate estimate of relative fossil abundance, the fossil data were transformed to a geometric scale, the rho scale proposed by Fox [1965]. The rho scale is defined as the logarithm to the base 2 of the number of specimens contained

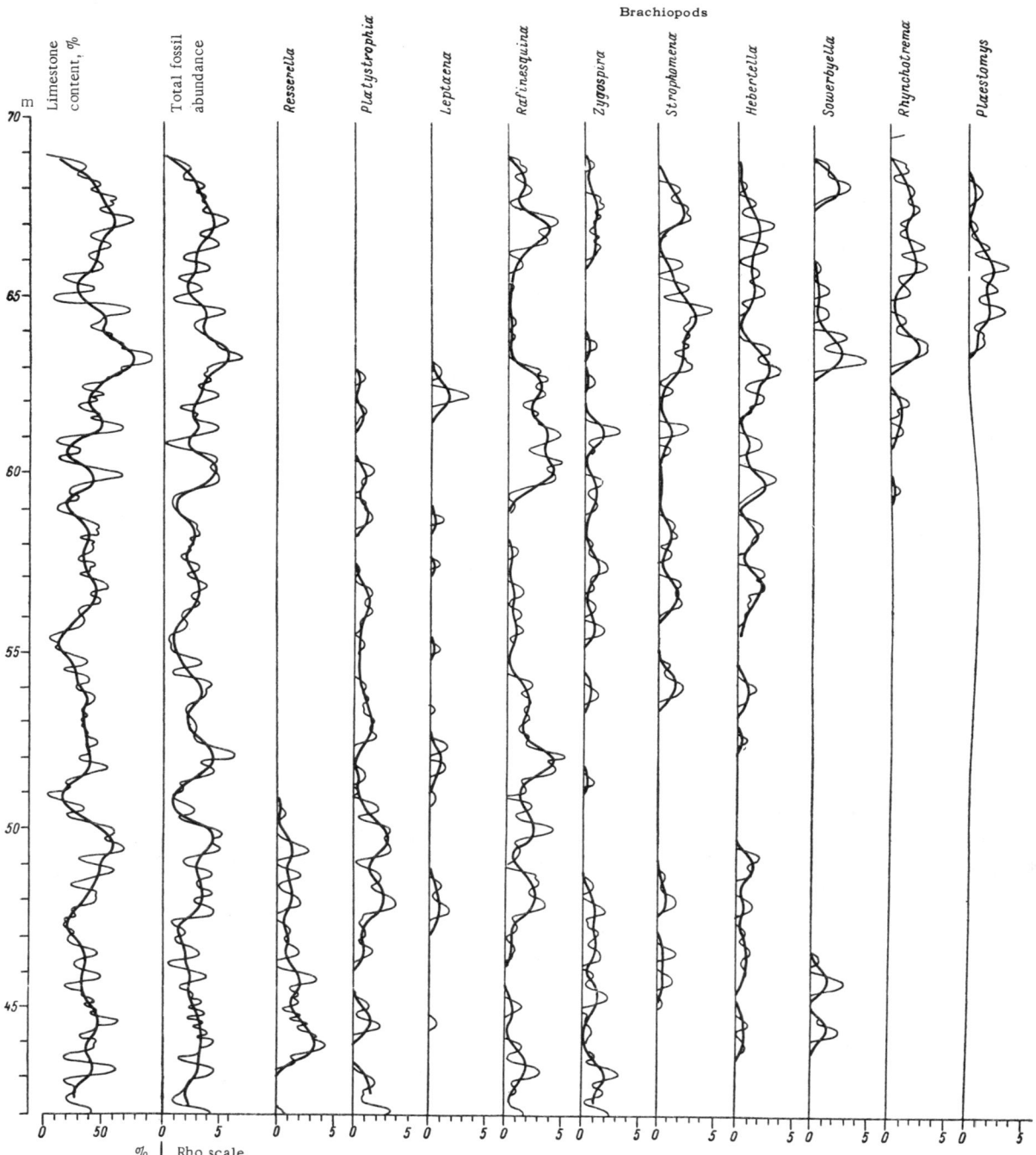

Fig. 2. Limestone percentage and brachiopod abundance curves. Rho scale is used for brachiopod abundance. Dark curve is based on Spencer's 21-term smoothing equation. Light curve is based on Sheppard's 7-term equation.

on an eight square decimeter bedding surface. In order to arrive at the rho scale, several absolute fossil counts were made by the method proposed by Fox [1962]. A cardboard cutout, which was 1 dm on a side, enclosing an area of 1 dm^2, was used in making the absolute count. The largest number of individuals of a single species encountered in several absolute counts was 65 specimens of the brachiopod, *Resserella meeki*. Since the logarithm to the base 2 of 64 is 6, it was decided to use 64 fossils per square decimeter as the upper limit. If this is transposed into the number of fossils per 8 square decimeters (64 × 8 = 512), it is possible to use a scale of 1-9, the rho scale, for recording relative fossil abundance. The chart below is used for comparing absolute fossil abundances per 8 square decimeters with the corresponding values on the rho scale.

Fossils per 8 dm^2	2	4	8	16	32	64	128	256	512
Rho scale, log$_2$	1	2	3	4	5	6	7	8	9

Time-trend curves computed for limestone percentage, relative fossil abundance, and relative abundance of each of the major brachiopod genera are given in Fig. 2. The rho scale from 1-9 is used to record relative fossil abundance. The darker curve on the chart was computed using Spencer's 21-term smoothing equation, (2), and the lighter curve was computed using Sheppard's 7-term equation, (4). A thickness of 10 cm, which is close to the average thickness for the limestone—shale pairs, is used for the smoothing interval. The curves represent a thickness of 28 m, from 42-70 m above the base of the section. The two curves are superimposed with the 21-term curve for major trends and the 7-term curve for minor oscillations. The major trends have regional significance and can be traced for several miles. The minor trends are more restricted and are the results of local fluctuations in the environment.

The relative abundance of the different brachiopod genera varies considerably within the designated interval. *Resserella* occurs only in the lower part of the interval between 42 and 51 m. *Platystrophia* and *Leptaena* extend from 42 up to 63 m. *Strophomena* and *Hebertella* are found up to 69 m but are absent between 50 and 54 m and are much more abundant in the upper part of the section. *Rafinesquina* and *Zygospira* are found sporadically throughout the entire interval. *Sowerbyella* occurs in the lower part of the interval with *Resserella*, but does not occur again until the 63-m level, where it is once more abundant. Along with *Sowerbyella*, *Rhynchotrema* and *Plaesiomys* are abundant between 63 and 69 m.

The relative abundance curves given in Fig. 2 allow for grouping of the brachiopod genera into assemblages which can be interpreted in terms of the depositional environment. The lowest assemblage occurs between 42 and 51 m, and is dominated by the genus *Resserella*. *Sowerbyella* occurs in the lower part of the interval, and *Platystrophia* and *Rafinesquina* are abundant near the top. *Zygospira*, *Leptaena*, *Strophomena*, and *Hebertella* are present in small numbers in the assemblage, but are not found in the upper 2 m of the interval. At the top of the interval, the percentage of limestone decreases significantly and *Resserella* drops out. From 51-63 m, the assemblage is marked by the presence of *Platystrophia* and *Leptaena*, along with *Strophomena*, *Hebertella*, *Rafinesquina*, *Zygospira*, and *Rhynchotrema*. *Rafinesquina* dominates near the base and at the top of the interval. The curves for *Strophomena* and *Hebertella* correspond quite closely, indicating that they commonly occur together. *Rhynchotrema* is found in only the upper 4 m of the interval. The top of the assemblage occurs at 63 m, where *Platystrophia* and *Leptaena* drop out together. Above 63 m, *Sowerbyella*, *Rhynchotrema*, and *Plaesiomys* form a new assemblage, along with *Rafinesquina*, *Zygospira*, *Strophomena*, and *Hebertella*. At the critical horizon of 63 m, where *Platystrophia* and *Leptaena* are last seen, and *Sowerbyella* and *Plaesiomys* are introduced, the limestone percentage curve reaches its highest value. At the top of the interval, the limestone percentage curve drops off to zero and a thick layer of shale is present.

To interpret the environmental factors which controlled the three assemblages shown in Fig. 2, additional information is needed from outcrop data and thin-section analysis. Similar curves for the same interval are being prepared for fossil condition and orientation. Curves based on thin-section data will also be used to show the distribution of microcrystalline calcite matrix and secondary sparry calcite cement. This material will be used along with the relative faunal-abundance curve to make the paleoecologic interpretations.

Conclusions

Time-trend curves are useful for showing the distribution in a vertical stratigraphic section or along a horizontal traverse of any variable which can be expressed numerically. The use of a high-speed computer to calculate and plot the time-trend curves allows the geologist several options which are difficult to obtain using simple computational equipment. By recording field and laboratory data in digital form, it is relatively easy to transfer the data to punched cards for processing on the high-speed computer. The same data may also be used with other computer programs to calculate correlation coefficients and to perform factor analysis.

In the example selected to demonstrate the power of the method, the relative abundance curves for ten brachiopod genera were calculated and plotted on the rho scale, along with curves for limestone percentage and total fossil abundance. The Spencer 21-term equation and Sheppard 7-term equation were used to compute the curves. The curves are superimposed to show the regional trends and local fluctuations in the environment. Three distinct fossil assemblages are recognized within the interval between 42 and 70 m. The stratigraphic position of the faunal assemblages is closely related to the percentage of limestone in an alternating sequence of limestone and shale layers. To make a complete environmental interpretation of the assemblages, additional curves are being prepared, using field data and thin-section information.

The use of time-trend curves should find wide application in paleontology and related fields of geology. In paleontology the curves can be used directly to show faunal distribution, and for stratigraphic correlation among adjacent measured sections. The curves can also be used indirectly along with factor analysis for environmental interpretation and detailed paleoecologic analysis. With the widespread application of computers to geology, time-trend analysis becomes a more readily accessible technique for geologic research.

REFERENCES

Cumings, E. R., "The stratigraphy and paleontology of the Cincinnati series of Indiana," Ind. Dept. of Geol. and Nat. Res., 32nd Ann. Report, pp. 607-1188 (1908).

Cumings, E. R., "Nomenclature and description of the geological formations of Indiana," in: Handbook of Indiana Geology, Ind. Dept. of Conserv. Publ. No. 21, pp. 403-570 (1922).

Cumings, E. R., and Galloway, J. J., "The stratigraphy and paleontology of the Tanners Creek section of the Cincinnati series of Indiana," Ind. Dept. of Geol. and Nat. Res., Vol. 37, pp. 353-479 (1912).

Foerste, A. E., "Preliminary notes on Cincinnatian and Lexington fossils," Denison Univ. Sci. Lab. Bull. 14, pp. 289-334 (1909).

Foerste, A. E., "Preliminary notes on Cincinnatian and Lexington fossils of Ohio, Indiana, Kentucky, and Tennessee," Denison Univ. Sci. Lab., Bull. 16, pp. 17-100 (1910).

Fox, W. T., "Stratigraphy and paleoecology of the Richmond Group in southeastern Indiana," Geol. Soc. Am. Bull., Vol. 73, pp. 621-642 (1962).

Fox, W. T., FORTRAN and FAP Program for Calculating and Plotting Time—Trend Curves using an IBM 7090 or 7094/1401 Computer System, Kansas Geo. Survey, Spec. Distrib. Pub., No. 12, pp. 1-24 (1964).

Fox, W. T., "The rho scale — A measure of relative population density for paleoecologic analysis," Geol. Soc. Am. Bull. (in preparation, 1965).

Fox, W. T., and Brown, J. A., "The use of time—trend analysis for environmental interpretation of limestone," J. Geol., Vol. 73, No. 3, pp. 510-518 (1965).

Milne, W. E., Numerical Calculus, Approximation, Interpolation, Finite Differences, Numerical Integration and Curve Fitting, Princeton University Press, Princeton, N. J. (1949).

Nickles, J. M., "The Richmond Group in Ohio and Indiana and its subdivisions with a note on the genus *Strophomena* and its type," Am. Geologist, Vol. 32, pp. 202-218 (1903).

Patton, J. B., Perry, T. G., and Wayne, W., Ordovician Stratigraphy, and the Physiography of Part of Southeastern Indiana, Ind. Geol. Survey Field Conf. Guidebook, No. 6 (1953), 29 pp.

Romanova, M. A., "Geology of the upper part of the red-bed deposits on the Cheleken Peninsula," Tr. Obshch. Estestvoisp., Vol. 49, No. 2, pp. 116-125 (1957).

Vistelius, A. B., "Notes on analytical geology," Dokl. Akad. Nauk SSSR, Vol. 44, No. 1, pp. 27-31 (1944).

Vistelius, A. B., "Sedimentation time—trend functions and their application for correlation of sedimentary deposits," J. Geol., Vol. 69, pp. 703-728 (1961).

Weiss, M. P., Edwards, W. R., Norman, C. E., and Sharp, E. R., The American Upper Ordovician Standard. VII. Stratigraphy and Petrology of the Cynthiana and Eden Formations of the Ohio Valley, Geol. Soc. Am., Special Paper No. 81 (1965), 76 pp.

Whittaker, E. T., and Robinson, G., The Calculus of Observations: A Treatise of Numerical Mathematics (2nd edition), London and Glasgow, Blackie and Son, Ltd. (1929).

V. Mapping Geological Characteristics

GEOMETRICAL PROPERTIES OF THE SURFACE OF THE ALEKSEEVKA UPLIFT IN THE KUIBYSHEV DISTRICT

M. D. Belonin

Laboratory of Mathematical Geology
V. A. Steklov Mathematical Institute
Academy of Sciences of the USSR
Leningrad, USSR

I. M. Zhukov

"Orenburgneft'" Association
Ministry of Geology of the USSR
Orenburg, USSR

Results of applying some concepts of differential geometry in describing deformation of a surface are discussed. Using an uplift of the Kuibyshev district as an example, we show the effects of the introduced characteristics areally and in section, their geological significance, and the possibility of using them to reconstruct the developmental history of the uplift. The problem of difficulties in computation, arising during attempted solution of the system of linear equations obtained by the method of least squares when finding an analytical representation of the surface, is examined.

In this paper an attempt has been made to study the geometrical properties of a surface in order to obtain supplemental information concerning aspects of evolution of the structural forms with time. As a subject for study, the Alekseevka uplift in the Kuibyshev district has been selected. Raw data for the study come from investigations of stratigraphic horizons specially made for the present paper.

To solve the problem set up, an analytical concept of structural surfaces has been used [Vistelius and Yanovskaya, 1963; Belonin, 1964], a necessary approach for studying the geometrical properties of such surfaces. All computations were made on a BESM-2 computer at the Computer Center of the Leningrad Branch of the Mathematical Institute according to programs developed by M. D. Belonin.

The Alekseevka uplift is in the eastern part of the Kuibyshev Oblast and lies on the Alekseevka salient of the crystalline basement in the Zhigulevsk system of anticlinal uplifts. It is clearly outlined by Devonian (Famennian*), Carboniferous, and Permian strata. It is a symmetrical fold of west-northwestern trend. It is characterized by relative coincidence in plan of the crestal zones for Carboniferous and Devonian rocks.

---

* We have no available information on older horizons.

Method of Investigation

In order to obtain supplementary information concerning the evolution of the structural forms, those geometrical properties of surfaces were studied that permit one to evaluate the degree and nature of bending of such surfaces. Primarily these include all types of curvature (total, mean, maximum, and minimum) and, also, the directions of maximum and minimum bending of the surface at a particular point [Rashevskii, 1950; Belonin, 1964].

Since determination of the indicated characteristics is extremely difficult in the absence of an analytical expression for the surface, structural marker horizons were first approximated by an exponential surface. This surface was defined by the relation

$$z(\varphi, \lambda)_k = \exp[f(\varphi, \lambda)],$$

where $z(\varphi, \lambda)_k$ is a structural marker at the k-th structural horizon; φ and λ are the coordinates of boreholes; and $f(\varphi, \lambda) = a_{11}\varphi^2 + a_{22}\lambda^2 + 2a_{12}\varphi\lambda + 2a_{13}\varphi + 2a_{23}\lambda + a_{33}$.

Further calculations consisted of determining the coefficients of the first and second differential forms of Gauss [Rashevskii, 1950] and computing all the indicated characteristics:

$$p = \frac{\partial z}{\partial \varphi}, \quad q = \frac{\partial z}{\partial \lambda}, \quad r = \frac{\partial^2 z}{\partial \varphi^2}, \quad s = \frac{\partial^2 z}{\partial \varphi \partial \lambda}, \quad t = \frac{\partial^2 z}{\partial \lambda^2}, \quad E = 1 + p^2, \quad F = pq, \quad G = 1 + q^2,$$

$$L = \frac{r}{\sqrt{1 + p^2 + q^2}}, \quad M = \frac{s}{\sqrt{1 + p^2 + q^2}}, \quad N = \frac{t}{\sqrt{1 + p^2 + q^2}},$$

$$K = \frac{LN - M^2}{EG - F^2}, \quad H = \frac{EN - 2FM + GL}{2(EG - F^2)}, \quad K = K_1 K_2, \quad H = \frac{K_1 + K_2}{2},$$

where E, F, G, L, M, and N are the coefficients of the first and second differential forms of Gauss; K_2 is maximum curvature, K_1 is minimum curvature, K is total (Gaussian) curvature, and H is mean curvature.

The directions of principal curvatures are found from solution of the quadratic equation

$$(LF - ME)\left(\frac{\partial \varphi}{\partial \lambda}\right)^2 + (LG - NE)\frac{\partial \varphi}{\partial \lambda} + (MG - NF) = 0.$$

The curvatures of normal sections corresponding to the principal directions of curvature are determined by the formula

$$K_{1,2} = \frac{L d\varphi^2 + 2M d\varphi d\lambda + N d\lambda^2}{E d\varphi^2 + 2F d\varphi d\lambda + G d\lambda^2}.$$

The characteristics of the degree of anisotropy in surface curvature about some point may be taken as the absolute value relating the principal curvatures (the coefficient of anisotropy of surface bending)

$$\beta = \left|\frac{K_2}{K_1}\right|.$$

The indicated geometrical characteristics of a surface possess a number of properties, of which the most important to us are the following.

1. The curvature, the principal directions of curvature, and the coefficient of anisotropy are local properties of the surface. Consequently, a study of these properties permits one to evaluate behavioral patterns of these characteristics for the entire surface.

2. Total curvature belongs to the internal geometry of the surface, and, consequently, it remains invariant during bending, being unaccompanied by elongation. A change in the value

of total curvature indicates extension (or compression) of the surface in the vicinity of the investigated point.

3. Change in the value of mean, maximum, or minimum curvature indicates extension (or compression) of individual sections of the surface and may serve as a characteristic of such extension (or compression).

4. Since the geometrical form of the surface in space is uniquely determined by the first and second differential forms of Gauss, all properties of the geometry of the surface are invariant relative to the choice of coordinates. This means that a value characterizing any particular geometrical property of the surface depends primarily on location of the point on the surface and does not depend on the position of the point in space.

In its application to geology, this last property means that investigated geometrical characteristics of structural forms will undergo no changes from mere tilting of the surface in space (with regional slope). In connection with the above discussion, we note that strike and dip of beds do not possess these properties. These change even with simple tilting of the surface, with no supplementary bending. This circumstance must be kept in mind when using attitudes of beds for defining tectonic movements.

It should be emphasized in conclusion that, since determination of the investigated characteristics is very difficult without any analytical expression of the surface, the use of an approximation device for this purpose requires high standards to achieve precision in the approximation. This permits us to judge to what extent the analytical representation of the investigated surface we have obtained may be suitable for further work with the surface as an analog of the actual form. We have ample grounds for believing that the surface corresponding to the required standard is exponential, if the standard deviation (s) between observed and theoretical values of the function is no greater than errors in the raw data. We have arbitrarily set this latter value at 3 m [Mashkovich, 1961; Ovanesov, 1962].

Computational Technique

Below we note several difficulties of computation encountered when solving the system of linear equations obtained by the method of least squares. In matrix form this system has the following notation:

$$\mathbf{L}\mathbf{a} = \mathbf{b},$$

where $\mathbf{b}$ is the given vector, $\mathbf{a}$ is the desired vector, and $\mathbf{L}$ is the assigned matrix in the form

$$\mathbf{L} = \sum_{i=1}^{m} \left(\begin{vmatrix} 1 \\ \varphi_i \\ \lambda_i \\ \varphi_i^2 \\ \lambda_i^2 \\ \varphi_i \lambda_i \end{vmatrix} \begin{vmatrix} 1 \varphi_i \lambda_i \varphi_i^2 \lambda_i^2 \varphi_i \lambda_i \end{vmatrix} \right),$$

where m is the number of observations.

Since all producible computations have but limited precision, we shall be interested primarily in stability of the characteristic vector depending on the round-off error and on the accuracy of representing the right sides of the linear systems.

Table 1. Effect of Origin and Scale of Measurement on Conditionality of the Matrix

Point of origin	Scale of measurement	Todd No., $P = \dfrac{\lambda_{max}}{\lambda_{min}}$
Observ. point is in 1st quadrant	1.0 km/division 10 km/division	$1 \cdot 10^5$ $3 \cdot 10^4$
Center of observation	1.0 km/division 0.10 km/division	48 $2.2 \cdot 10^5$

Note. λ_{max} and λ_{min} are the maximum and minimum characteristic values of the matrix **L**.

Let us examine the effect of the round-off errors on the results obtained on the assumption that the right sides of the linear systems are accurately represented.

We know from algebra that any matrix equalities corresponding to systems of linear equations are invariant, relatively arbitrary, linear transformations [Faddeev and Faddeeva, 1963]. This means that the form of the function does not depend on choice of origin or on choice of the scale used in measurement.* In practice, when finding a numerical solution to the system of equations, different results are obtained, somewhat at variance with theoretical values. It is shown that the form of the function does not possess strict invariance relative to the linear transformations. Here, the degree of deviation from the theoretical scheme depends to a considerable extent on the choice of scale used in measuring and on the choice of origin [Lantsosh, 1960]. Unfortunate choice of these parameters may lead to considerable deviation between obtained results and true values (especially when the mantissas of the numbers are large†). This should be kept in mind especially when a large number of observations are made. The cause of the indicated deviations lies in the limited precision of calculating, because of which signs are lost. Naturally, when the matrices are poorly founded, the variation limits of the characteristic vector may be rather broad. It therefore becomes necessary to find a characteristic vector by solving a system that corresponds to the best-founded matrix. This solution must then be considered optimal.

If one succeeds in selecting a measuring scale and point of origin such that, during formation of the system of equations and the subsequent solution, there is practically no loss of signs (for a given precision of calculation), it is not then necessary to strive for a highly conditional matrix. This case corresponds to the theoretical scheme analyzed above, and it is possible only when the number of observations is small.

In order to find the optimal solution under other conditions, as shown by experiments, it is necessary to place the origin at the center of observation, and the scale must be so chosen that, as much as possible, the matrix is orthogonalized by a very oblique system of equations [Lantsosh, 1960].

It is clear that, having aimed for the maximum possible conditionality of the matrix under the given conditions, it is easy to go on to an evaluation of the stability of the characteristic vector, depending on the precision of representing the right sides of the linear system [Faddeev and Faddeeva, 1963; Lantsosh, 1960].

For the Alekseevka uplift, within which structural marker horizons were approximated, it was possible to choose a measuring scale and point of origin such that the scheme of computation proved to be similar to the theoretical. Here, even in case of variably conditional matrices (Table 1), variation of the characteristic vector proved to be so insignificant that it could be neglected.

---

* In the problem here considered the choices are completely arbitrary.
† For this reason we do not recommend the use of a degree net of coordinates where point locations are defined by seconds or parts of a second.

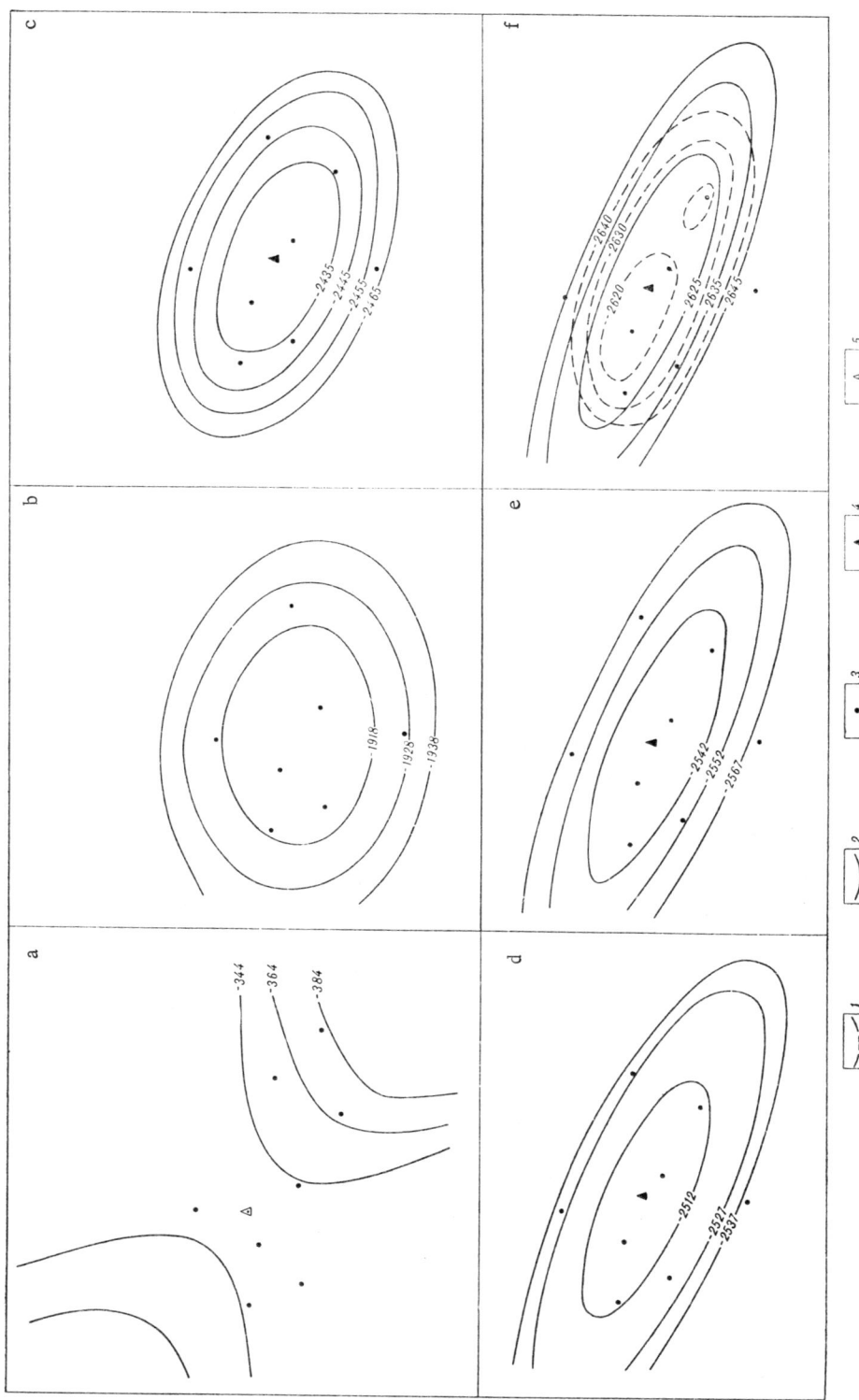

Fig. 1. Structure maps of the Alekseevka uplift. On top of: a) the Kalinovka Series; b) Bashkirian Series; c) Tula Formation; d) coal horizon; e) Tournaisian Series; f) Famennian Series. 1) Structure contours, drawn with linear interpolation; 2) computed structure contours; 3) borehole, data from which was used to approximate structure datum; 4) position of computed center of elliptical surface; 5) position of saddle point on hyperbolic surface.

Table 2. Statistics of Surface Parameters of the Alekseevka Uplift
(for 8 observations)

Structure surface	a_{33}	$2a_{13} \cdot 10^{-2}$	$2a_{23}$	$a_{11} \cdot 10^{-2}$	$a_{22} \cdot 10^{-2}$
Top of Kalinovka Series (VII)	−0.9841	−8.9148	−0.1247	−0.3944	−0.6846
Surface of Vereyan Formation (VI)	−1.5964	−7.1722	−0.1055	−0.9942	−1.9246
Top of Bashkirian Series (V)	−1.6256	7.5673	0.1443	−1.2290	−2.578
Top of Tula Formation (IV)	−2.1204	19.350	0.3984	−2.0730	−6.206
Top of coal horizon (III)	−2.2518	21.281	0.4616	−1.6207	−6.471
Top of Tournaisian Series (II)	−2.3605	24.1137	0.5148	−1.791	−7.036
Top of Famennian Series (I)	−2.452	25.1983	0.5711	−1.825	−7.928

$2a_{12} \cdot 10^{-1}$	S, м	$I_2 \cdot 10^{-3}$	Strike, α	φ_0	λ_0	$\gamma = a : b$
0.3518	2.9	−0.336	N. 37° W.	2.3522	3.0610	—
0.0408	1.8	0.187	N. 82° E.	3.1147	2.410	1.44
−0.0312	1.8	0.314	N. 83° W.	2.7442	2.633	1.46
−0.299	2.1	0.980	N. 67° W.	2.790	2.612	2.20
−0.481	1.3	0.380	N. 67° W	2.795	2.574	3.70
−0.554	0.8	0.494	N. 67° W.	2.746	2.578	3.70
−0.582	1.5	0.601	N. 68° W.	2.800	2.575	3.71

Note. $I_2 = \begin{vmatrix} a_{11} & a_{12} \\ a_{12} & a_{22} \end{vmatrix}$; a and b are the long and short semiaxes of the ellipse, respectively.

Results of Investigation

To explain the geometrical features of the structural forms within the Alekseevka uplift, datum planes on seven horizons of different ages were approximated (Table 2). The results furnish grounds for believing that the analytical representation of the investigated structural forms may be accepted as perfectly satisfactory. From the data supplied, it is clear that, for the Alekseevka uplift, the approximation of structural datum planes is characterized by a family of surfaces, yielding ellipses in projections of the Carboniferous strata, hyperbolas in projections of the Permian (Fig. 1).

To solve the problem we have set, it was necessary to explain how the geometrical characteristics change (1) within each of the investigated structural forms, and (2) on going from surfaces of older beds to surfaces of younger beds.

In keeping with this requirement, the results of our investigation are discussed below.

Change of Geometrical Characteristics within the Structural Forms

The approximation scheme here considered permits us to obtain a better approximation only for symmetrical forms. Of these, the most interesting are structures represented by closed contours, of which exponential surfaces are analogs, in projections giving ellipses. We therefore turned our attention chiefly to these surfaces. For the structural forms of this type

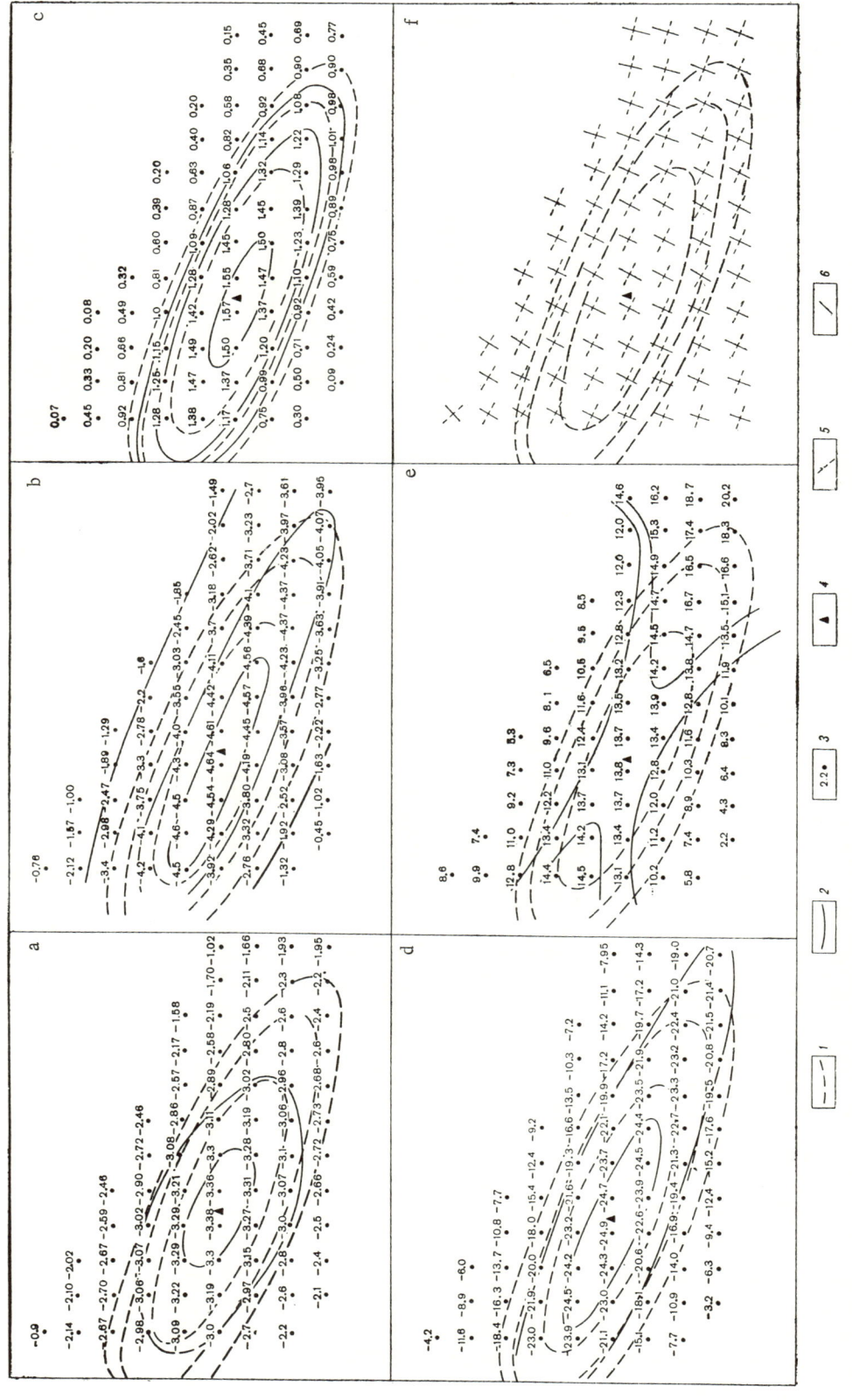

Fig. 2. Maps of curvature, coefficient of anisotropy (β), and orientation of the principal surface trends on the top of the Famennian Series of the Alekseevka uplift. Curvature maps: a) minimum ($K_1 \cdot 10^{-3}$); b) maximum ($K_2 \cdot 10^{-2}$); c) total (Gaussian) ($K \cdot 10^{-4}$); d) mean ($H \cdot 10^{-3}$); e) map of anisotropy of surface bending; f) map of orientation of principal trends. 1) Structure contour on top of the Famennian Series; 2) line of equal curvature (of corresponding type) or of equal coefficient of anisotropy and surface bending; 3) value of corresponding index at the point of observation; 4) location of computational center; 5) orientation of principal direction corresponding to maximum curvature; 6) orientation of principal direction corresponding to minimum curvature.

that we studied,* a single picture is characteristic of the behavior of all types of curvature, principal directions, and coefficients of anisotropy. Differences are found only in quantitative values, and these appear to be substantial when tracing the evolution of the structural surfaces with time.

As an example, let us look at the surface at the top of the Famennian Series in the Alekseevka area (Table 2, Fig. 2). From the maps representing all types of curvature and coefficients of anisotropy, it may be seen that the surfaces they describe have the same trend as the corresponding structural form.† As for the structure contours on each surface, some differences are observed here. For example, the distribution of values of Gaussian curvature agree completely with structure contours on the surface (Fig. 2c), but this does not hold for the mean and principal curvatures. The isopleths for equal values of minimum curvature outline a surface not so elongate as the structural form itself (Fig. 2a). The surfaces corresponding to the other types of curvature (Fig. 2b, d) are more elongate than the surface on top of the Famennian Series. A completely different picture is observed for the coefficient of anisotropy (Fig. 2e). The corresponding surface in the projection has the form of a hyperbola with a center coinciding with that of the structural form. The greatest value of anisotropy is found in the central (quaquaversal) parts of the structure, the least on the limbs. Values are intermediate near anticlinal axes. The principal directions of the surfaces exhibit a close connection with the type of fold and are oriented (in projection) across and along the strike of the structure surface, respectively (Fig. 2f). There is no dependence on strike and dip of the bed at the point of investigation.

The behavior of the geometrical characteristics permits us to draw the following conclusions concerning tectonic movements that lead to the formation of structural forms of the described type.

1. The irregular distribution of values of the investigated characteristics indicates that the surface underwent irregular bending at each point during the process of deformation.

2. According to the values of total (Gaussian) curvature, the greatest stretching of the surface took place on the crest, the least on the limbs and flanks.

3. The distribution of the coefficients of anisotropy attests to the fact that the observed bending at each point took place by irregular bending of the surface in different directions. Folds developed when the axial zones of anticlines and apical zones of domal structures were deformed very differently in two orthogonal directions.

4. The indicated orientations of the principal directions are possibly related to the appearance of longitudinal and transverse faults in transverse folds, the catagory of most platform structures [Gzovskii, 1963]. Attention should also be turned to the similarity of the orientation of the principal normal stresses associated with the growth of transverse folds [Gzovskii, 1963], to the orientation of the principal directions of curvature and the normals to the surface of an anticline. This fact may be readily explained by pointing out that the orientation of the principal directions are closely related to stresses arising because of the growth of folds.

Change in the Geometric Characteristics of Approximated
Surfaces Upward in the Section

From the data we have cited on approximation of structural datum planes in the Alekseevka uplift (Table 2), it may be seen that no appreciable displacement of anticlinal crests occurs from older to younger strata. Differences among surfaces I-VI involve merely a de-

---

* We have in mind uplifts of platform type, with dips on the flanks no greater than 45°.
† Lack of coincidence is possible only for tilted surfaces, but these are not considered in the present paper.

crease (from older to younger strata) in degree of elongation (γ) (Table 2) and a change in amplitude of the structural form. The first of the indicated parameters for surfaces I-III (Table 2) are more than twice the corresponding indices for surfaces V and VI (1.44 for VI and 3.71 for I). At the same time a change in trend of the uplift may be observed, the trend becoming nearly east-west for the less-elongated structures. It should be noted that uplifts of the Devonian and Carboniferous strata correspond to a conjunction of two structural forms in the Permian strata.

The changes in values (upward in the section) reflecting any particular property of surface geometry are much more complex. Attention should be turned to the persistence of trend (in projection) of the principal directions, characterized by northwesterly strike (N. 83-55° W.) for the minimum direction and northeasterly (N. 7-35° E.) for the maximum bending of the surface (Figs. 2f and 3c). This cannot be said of all types of curvature or the coefficients of anisotropy. Surfaces I-III exhibit the closest values of these characteristics. But even here some differences are observed, as indicated below.

The value of total (Gaussian) curvature for corresponding points proves to be greatest for surface I, least for surface III. However, the diminution in value of total curvature upward in the section is achieved at the expense of irregular bending of the surfaces in different directions. In this connection, the difference in values of principal curvatures is determined by the position of the point on the surface, a fact confirmed by the distribution of values of the coefficient of anisotropy of surface bending (Figs. 2e and 3j). It is shown that, for the axial segments of the structures, decrease in the indicated coefficients upward in the section is characteristic. Their values for surfaces I and II are closer together than the values for II and III. For the structural form at the top of the coal horizon, this coefficient at the crest of the uplift reaches a value of 11.5, as against 13.6 for II and 13.7 for I. On the limbs of the structures, a specific characteristic is that the investigated index is greater for surface II than for surface I (Figs. 2e and 3j). The structural form at the top of the coal horizon is also characterized here by lower values of the coefficients of anisotropy.

The difference between the characteristics of surfaces IV and III is much more substantial. Whereas the values of the coefficients a_{22}, characterizing the curvature of the surface along a north-south line, are less for surface IV than for surface III (−6.2 and −6.5, respectively), the value of the coefficient a_{11} is greater for surface IV than for surface III (−2.1 and −1.6, respectively). The difference in the values of the coefficients for a_{11} proves to be more significant than for a_{22}. In comparing curvature maps of surfaces IV and III, one's attention is caught by the greater values of total and minimum curvatures for surface IV than for surface III (Figs. 2a, c, and 3g, h). For the maximum and mean curvatures and the coefficient of anisotropy, the relations are characteristically just the opposite (Figs. 2b, e and 3f, i).

The surface on the top of the Bashkirian Series (V) differs markedly from surface IV in values of the statistics of exponential surfaces a_{11} and a_{22} (−1.2 and −2.1 for a_{11} and −2.6 and −6.2 for a_{22}, respectively) for surfaces V and IV. The curvature maps for the given structural form are characterized by lower values of total, mean, and principal curvature than the maps for surface IV. We should note the similarity in values of maximum and minimum curvature for surface V and the slight variations in values of the coefficient of anisotropy within the investigated structural form.

The surface at the top of the Vereyan Formation is characterized by similarity in values of all its indices with those for surface V. They are everywhere less than those for the underlying structural forms.

The surface on top of the Kalinovka Series is distinguished from all lower forms primarily by belonging to a different type of exponential form. It is characterized by a lower value (as compared to surfaces I-VI) of maximum curvature (in absolute value) and a greater value of minimum curvature. The degree of anisotropy of surface curvature at the top of the Kalinovka Series is appreciably less than for the underlying structural forms (Figs. 2e and 3d, i, j). It is least at the junctions of the structural forms and systematically increases from here toward the indicated uplifts and the warps at the top of surface VII (Fig. 3d). It may be noted that the sign of maximum curvature for the given structural form is positive (Fig. 3a).

Geologic Interpretation of the Obtained Data

The geologic interpretation of the obtained data, given below, is based on the above discussions and on earlier work of one of the present authors [Belonin, 1964].

An analysis of all indices of the approximated surfaces permits us to state that the Alekseevka uplift in post-Frasnian time grew in several stages. The first of these corresponds to development of an uplift at the top of the Famennian, Tournaisian, and coal deposits, and it belongs to the interval of time from the end of the Devonian to the end of coal accumulation. This is attested by the rather similar values of the characteristics of surfaces I and III (Table 2). The peculiarities in behavior of H, K, K_1, K_2, and β, noted above, are possibly related to the specific aspects of erosion of the Famennian and Tournaisian deposits [Resolutions . . . , 1962; Peshtich, Preobrazhenskaya, and Ivanova, 1963; History . . . , 1964]. The rather similar values of the indicated characteristics for surfaces I and II furnish reliable grounds for stating that the Tournaisian deposits were only slightly eroded. For this reason no sharply defined erosional forms could have developed within the investigated area. It is impossible to determine the amount of erosion of the Famennian deposits because of absence of information concerning older deposits. We noted above that the uplifts along the top of the Famennian and Tournaisian strata exhibit great similarity. It is difficult to picture the origin of such close similarity in structural forms I and II where erosional prominences of the Famennian deposits are obscured (if we admit their existence at the beginning of the Tournaisian cycle of deposition) and where subsequent erosion conforms very closely to the shape of the buried erosional remnant. It is more natural to assume that no appreciable elevation (or remnant) existed along the top of Fammenian strata during the period of Tournaisian sedimentation, and that contours of such an elevation simultaneously on surfaces I and II rather clearly mark the beginning of erosion of the Tournaisian deposits. This erosion affected insignificantly only the peripheral parts of the developing uplift. All this is in complete agreement with the above-described behavior of the characteristics of surfaces I and II.

The observed insignificant differences between elevations on the tops of III and II (but stronger than between I and II) attest to comparatively weakly expressed uplift at this time along the top of the Famennian and Tournaisian strata. This is even more plausible when we consider the terrigenous composition of the coal horizon.

The time interval from the end of the coal episode to the end of the Tulian stage corresponds to a new stage of uplift. It is characterized by a change in trend of tectonic movement, as a consequence of which, the top of the coal bed was bent positively in a northeasterly direction and negatively in a northwesterly direction, i.e., along and across the trend of the present structural form, respectively. Support of this view is found in the greater values of the coefficient a_{11} for surface V than for IV and a smaller value of a_{22} for V than IV. Support is also found in the behavior of the values of total and minimum curvature. The higher value of Gaussian curvature for IV indicates that, in the final analysis, it experienced greater elongation than surfaces I-III. This latter took place also because of more intense bending of the surface along the trend of the structural form. Movements of opposite sign in a northwestern direction led to destruction of the indicated uplift on top of the Tournaisian and Famennian deposits. It is interesting to note that the direction of greatest and least bending of the surfaces at this stage were different from now: intersection with the maximum curvature was northwesterly, with

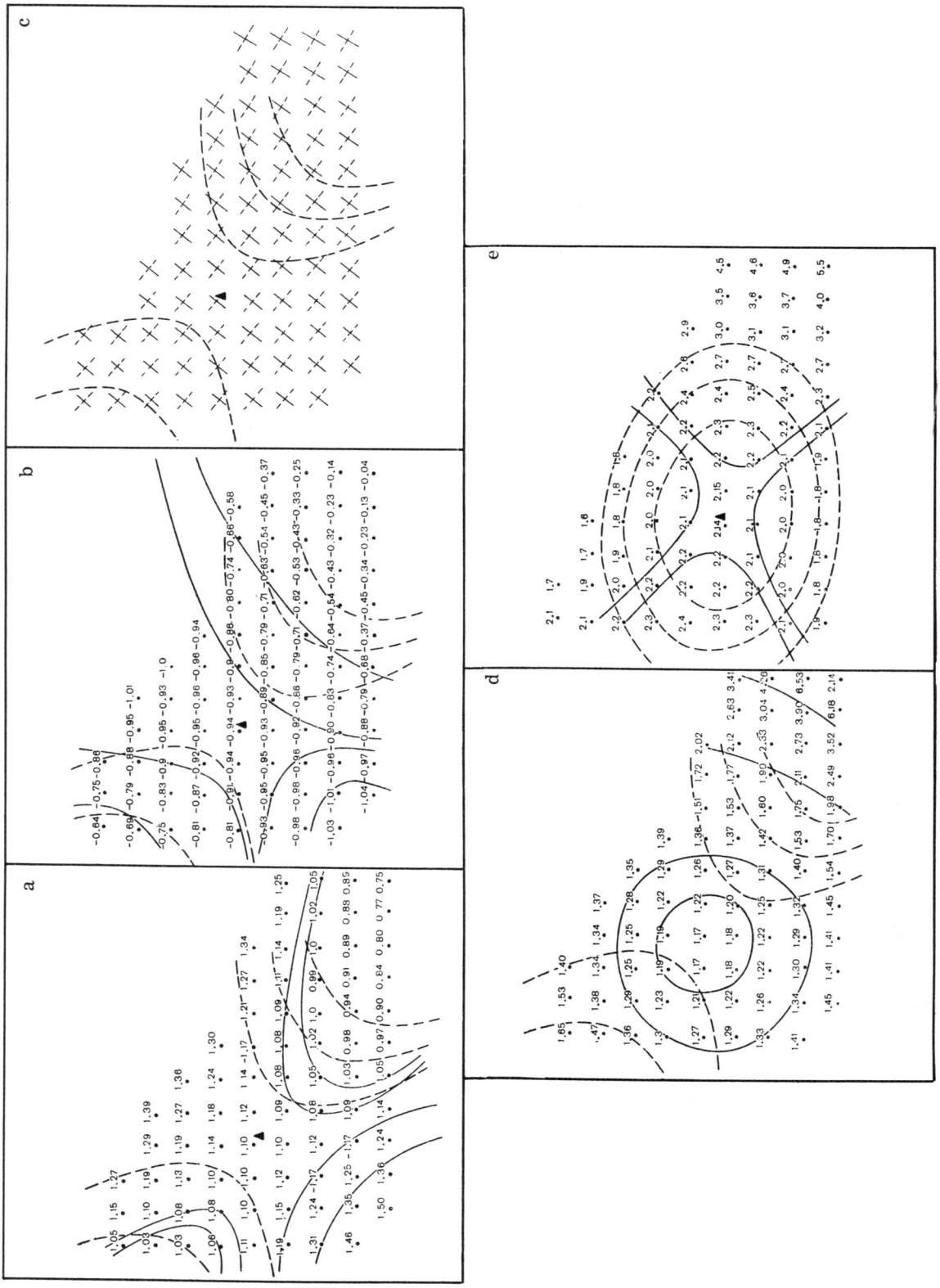

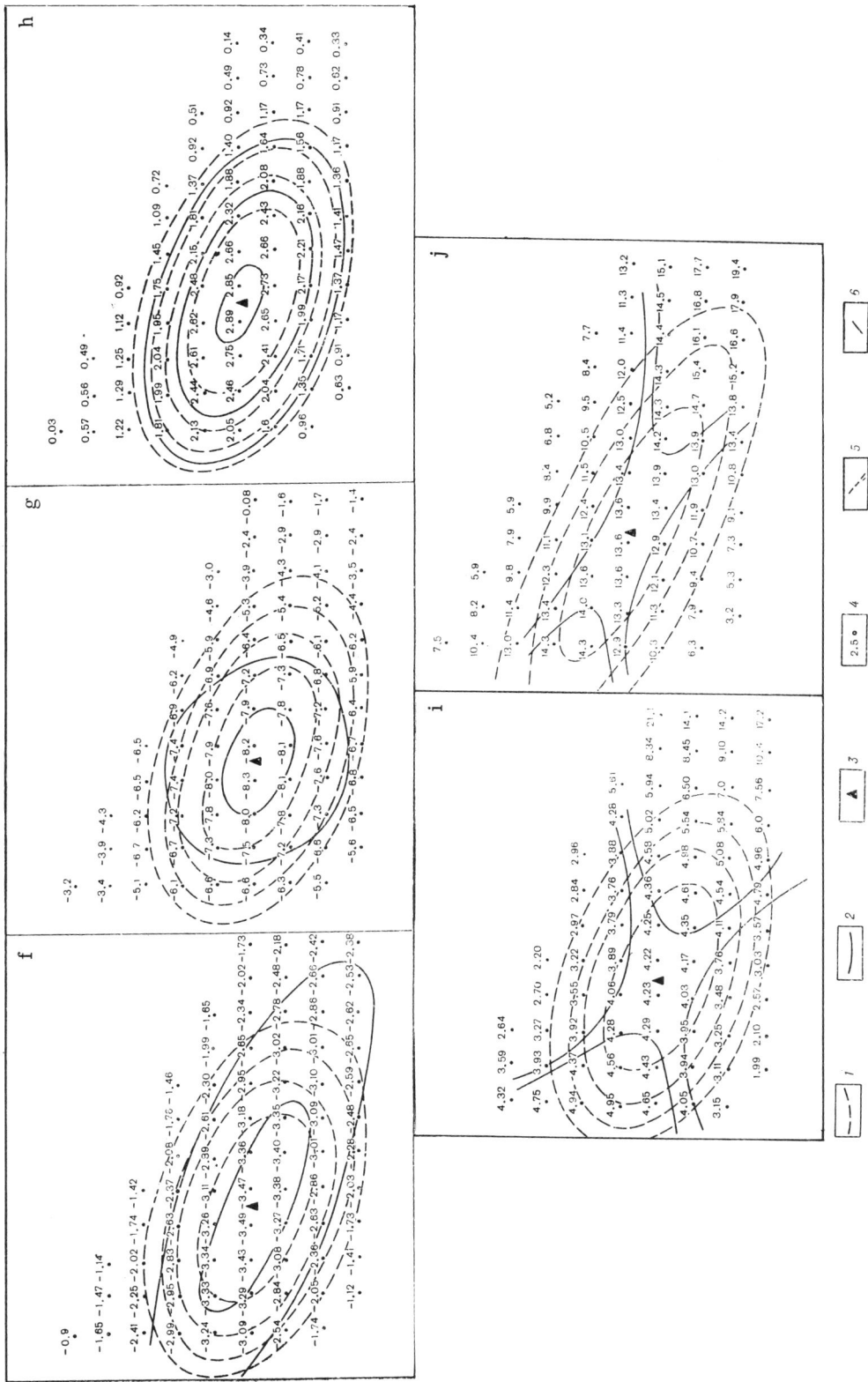

Fig. 3. Maps of curvature, coefficient of anisotropy, and orientation of the principal directions of different horizons on the Alekseevka uplift. On top of the Kalinovka Series: a) map of maximum curvature ($K_2 \cdot 10^{-2}$); b) map of minimum curvature ($K_1 \cdot 10^{-2}$); c) orientation of principal directions; d) bending anisotropy of the surface. On top of the Bashkirian Series: e) bending anisotropy of the surface. On top of the Tula Formation: f) maximum curvature ($K_2 \cdot 10^{-2}$); g) minimum curvature ($K_1 \cdot 10^{-3}$); h) total curvature ($K \cdot 10^{-4}$); i) bending anisotropy of the surface. On top of the Tournaisian Series: j) bending anisotropy of the surface. 1) Structure contour on corresponding surface; 2) isopleth for equal values of curvature and coefficients of anisotropy; 3) center and point of junction of structural forms; 4) point at which the corresponding index was determined; 5) orientation of principal direction corresponding to maximum curvature; 6) orientation of principal direction corresponding to minimum curvature.

the minimum northeasterly. The most intense movements took place in the period corresponding to the interval from the end of Tula time to the end of the Bashkirian epoch, as attested by the great difference in values of all characteristics of the exponential surfaces IV and V (Figs. 1-3, Table 2). Smaller values of these characteristics for surface V than for the underlying surfaces are grounds for stating that an uplift occurred at the top of the Tula formation. Uplifts were again developed on surfaces I, II, and III. A feature of the structure-forming movements at this stage was sharply different bending of the surfaces in northwesterly and northeasterly directions. It was strongest in a northeasterly direction across the trend of the growing uplift at the top of I-IV. The rising structural forms on all four horizons were characterized by a common northwesterly trend (N. 67° W.) and common position of the crests, but the degree of elongation (Table 2) and the curvature (bending) were different. This is especially clear when we compare surface IV with any of the surfaces I-III. The degree of anisotropy and the maximum curvature at corresponding points of the compared surfaces prove to be least for IV, whereas the opposite picture holds for total and minimal curvatures. This means that the surface bent under conditions of more uniform and intense elongation (stretching) of the surface as a whole than took place with surfaces I-III. For the indicated reasons the surface is characterized by contours not so elongated as those of the underlying surfaces. The sign of movement remained the same during the period from the end of the Bashkirian to the end of the Vereyan. The direction and amount of maximum curvature of the surfaces, however, underwent some modification from the preceding stage. This included the greatest shift observed of the crest of the rising uplift on top of the Bashkirian Series.

Further study of the evolution of uplifting is considerably hindered not only by the absence of sufficiently reliable marker beds in deposits of Late Carboniferous age but also by the presence of numerous episodes of erosion that occurred during the Early and Late Permian. It may be stated that the surface on the Kalinovka Series corresponds better to the requirements of paleostructural analysis. Thus, following the investigated stages of uplift growth is a stage corresponding to the time interval from the end of the Vereyan to the end of the Kalinovka epoch. It corresponds to the development at the top of the Vereyan horizon of an uplift of northwesterly trend, having more rounded outlines than the corresponding surface on top of the Bashkirian (Table 2), with some shift (as compared with older horizons) on the southeastern crest (Table 2). The degree of anisotropy of bending attests to deformation of the surface, under conditions of uniform warping, after the investigated time interval. The intensity of the tectonic movements, judging from the comparative change in value of curvature at the given and preceding stages, was considerably greater than during the period from the end of the Bashkirian to the end of the Vereyan.

In the history of the Alekseevka uplift, movements of post-Kalinovka time played an appreciable but not fundamental role. They could not have had any substantial effect on the configuration of the uplift on the underlying horizons, although the degree of curvature of the latter changed. The formation of structural form VII, approximating the surface given in the projection of a hyperbola (Figs. 3a,b,c and Table 2), corresponds to this period. Development of this form was accompanied by warping of the Kalinovka Series in a northeasterly direction to a much lesser degree than warping of the corresponding surfaces that took place at all stages, except for the first and second. It is interesting that the principal directions were preserved, not only in orientation but in kind (Fig. 3c).

Conclusions

In the present work we have attempted to obtain supplementary information concerning features in the evolution of structural surfaces. Special attention was given to a study of the geometric properties of the surfaces, since they permit us to evaluate the conditions under which the surfaces were warped and, consequently, to determine indirectly the nature of the tectonic movements.

In coming to a summary of all we have discussed concerning the evolution of the Alekseevka uplift, the following may be said.

1. In the developmental history of the uplift, several stages of growth may be noted, characterized by specific features. The growth was irregular: periods of intensified development alternated with periods of relatively slow growth, even partial destruction of the already formed uplifts. Maximum movements took place from the end of the Tulian to the end of the Bashkirian epoch.

2. The development of the uplift took place under conditions of irregular warping of the surfaces in the principal directions. These latter preserved their orientation almost unchanged throughout the entire history of uplift development. The types (names) of these directions also remained unchanged. They changed locally only during the period corresponding to the destruction of the uplift on surfaces I-III.

3. The trend of the developing structural forms throughout the entire history of growth of the uplift underwent no marked changes, remaining northwesterly for the more elongated forms and sublateral for the less-elongated forms.

4. Position of the crests of the uplifts on all the investigated stratigraphic horizons was distinguished by great stability, especially during formation of positive structures at the top of the Devonian and Carboniferous strata.

5. Although erosion of Famennian and Tournaisian deposits took place in the investigated area, it did not lead to development of any sharply defined erosional forms or to the appearance of any conflict in structural arrangements of the horizons.

REFERENCES

Belonin, M. D., "Basic outlines of the evolution of the Sokolovy Gory and Bagaevka structures in the Saratov part of the Volga region," Sov. Geol., No. 12, pp. 90-109 (1964).

Faddeev, D. K., and Faddeeva, V. N., Digital Methods of Linear Algebra, Fizmatgiz, Moscow-Leningrad (1963).

Gzovskii, M. V., Basic Questions on the Tectonophysics and Tectonics of the Baidzhansai Anticlinorium, Parts III, IV, Izd. Akad. Nauk SSSR, Moscow (1963).

History of the Geologic Development of the Russian Platform and Its Framework, Izd. Nedra, Moscow (1964).

Lantsosh, K., Practical Methods of Applied Analysis, Fizmatgiz, Moscow (1960).

Mashkovich, K. A., Methods of Prospecting and Exploration for Oil and Gas in the Saratov District of the Volga Region, Gostoptekhizdat, Moscow (1961).

Ovanesov, G. P., The Formation of Oil and Gas Deposits in Bashkiria, Their Classification, and Methods of Prospecting for Them, Gostoptekhizdat, Moscow (1962).

Peshtich, E. L., Preobrazhenskaya, G. S., and K. P. Ivanova, Investigation of the Conditions under which Oil Deposits Formed in the Southeastern Volga—Ural Region, Tr. Vses. Nauchn.-Issled. Geologorazved. Inst., No. 126, Gostoptekhizdat, Leningrad (1963).

Rashevskii, P. K., A Course in Differential Geometry, GITTL, Moscow (1950).

Resolutions of the Committee on Refining the Unified Scheme of the Upper Proterozoic and Upper Paleozoic of the Volga—Ural Oil and Gas Province, Gostoptekhizdat, Moscow (1962).

Vistelius, A. B., and Yanovskaya, T. B., "Programming problems of geology and geochemistry for use with all-purpose electronic computers," Geol. Rudn. Mestorozhd., No. 3, pp. 34-48 (1963).

SORTING OF CLASTIC MATERIAL IN EOLIAN DEPOSITS OF CENTRAL KARA KUM

M. A. Romanova

Laboratory of Mathematical Geology
V. A. Steklov Mathematical Institute
Academy of Sciences of the USSR
Leningrad, USSR

Eolian sorting of clastic material in recent sands of the Kara Kum Desert is analyzed. The mobility of the components is determined by the value of highest gradient of trend surfaces. The causes relating to formation of mineral associations in the eolian sands are examined on the basis of a linear model.

It is well known that currents of windborne sands differ markedly in their regimes, giving rise to extremely irregular areal distributions of the accumulating clastic components. Under similar conditions, general patterns of variation in areal characteristics of the sediments may be masked by random fluctuations, which may at times be overrated through subjective considerations. In order to bring to light these common patterns in our search for a solution to the problem, special mathematical methods have been used. Computations were made on BESM-2 and M-20 computers by programs of the Laboratory of Mathematical Geology, prepared by I. N. Golynko and M. D. Belonin.

The area chosen for analysis is the Central Kara Kum Desert, within a rather narrow zone separating the zones of recent deposits of the Zaunguzskii (Trans-Unguz) and Nizmennyi (Low) Kara Kum (Fig. 1).

Results of a spectrophotometric survey of the sand surface made earlier in the Kara Kum have permitted us to distinguish an east-west zone here, differing in brightness properties from sands to the north and south and consisting of mixed sands of undoubted eolian origin [Romanova, 1964a, 1946b]. To the north of the investigated zone are the Zaunguzskii (Trans-Unguz) sands (Table 1) with high reflectivity, consisting chiefly of quartz and almost free of heavy minerals. Within the Zaunguzskii Kara Kum these sands rest on Sarmatian and older rocks. Their origin is complex: alluvial and deltaic, from fresh or saline continental basins [Luppov, 1963]. To the south of the investigated zone occur sands with low reflectivity, with half the quartz content of the northern sands, rich in heavy minerals (Kara Kum sands, Table 1), belonging to the ancient alluvial, partly marine deposits of Khazarian—Bakinian age. The northern boundary of the

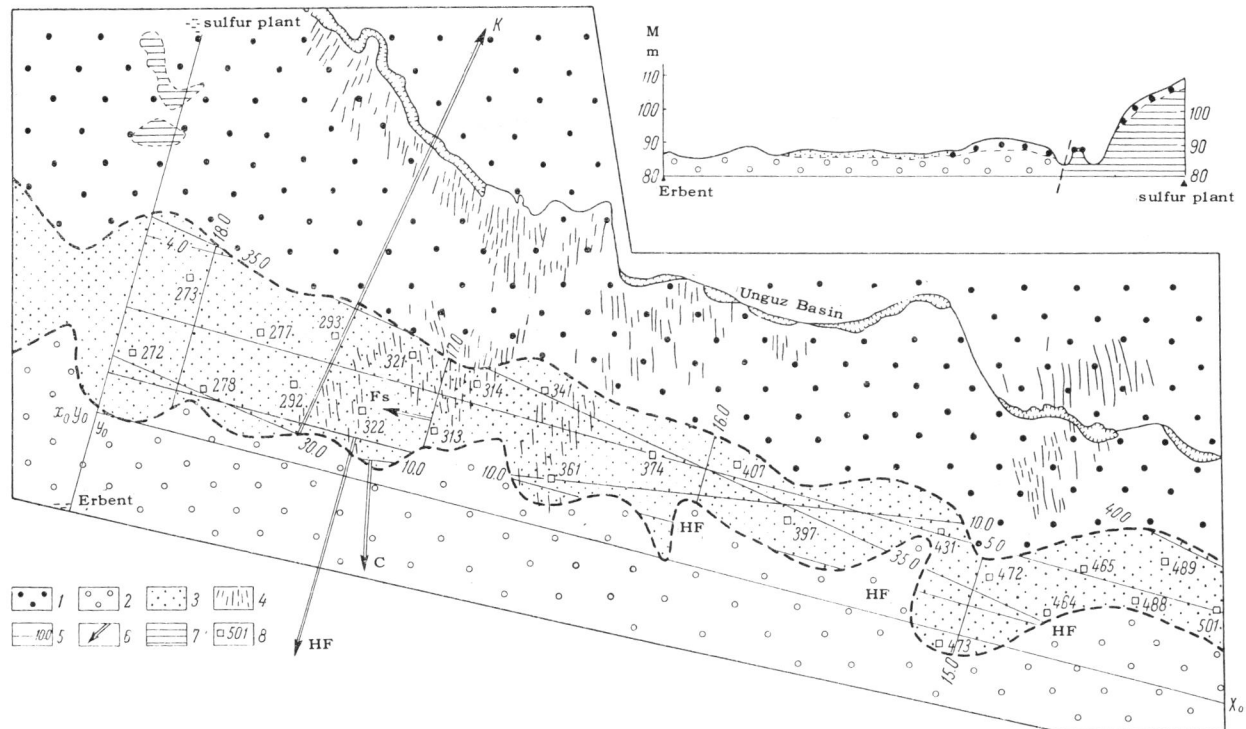

Fig. 1. Location of analyzed zone of eolian sands. Distribution: 1) essentially quartz sand (Zaunguzskii); 2) graywacke (Kara Kum) sands; 3) sands of mixed composition; 4) orientation of ridge forms; 5) location of trend lines (approximate values); 6) gradient vector: K for quartz, Fs for feldspar, C for carbonate rocks, and HF for summed heavy fraction; 7) Sarmation deposits ($N_1^3 s$); 8) outcrop number.

Kara Kum sands, determined by exposures of bedrock and by borehole data, passes along the Unguz Basin (Fig. 1). South of the Unguz Basin (transitional zone), the Kara Kum sands are covered by eolian sediments, similar in composition to the Zaunguzskii sands. The distribution of Zaunguzskii sands south of the Unguz basin may be explained by systematic redistribution by wind currents of prevailing direction. Above the Kara Kum, which lies within the northern belt of subtropical deserts, strong jet streams of air with persistent periodicity of direction are active: from north to south in summer, south to north in winter. The northerly winds, blowing in the driest part of the year, are effective for transporting clastic material. The southerly winds are less effective, since the sands are moist in winter, and such sands are not readily moved. On this basis, one should expect a gradual movement of sands from north to south. This movement is facilitated by the greater elevation of the Zaunguzskii Kara Kum, which rises above the Nizmennyi Kara Kum by approximately 50-70 m.

Thus, on the basis of the physical-geographic conditions of the investigated region, we propose that the model of the process include the systematic shifting of sands from the surface of the Zaunguzskii Kara Kum to the south, toward the Nizmennyi Kara Kum. The objectivity of this assumption is confirmed by the nature of microrelief on the surface. Within the investigated region the accumulated deposits have ridge forms trending northeast (N. 15-20° E.) (Fig. 1). The length of individual ridges ranges from several tens of meters to 5-8 km, and the width from a few to 20-30 m. The crest of a ridge is generally flat and the eastern slope is steeper than the western (evidence of the shift of sand to the east also, but which we neglect

Table 1. Lithologic Characteristics of the Investigated Sands (in %)

Component	Zaunguzskii, 55		Nizmennyi, 124		Eolian, 25	
	$\bar{x}_I$	s_I	$\bar{x}_{II}$	s_{II}	$\bar{x}_{III}$	s_{III}
Grain-size fractions, mm						
> 0.21	24.63	12.07	23.72	13.12	20.19	9.92
$0.15 \leqslant 0.21$	30.91	8.74	36.70	7.31	37.87	6.66
$0.105 \leqslant 0.15$	30.44	10.99	27.22	9.73	28.68	7.57
$0.074 \leqslant 0.105$	11.76	6.72	12.00	23.18	11.13	5.17
$0.056 \leqslant 0.074$	1.36	1.19	1.94	2.13	2.02	1.50
< 0.056	0.77	0.80	0.72	3.66	0.46	0.49
Minerals and rocks						
Quartz	49.38	13.37	24.38	7.20	35.75	9.53
Feldspar	18.65	7.60	15.83	5.72	18.20	5.92
Carbonate rock	6.91	4.32	12.68	3.82	10.10	3.59
Magmatic rock	23.30	8.50	37.86	7.60	24.20	5.30
Heavy fraction	1.76	0.72	9.30	2.38	5.27	1.65
Minerals in heavy fraction (in % · 100)						
Hematite	4.08	5.25	8.39	36.06	3.80	2.69
Limonite	9.62	6.13	20.52	10.55	15.91	6.89
Ilmenite	13.82	9.89	15.74	16.63	10.40	7.22
Rutile	0.94	0.72	1.37	1.45	1.04	1.01
Leucoxene	4.91	2.74	2.66	3.06	3.15	2.25
Magnetitie	1.00	1.01	10.74	7.01	4.34	3.03
Kyanite	6.25	5.56	58.50	37.84	43.88	23.11
Staurolite	2.44	1.74	5.81	4.83	3.96	3.37
Zircon	2.83	1.77	3.97	3.95	2.56	2.23
Amphibole	29.92	18.88	244.74	86.68	130.18	52.57
Pyroxene	1.00	1.36	25.86	18.25	11.65	6.65
Garnet	12.54	10.90	47.22	22.95	28.62	17.49
Epidote	62.87	34.27	318.63	104.82	160.71	63.21
Tourmaline	5.06	3.12	5.71	5.78	6.70	4.77
Chlorite + mica	3.33	4.60	63.19	34.98	30.44	3.87
Sphene	1.19	1.70	3.58	3.80	2.03	2.17

Note. Figures at head of tables indicate number of observations.

because it is strongly subordinate to the southerly trend). The accretion process of eolian forms in the direction of principal wind direction, with the formation of ridgelike relief, has been studied by many investigators, and it has been shown that the size of a ridge depends on the strength of wind currents, duration of wind action, as well as features of surface relief [Exner, 1920; Fedorovich, 1940, 1950; Ermilov, 1959; Hörner, 1957; Priestley, 1964; Bondarev, 1966].

It is further assumed that the mobility of the clastic material is determined by the physical properties (size, shape, and specific gravity) of the clastic minerals. This assumption was based on the systematic patterns established by many investigators. These patterns are summarized in very condensed form below.

1. The clastic material of sands is moved chiefly at a height of 0-10 cm from the surface [Znamenskii, 1942; Petrov, 1950]. According to Chepil [1945], the height is 5-8 cm, according to Williams [1963] about 9 cm. It has been established that grains move by saltation (rebound from the surface) [Bagnold, 1931; Petrov, 1950; Chepil, 1961] or by rolling [Senkevich, 1962].

2. The mobility of grains depends on their rounding. Rounded grains begin to move earlier and are thrown higher than angular grains [Chepil, 1961]. Williams [1964] has shown that in a weak wind predominantly the more angular grains are moved, whereas the more rounded grains are moved chiefly in moderate and strong winds; but at any particular wind velocity the amount of material transported is greater the better rounded the grains in general.

3. Bagnold [1941], in determining the capacity of grains to be moved, introduced a special characteristic, the susceptibility, equal to the ratio of the force of wind acting on the grain to the force of gravity of the transported grain. This relationship explains very well the sorting of clastic material in eolian sediments and, in particular, the predominance of grain sizes 0.15-0.30 mm in eolian sand.

4. Many years of observation on movement of the Kara Kum sands has established the fact that the sands begin to move at a surface wind velocity of 4-5 m/sec [Orlov, 1960]. It has been noted that coarse-grained sand, rich in heavy minerals, accumulates in deflation areas [Sidorenko, 1956] and that sand accumulates if the amount of suspended material in the surface layer of wind 1 cm thick is twice the average content per centimeter in the interval 0-10 cm from the surface [Znamenskii, 1959].

5. The dependence of the mobility of sand grains on specific gravity was studied by Vistelius and Sarsadskii [1952] in relation to conditions in an aqueous medium. The slope of the straight line determining the velocity of a water-transported mineral depending on its specific gravity, as used in this work, may be considered the gradient of sediment precipitation from the current of water-borne sand.

Under conditions of eolian transport of sediment, we also start with the assumption that the mobility of the clastic material is defined by the gradient of grain-settling from the stream of airborne sand. We assume that: 1) common systematic areal patterns of change in contents of components of the sediment are rather precisely approximated by a trend surface, and, 2) the velocity and direction of sand-bearing winds are approximately uniform for small segments of landscape irregularities.

Methods of Investigation

The sorting of clastic material in sediments is examined within a zone of mixed sand over an area of 20 × 160 km, where observations were made on a 10 × 10 km grid and 25 samples of sand were analyzed. The samples were collected from the flat crests of accumulation forms. Grain-size, petrographic and mineralogic studies were made by ordinary methods. The characteristics obtained were analyzed by means of two-dimensional regression. In view of the fact that we were confronted with methodological difficulties when computing trend surfaces, we shall dwell on this method in somewhat more detail.

The computed values of gradients were interpreted on the basis of geologic considerations. Assumptions concerning the effect of any particular characteristic on the gradient were made on the basis of previously established patterns of distribution of the clastic material and were tested by correlation or regression analysis.

The effect of grain size on mobility of the sediment was determined by simple comparison of average contents of separate grain-size fractions in the mixed sands with the values of the highest gradients of these fractions in the eolian sands. In addition, the linear correlation coefficients between contents of individual fractions in the sand and contents of fragmental minerals in these fractions were computed.

The effect of specific gravity of the fragmental minerals on mobility of sand grains was tested by regression analysis (correlation analysis in this case is inapplicable, since specific gravity of minerals is not a random value).

Evaluation of the statistics obtained during correlation and regression analysis has permitted us to demonstrate that no contradiction exists between the assumptions we have made.

The assumption that the average composition of the mixed sands may have an influence on the highest gradient was tested by computation of paired linear correlation coefficients between the normalized difference of average contents of sand components and the value of their highest gradients. Linear correlation in this case is permissible, since the compared characteristics are random and independent variables.

The assumption that grain shape affects sand mobility has not been tested. Such dependence might be established by using factor analysis, which we hope to do once we have developed the proper model.

Analysis of the Trend Surface

The method of analyzing observed values of a trend surface has become widely used in geology, especially since introduction of electronic computers in practical work. An extensive literature exists on this question. The method, as everyone knows, is based on approximations of investigated values of a trend surface, which is computed by the method of least squares.

The choice of approximation function (trend surface) depends on the properties of the investigated characteristics; in particular, for geologic characteristics, Vistelius [Vistelius and Yanovskaya, 1963] proposed that an exponential function of the following type be used:

$$z'_{ij} = \exp[P_m(\varphi_i; \lambda_j)], \qquad (1)$$

where m is the degree of the polynomial, being the exponent by which the selection of $P_m(\varphi, \lambda)$ is made according to logarithms of z_{ij}.

Advantages of the exponential function are the following.

The exponent cannot take on negative values, a trait that corresponds to the sense of the characteristics we are investigating.

The taking of logarithms of observed values with large positive distributional asymmetry smooths these values in itself, i.e., observations strongly deviating from average values must not have a distorting effect on the function.

The exponent is readily linearized and, consequently, all tools of the method of least squares may be applied to the observed values. However, difficulties associated with choice of the degree of the polynomial of the smoothing function may arise when analyzing the trend surface.

As we know, exponential trend surfaces may have degrees of the polynomial of different orders. It is natural that polynomials of low order give poorer approximations to observed values than high-order polynomials. The use of polynomials of excessively high order is harmful, however, since these polynomials deviate farther from the value of the function between points of interpolation the higher the order of the polynomial. Furthermore, when using polynomials of a high order, the awkwardness of the computation increases. It is thus necessary to make the right choice of degree of the polynomial, represented by the exponent, when solving this problem.

We have found empirically that coefficients of polynomials of the second and third degree give a good approximation in the sense of a minimum of squared deviation, but unstable solutions may result.

It is known that solution of a normal system, when using the method of least squares, reduces to a solution of the system

$$X = A^{-1}Y,$$

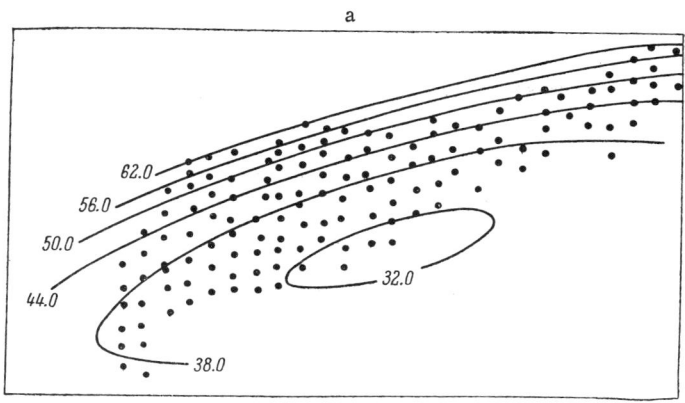

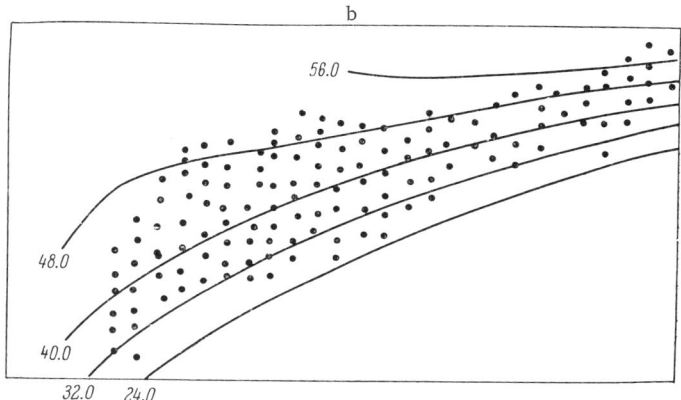

Fig. 2. Trend surfaces of quartz content in eolian sands. a) Computations for standard coordinates of observation points; b) for conditional coordinates.

where values of the vector **X** are found from the inverse of the matrix of the coefficients a_{ij} and from the values of the free terms of y_i. It is possible that small changes in the elements of the inverse matrix $\mathbf{A}^{-1}$ will lead to a marked change in elements of the characteristic vector **X**, and that this will, in turn, strongly affect the geometric aspects of the functional as a whole.

In other words, small changes in position of the observation point $Z_{i,j}$, when the matrix **A** contains only data on the position of observation points, will lead to completely different results of approximating observed characteristics. This phenomenon is referred to as poor conditionality of the matrix.

To evaluate the conditionality of the matrix, we have used the Todd number. This number, as everyone knows, is equal to the ratio of the largest characteristic number of the matrix to the smallest, and the smaller the value the greater the conditionality of the matrix. The Todd numbers thus obtained for approximation, on the basis of polynomials of the first three degrees, show that the conditionality for a polynomial of the first degree is comparatively satisfactory. For polynomials of the second and third degree the conditionality worsens sharply and it becomes doubtful if polynomials of higher degree can be used.

To test the effect of poor conditionality on the behavior of isopleths representing mineral contents, special studies were made. We shall examine one of these. Below we furnish a matrix of the system considering only the coordinates of observation points for $P_2(\varphi, \lambda)$ (with coordinate points computed from Greenwich)

$$\begin{pmatrix} 138.0000000 & 141.8558280 & 145.8483910 & 95.6175964 & 66.2545608 & 98.2822537 \\ 141.8558280 & 145.8483910 & 149.9832990 & 98.2822537 & 68.0961202 & 101.0411900 \\ 145.8483910 & 149.9832990 & 154.2664150 & 101.0411900 & 70.0026883 & 103.8982670 \\ 96.6175964 & 98.2822537 & 101.0411900 & 66.2545608 & 45.9105963 & 68.0961202 \\ 66.2545608 & 68.0961202 & 70.0026883 & 45.9105963 & 31.8148046 & 47.1833691 \\ 98.2822537 & 101.0411900 & 103.8982670 & 68.0961202 & 47.1833691 & 70.0026883 \end{pmatrix}$$

The Todd number of this matrix is approximately $1.01 \cdot 10^9$, i.e., the conditionality of the system is very poor. By substituting the vector of the free terms for quartz content, given below, we obtain a solution indicating $P_2(\varphi, \lambda)$ to be the surface of an ellipsoid: isopleths of the appropriate exponent are shown in Fig. 2a.

System solutions	Free terms
2784.618170	609.721846
−1915.663810	626.814767
389.108106	644.515662
−5222.710140	422.480888
2584.695130	292.753252
1622.519260	434.293623

In view of the very poor conditionality, we tested the effect on the system of a change in point of origin, transferring it to the point $\varphi_0 = 35°$ N. Lat., $\lambda = 55°$ E. Long. The matrix of this system is given below:

$$\begin{pmatrix} 138.000000000 & 9.4057005400 & 0.67018386900 & 11.310029200 & 0.92985384100 & 0.76399582500 \\ 9.405700540 & 0.6701838690 & 0.04980651570 & 0.7639958250 & 0.06226937150 & 0.05392288560 \\ 0.670183869 & 0.0498065157 & 0.00384706764 & 0.0539228856 & 0.00435413425 & 0.00396933272 \\ 11.310029200 & 0.7639958250 & 0.05392288560 & 0.9298538410 & 0.07668427900 & 0.06226937150 \\ 0.929853841 & 0.0622693715 & 0.00435413425 & 0.0766844790 & 0.00634329529 & 0.00509246301 \\ 0.763995825 & 0.0539228856 & 0.00396933272 & 0.0622693715 & 0.00509246301 & 0.00435413425 \end{pmatrix}$$

The Todd number of this matrix is approximately $2.4 \cdot 10^9$, i.e., the conditionality of this system is also poor. In finding the solution for quartz one obtains a hyperboloid in this case; the isopleths for the exponent are shown in Fig. 2b.

System solutions	Free terms
−9.91399995	609.72184600
9.41993032	41.617544720
170.96738100	2.96998932
304.28632000	49.98642430
−1550.00080000	4.11099244
−295.21074100	3.381330610

From the two parts of Fig. 2 it may be seen that within the designed range of the function the isopleths are almost coincident, but at the margins the isopleths give entirely different pictures, possibly leading to different geologic interpretations.

Thus, on the basis of the method of least squares, it proved impossible in our case to use an approximation of high degree without special precautions. On the basis of the above discussion, we limited ourselves to the study of a small zone and we used a trend surface with a linear function as the exponent (m = 1). The corresponding matrix of the system

$$\begin{pmatrix} 25.0000 & 360.8000 & 70.4500 \\ 360.8000 & 8109.3000 & 948.4525 \\ 70.4500 & 948.4525 & 241.5175 \end{pmatrix}$$

has a Todd number of 3045.4692, which, considering the high precision of machine computation, is rather satisfactory.

Table 2. Solution of the System

Component	a_0	a_1	a_2
Grain-size fraction, mm			
< 0.056	−1.0195	−0.02194	0.07885
$0.056 \leqslant 0.074$	−0.1896	0.02523	0.09776
$0.074 \leqslant 0.105$	2.3276	−0.01061	0.04570
$0.105 \leqslant 0.150$	3.3167	−0.004188	0.02090
$0.150 \leqslant 0.21$	3.6907	−0.006916	0.01027
> 0.21	2.7320	0.02296	−0.07388
Minerals and rocks			
Quartz	3.3427	0.006402	0.03342
Carbonate rock	2.2819	0.004518	−0.02866
Magmatic rock	3.1503	−0.001156	0.009597
Amphibole	5.2161	−0.006135	−0.1084
Epidote	5.4049	−0.01441	−0.06884
Feldspar	2.96097	−0.007537	−0.0000439
Garnet	3.4435	0.02501	−0.1988
Zircon	0.2724	0.07502	−0.3308
Limonite	3.2092	−0.005612	−0.1438
Kyanite	3.9019	−0.01457	−0.02721
Mica	3.4049	−0.01746	−0.0347
Magnetite	1.5127	0.01817	−0.1594
Leucoxene	−0.3458	0.03196	0.1929
Staurolite	1.6033	−0.004056	−0.5356
Rutile	−0.5903	0.04829	−0.3467
Hematite	0.06409	0.02059	0.07190
Sphene	−0.4776	0.01329	−0.1246
Pyroxene	2.4385	0.01290	−0.09730
Ilmenite	2.2283	−0.01699	−0.04314
Tourmaline	1.0537	0.02790	−0.01689
Heavy fraction	1.9520	−0.002458	−0.09606

The accepted solution required analysis from the geologic point of view. Actually, the investigated zone of eolian deposits is bounded in the eastern part, along a considerable distance, by a relatively straight morphological bench, oriented approximately along an east-west line. This line terminates in the region of the Sernyi Hills (Sulfur Hills); and farther to the west the relief changes. Starting with the assumption that the sands moved from north to south because of steady winds, we may naturally expect the sand-laden winds in the eastern part of the zone to give a distribution of sand components that permits approximation by an exponential curve with a linear function as the exponent. In the western part of the zone, where the sand-laden winds change direction because of more complex relief, the isopleths of mineral content exhibit variable bending. These bendings cannot be reflected by an exponential curve with a linear function as the exponent. Thus, in order to preserve the geologic correctness of the problem when using a linear exponent in (1), we had to restrict our analysis to the eastern part of the zone of eolian sands bordered on the south by the rectilinear segment of Unguz, which was described at the beginning of the section Methods of Investigation. In order to simplify the computations, Cartesian coordinates were introduced with the origin at the points X_0, Y_0 (Fig. 1) and the horizontal axis oriented approximately parallel to the morphological bench of the Unguz, trending N. 75° W.

As a result, the value of the approximate characteristics were computed by the equation

$$z'_i = \exp[a_0 + a_1 x_i + a_2 y_i]. \tag{2}$$

The trend surfaces for 27 lithologic characteristics of the eolian sands were computed by this equation. The values of a_0, a_1, and a_2 are given in Table 2.

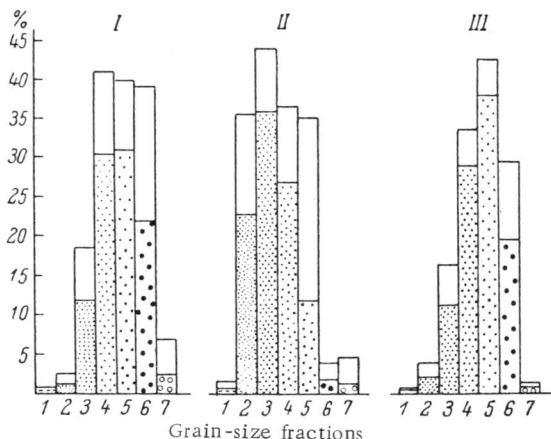

Fig. 3. Grain-size distribution in the sands. I) Zaunguzskii (55 observations); II) Kara Kum (124 observations); III) mixed composition (25 observations). Ordinate axis: frequency of average contents of fraction and standard (part of columns without pattern). Abscissa axis, fractions (mm): 1) < 0.056; 2) $0.056 \leq 0.074$; 3) $0.074 \leq 0.105$; 4) $0.105 \leq 0.150$; 5) $0.15 \leq 0.21$; 6) $0.21 \leq 0.29$; 7) > 0.29.

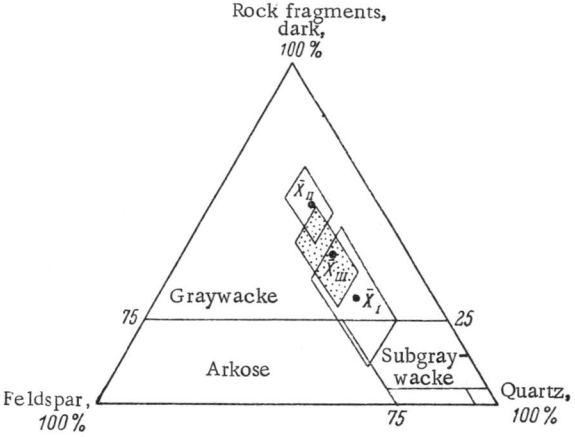

Fig. 4. Average petrographic composition of sands. Sands: X_I) Zaunguzskii; X_{II}) Kara Kum; X_{III}) mixed. Outlines of the fields are drawn according to standard deviations of the means.

The values of the highest gradients of the trend surfaces were computed according to the equation

$$\text{grad}_i = \exp(a_0 + a_1 x_i + a_2 y_i) \sqrt{a_1^2 + a_2^2}, \quad (3)$$

where grad_i is the gradient, or the measure of greatest increase of the trend surface of the investigated characteristic at a given point.

The gradient modulus is also the value by which the degree of mobility of the fragmental particles of sand was determined.

Geologic Interpretation of Gradients of Sand Characteristics

The grain-size distribution of eolian sands in the analyzed zone is similar to that of the mixed sands (Table 1, Fig. 3). One difference is found in the somewhat greater and stable content of the fraction $0.15 \leq 0.21$ mm. In content of the fine-grained fraction they are similar to the sands from the source area (Zaunguzskii), but they are considerably poorer in these fractions than the sands in the zone of accumulation (Kara Kum). Thus, the eolian sands have better sorting of grain sizes even than the rather well-sorted, substantially quartz sands of the Zaunguzskii Kara Kum.

The eolian sands proved to have a composition represented by the average of the compositions of the mixed sands, as should be expected (Table 1, Fig. 4). Difference in content of individual components, determined by different mobilities of grains during eolian transport, is reflected in the values of the highest gradients of the fragmental minerals (Table 3). Let us examine the gradients of the most widespread clastic minerals (in order of decreasing values).

1. Quartz. Most quartz grains are in the fraction $0.105 \leq 0.150$ mm. The grains are rather well rounded with smooth polished surfaces, locally coated with a film of carbonate or desert varnish. The trend lines of quartz content are approximately parallel to the zone of eolian sands. As seen from Table 3, the gradient is oriented N. 25° E., i.e., the content of quartz in the direction of sand migration sharply declines.

Table 3. Maximum Gradients of the Compositional Characteristics of Eolian Sands in the Central Kara Kum

Component	Gradient modulus, %/5 km	Orientation of gradient vector	Paired correlation coefficients between fraction and clastic mineral			
			r		− r	
Grain-size fraction, mm						
< 0.056	0.0492	N. 1° W.	Hematite	0.88	Epidote	0.20
			Leucoxene	0.61		
			Zircon	0.88		
			Feldspar	0.24		
0.056 ≤ 0.074	0.3002	N. 29° E.	Hematite	0.46	Epidote	0.27
			Leucoxene	0.44		
			Zircon	0.50		
			Feldspar	0.22		
0.074 ≤ 0.105	0.6445	N. 2° E.			Hematite	0.23
					Zircon	0.22
					Garnet	0.21
0.105 ≤ 0.150	0.6713	N. 4° E.	Quartz	0.17	Hematite	0.32
					Leucoxene	0.31
					Magnetite	0.28
					Zircon	0.31
					Kyanite	0.30
					Amphibole	0.23
					Mica	0.40
0.150 ≤ 0.21	0.5317	N. 29° W.	Limonite	0.26	Leucoxene	0.27
			Kyanite	0.23	Zircon	0.36
			Amphibole	0.30	Feldspar	0.21
			Epidote	0.36		
			Garnet	0.25		
> 0.21	2.2271	S. 2° E.	Magnetite	0.34	Limestone	0.23
			Kyanite	0.23		
			Mica	0.33		
Minerals and rocks						
1. Quartz	1.390	N. 25° E.			Amphibole	0.23
					Magnetite	0.20
2. Carbonate rock	0.601	S. 6° W.	Kyanite	0.32		
			Amphibole	0.35		
			Epidote	0.28		
3. Amphibole	0.218	S. 19° W.	Limonite	0.43	Quartz	0.23
			Magnetite	0.82		
			Kyanite	0.71		
			Garnet	0.61		
			Epidote	0.75		
			Mica	0.66		
4. Epidote	0.156	S. 27° W.	Limonite	0.39		
			Magnetite	0.59		
			Kyanite	0.41		
			Garnet	0.61		
			Mica	0.39		
5. Feldspar	0.145	N. 75° W.	Hematite	0.22		
			Leucoxene	0.33		
			Mica	0.20		

Table 3 (continued)

Component	Gradient modulus, %/5 km	Orientation of gradient vector	Paired correlation coefficients between fraction and clastic mineral			
			r		− r	
6. Garnet	0.121	S. 8° W.	Limonite	0.46		
			Ilmenite	0.44		
			Magnetite	0.65		
			Amphibole	0.61		
			Epidote	0.61		
			Mica	0.41		
7. Zircon	0.039	S. 2° W.	Hematite	0.91		
			Leucoxene	0.66		
8. Limonite	0.035	S. 17° W.	Magnetite	0.45		
			Kyanite	0.33		
			Amphibole	0.43		
			Garnet	0.46		
			Epidote	0.39		
			Mica	0.55		
9. Kyanite	0.015	S. 43° W.	Magnetite	0.67	Quartz	0.19
			Amphibole	0.71		
			Epidote	0.41		
			Carbonate	0.32		
			Mica	0.61		
10. Mica	0.012	S. 42° W.	Limonite	0.55	Quartz	0.16
			Magnetite	0.69		
			Kyanite	0.61		
			Staurolite	0.33		
			Amphibole	0.66		
			Pyroxene	0.41		
			Garnet	0.41		
			Epidote	0.39		
			Feldspar	0.20		
11. Magnetite	0.011	S. 8° W.	Limonite	0.45	Quartz	0.20
			Kyanite	0.67		
			Amphibole	0.82		
			Garnet	0.65		
			Epidote	0.59		
			Mica	0.69		
12. Leucoxene	0.0099	N. 24° E.	Hematite	0.60		
			Zircon	0.66		
			Feldspar	0.33		
13. Staurolite	0.0054	S. 16° W.	Hematite	0.45		
			Leucoxene	0.43		
			Magnetite	0.36		
			Kyanite	0.30		
			Zircon	0.42		
			Amphibole	0.29		
			Garnet	0.34		
			Epidote	0.29		
14. Rutile	0.0035	S. 8° W.	Tourmaline	0.78		

Table 3 (continued)

Component	Gradient modulus, %/5 km	Orientation of gradient vector	Paired correlation coefficients between fraction and clastic mineral			
			r		−r	
15. Hematite	0.0022	N. 31° E.	Leucoxene	0.60		
			Staurolite	0.45		
			Zircon	0.91		
			Feldspar	0.22		
16. Sphene	0.0013	S. 8° W.	Limonite	0.33		
			Magnetite	0.35		
			Amphibole	0.26		
			Pyroxene	0.24		
			Garnet	0.35		
			Epidote	0.36		
			Mica	0.36		
17. Pyroxene	0.0010	S. 7° W.	Limonite	0.29	Quartz	0.20
			Ilmenite	0.29		
			Magnetite	0.63		
			Kyanite	0.44		
			Amphibole	0.67		
			Garnet	0.57		
			Epidote	0.59		
			Mica	0.41		
18. Ilmenite	0.0005	S. 35° W.	Staurolite	0.20		
			Pyroxene	0.29		
			Garnet	0.44		

Note. In computing the correlation coefficients, 119 observations within the fore-Unguz region were considered.

2. **Carbonate Rock Fragments (Limestone).** The principal source of this material is found in outcrops of Neogene bedrock in basins of the Unguz. The grains are rather unstable, easily broken, abundant at first in the fine-grained fraction, then in the silty fraction, being removed by wind from the zone of accumulation of the sandy material. Mobility of the grains is rather high because of their light weight and generally advanced degree of rounding. The content gradient is oriented toward the south, the maximum value being observed along the southern border of the zone of eolian sands.

3. **Amphibole.** This mineral forms chiefly large, slightly rounded grains in the Kara Kum sands. Its content increases sharply from north to south, reaching a maximum along the southern border of the zone. Amphibole grains, along with kyanite, are concentrated in the coarse-grained fraction.

4. **Epidote.** Epidote grains are generally smaller than amphibole grains and are concentrated in the medium-grained, rarely the fine-grained fraction. They are associated with garnet and magnetite, and they show a negative correlation with feldspar. The content gradient is oriented toward the southwest.

5. **Feldspar.** Grains of this mineral are concentrated in the finest fraction. They show a negative correlation with the fraction $0.15 \leq 0.21$ mm. Rather insignificant concentrations of feldspar grains are noted on the sharp crests of the accumulation forms in association with mica and leucoxene. The content gradient of feldspar is oriented along the axis of the zone of eolian sand (N. 75° W.). This gradient must be excluded from determination of mobility of the clastic grains in their movement from north to south.

6. **Garnet.** This mineral occurs as light-rose angular grains or, because of secondary splitting, as hackly grains, and is concentrated in the most widespread sand fraction in association with magnetite, amphibole, epidote, and other minerals. The content of garnet in the eolian sands increases toward the south.

7. **Zircon.** Most zircon grains are concentrated in the finest fractions, but some may be found in coarse-grained fractions in association with garnet. Fine-grained, well-rounded zircon is invariably associated with hematite and leucoxene. A mixing of zircons from two source areas occurs in the investigated sands: the fine-grained well-rounded forms belong to the Zaunguzskii sands; the coarse, slightly rounded zircons belong to the Kara Kum sands. As a consequence, the gradient of total zircon content obtained in our study is unsuitable for determining mobility. The total content of zircon in the eolian sands is less than in any of the mixed types, but it increases somewhat toward the south.

8. **Limonite.** Rounded spongy grains or films and tubes of limonite are found chiefly in the coarse-grained fraction in association with mica and garnet. The content gradient generally increases toward the southwest.

9. **Kyanite.** Large platy grains of this mineral, rarely small hackly forms, transparent, or limonitized along cleavage traces, are concentrated in the coarsest sand fractions. They are associated with amphibole, magnetite, and mica, and are negatively correlated with quartz. The content gradient of kyanite is slight, oriented toward the southwest. The low gradient within the analyzed field is explained by the very great mobility of kyanite grains. These grains are normally carried far beyond the accumulation site of the examined deposit. Within the investigated area only very large grains remain, settling with the minerals indicated above.

10. **Mica and Chlorite.** These minerals are combined in a single group because of commonly encountered transitions, generally grains of biotite in some stage of chloritization, and because of chlorite pseudomorphs after biotite. These minerals are most highly concentrated in the coarser fractions, but almost nowhere in fractions rich in quartz. The negative correlation between this group and quartz is $r = -0.40$. Strong positive correlation with amphibole and kyanite is characteristic. A low content gradient for mica—chlorite within the investigated zone of eolian sands also follows from the great capacity of these minerals to be transported by wind. The grains of mica and chlorite are generally carried far beyond the limits of the zone of accumulation of the eolian sands, being concentrated on the crests of the large ridge forms.

11. **Magnetite.** Two types of magnetite grains are present: small, well-rounded grains with a lustrous surface, characteristic of the Zaunguzskii sands, and large angular grains, commonly in intergrowths with amphibole and mica, characteristic of the Kara Kum sands. The mineral associations differ for the two types: either zircon and tourmaline (Zaunguzskii sands), or amphibole and mica (Kara Kum sands). The total magnetite content increases in the direction of sand movement from north to south, but the low gradient must be considered a result of variable mobility of the magnetite grains: relatively rapid transport of fine rounded grains, slower transport for large angular grains.

12. **Leucoxene.** This mineral forms large, equant grains, the content of which declines with movement of the sands from north to south. It is concentrated in the finest fractions, being easily destroyed during transportation. It is characteristically associated with feldspar.

13. **Staurolite.** Rounded and subrounded grains of this mineral occur in association with hematite, leucoxene, zircon, and other minerals. It is concentrated in the finest fractions. The gradient is low and is directed toward the southwest.

Table 4. Linear Regression (η_i) of the Logarithm of Maximum Gradient (ξ_i) on Specific Gravity (x_i) of Clastic Minerals in the Kara Kum Sands

Ser. No.	Component	x_i	ξ_i	η_i	Ser. No.	Component	x_i	ξ_i	η_i
1	Quartz	2.6	0.143	−0.734	8	Kyanite	3.6	−1.818	−1.821
2	Carbonate	2.7	−0.221	−0.842	9	Staurolite	3.7	−2.271	−1.929
3	Mica	2.9	−1.936	−1.060	10	Limonite	3.8	−1.452	−1.940
4	Amphibole	3.2	−0.662	−1.386	11	Rutile	4.2	−2.456	−2.473
5	Epidote	3.3	−0.807	−1.495	12	Ilmenite	4.8	−3.334	−3.125
6	Sphene	3.4	−2.903	−1.603	13	Hematite	5.0	−2.662	−3.342
7	Pyroxene	3.4	−3.008	−1.603					

$$\bar{x} = 3.58 \quad \bar{\xi} = 1.799$$

$$\Sigma (x_i - \bar{x})^2 = 6.401; \quad a = \frac{\Sigma \xi_i (x_i - \bar{x})}{\Sigma (x_i - \bar{x})^2} = -1.087;$$

$$\eta_i = -1.799 + [-1.087 \cdot (x_i - \bar{x})].$$

The lines of regression and the confidence interval were computed by L. N. Bol'shev and N. V. Smirnov [1965].

14. Rutile. This mineral occurs in bright red, highly polished, elongated grains. It is found only in association with tourmaline and is concentrated chiefly in sands at the margin of the basin. Despite the low rutile content, its gradient is clearly defined, directed almost due south.

15. Hematite. Rounded, subrounded, and platy grains are concentrated in the finest fractions with zircon, leucoxene, and staurolite. A positive correlation with feldspar is characteristic. The amount of hematite increases toward the northeast, but the gradient is very low. The total hematite content in the eolian sands is considerably smaller than in the sands in the zone of accumulation, a fact that may be explained by the low mobility and low stability during eolian transport.

16. Sphene. Light-colored semitransparent grains are found in small quantities in association with epidote, mica, and other minerals. The gradient is low, directed along the path of sand movement.

17. Pyroxene. Prismatic, slightly rounded grains are concentrated in the fraction $0.15 \leq 0.21$ mm and are associated with amphibole, magnetite, and epidote; the correlation with quartz is negative. The gradient is oriented along the direction of sand transport.

18. Ilmenite. According to its mobility, this mineral occupies one of the last positions. Rounded, polished grains, rarely with remains of leucoxene films, are found. The amount of ilmenite in the eolian sands is less than in any of the mixed sands. The gradient is low and is directed to the southwest. Correlation with garnet and pyroxene is positive.

The indicated values of maximum gradients of sand components permit us to draw the following conclusions concerning the mobility of the Kara Kum sands.

1. The mobility of the sands depends on grain size. The fraction coarser than 0.21 mm falls from the sand-bearing winds most rapidly, forming a maximum on the forward slope of the advancing sands. The low gradient of the fraction < 0.056 mm is explained by the removal of fine grains from the zone of accumulation of the sand fraction. As a consequence, the eolian sands are well sorted, with a steady dominance of the fraction $0.15 \leq 0.21$ mm.

2. The mobility of the sands depends on the physical properties of the clastic grains: equant grains of the light fraction are most mobile (quartz, carbonates); minerals of inter-

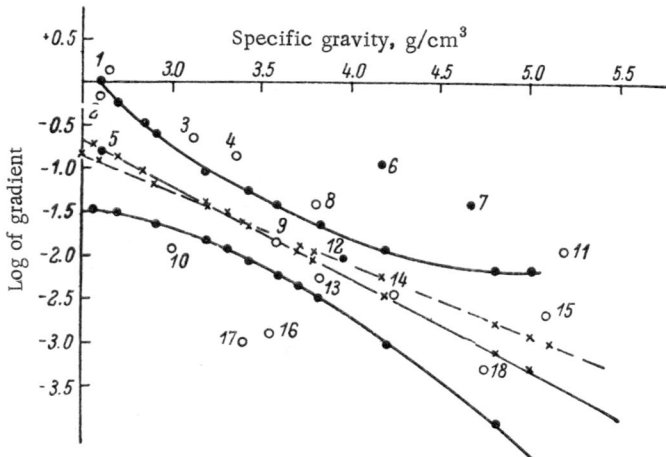

Fig. 5. Linear regression of the dependence of maximum gradient on specific gravity of clastic minerals. Numerals near points correspond to mineral numbers as given in Table 3; open circles represent minerals for which regression was computed; crosses indicate computed regression; dashed line is for computed regression with inclusion of magnetite.

mediate specific gravity have fair mobility (amphibole, epidote); garnet and limonite have lower mobility; and hematite and ilmenite grains are least mobile. A special group of minerals is distinguished, the grains of which are tabular, and which, though the minerals are rather heavy (kyanite, mica, chlorite), are removed from the zone of accumulation of the investigated sands. This means that the content gradient of these minerals is small in the investigated area.

3. Dependence of mobility of clastic material on specific gravity of the minerals was tested by computing linear regression (Table 4). In computing the slope of the linear regression, only components whose gradients were oriented approximately parallel to the direction of sand transport were used. For this reason, feldspar (whose gradient is oriented almost at right angles to the direction of sand transport) and tourmaline (whose gradient is oriented obliquely to the front of advancing sand) were excluded from the analysis.

Grains of magnetite, which commonly occur in intergrowths of dark minerals, and zircon, which belongs to two sharply different types and different source areas, were also excluded in computing the regressions.

The slope of the linear regression of the logarithm of the ratio of maximum gradient to specific gravity of the clastic minerals was found to be −1.087 (including magnetite it would be −0.835). The confidence interval for the linear function $u(x_i, \xi_i)$ for the range of specific gravities from 2.6 to 5.0 is illustrated in Fig. 5.

4. The assumption concerning the effect of the original material of the mixed sands on grain mobility was tested by computing the linear correlation coefficient between the standardized difference of average contents of individual components (Δ_i) and the value of maximum gradients. For this, Δ_i was computed from the relation:

$$\Delta_{ij} = \frac{|X_I - X_{II}|}{X_{max}},$$

where $\overline{X}_I$ is the average content of a component in the Zaunguzskii sands, and X_{II} is the average content of the component in the Kara Kum Sands.

The correlation coefficients thus obtained are insignificant ($r = +0.22$ at $t = 0.8$).

Results of the investigation permit us to draw two fundamental conclusions.

1. In approximating observed values of the exponent in the degree of a polynomial of high order, an unstable solution may be obtained, associated with poor conditionality of the matrix. To put it differently, results of approximation by using polynomials of high order require supplementary testing. But the maximum gradients of trend surfaces of a compositional component reliably describe the mobility of the components of the investigated sands.

2. Grain size and specific gravity of a mineral affect the gradient of clastic grains. In regard to this, eolian sorting of the clastic minerals of the Kara Kum sands tended toward enrichment in grains in the fraction $0.15 \leq 0.21$ mm. Greatest mobility was observed for quartz grains and carbonate rocks, normally having rounded shapes. Ilmenite exhibits the lowest mobility.

The role of grain shape remains untested quantitatively, although it is obvious that platy grains (mica, chlorite, kyanite) are removed from the zone of accumulation of the investigated sands.

REFERENCES

Bagnold, R. A., The Physics of Blown Sand and Desert Dunes, London (1941).

Bondarev, L. G., "Eolian sands along the upper reaches of the Bol'shoi Naryn," in: Physical Geography of Tien Shan, Izd. Akad. Nauk KirgSSR (1966), pp. 12-19.

Bol'shev, L. N., and Smirnov, N. V., Tables of Mathematical Statistics, Izd. Nauka, Moscow (1965).

Chepil, W. S., "Dynamics of wind erosion: The nature of movement of soil by wind," Soil Sci. Soc. Am. Proc., Vol. 60, pp. 305-320 (1945).

Chepil, W. S., "The use of spheres to measure lift and drag on wind-eroded soil grains," Soil. Sci. Soc. Am. Proc., Vol. 25, pp. 343-345 (1961).

Ermilov, I. Ya., "Eolian accumulation forms in the saline deserts of western Turkmenistan," Izv. Vsesoyuzn. Geogr. Obshch., Vol. 81, No. 3, pp. 327-333 (1949).

Exner, F. M., "Zur Physik der Dunen," Sitzber. Akad. Wiss., Pt. II, Vol. 129, Wien, 929-959 (1920).

Fedorovich, B. A., The Role of Wind in the Formation of the Sandy Landscape of the Desert, Tr. Inst. Geogr., Vol. 36, Izd. Akad. Nauk SSSR, Moscow (1940).

Fedorovich, B. A., "Origin and development of sandy strata in the desert," Azii. Mat. po Chetvert. per. SSSR, No. 2, pp. 221-233 (1950).

Horner, G. N., "Some notes and data concerning dunes and sand drift in the Gobi Desert," Sci. Exped. Northwest Provinces China, Sino-Swedish Exped., Publ. 40, Vol. III, Geol., No. 5, p. 40 (1957).

Luppov, N. P., "The Middle Pliocene stage in the geologic history of the trans-Caspian region," in: Problems of Oil and Gas Occurrence in Central Asia, No. 14, pp. 11-37, Gostoptekhizdat, Leningrad (1963).

Orlov, B. P., "Mechanism of advance of individual sand grains and sandy accumulations," Zemlevedenie, Nov. Ser., Vol. 5, pp. 25-41 (1960).

Petrov, M. P., Moving Sands of the Soviet Union and the Fight Against Them, Izd. Geogr. Lit., Moscow (1950).

Priestly, C. H. B., Turbulent Transfer in the Lower Atmosphere, University of Chicago Press, Chicago (1959).

Romanova, M. A., "Separating the sands of the central Kara Kum into districts according to spectral brightness," Dokl. Akad. Nauk SSSR, Vol. 156, No. 5, pp. 1095-1098 (1964a).

Romanova, M. A., "Recent deposits of sand in central Kara Kum and the problem of studying subsurface structures," Sov. Geol., No. 12, pp. 70-89 (1964b).

Senkevich, B. I., "Experimental investigations of eolian erosion of sands," Izv. Akad. Nauk TurkmSSR, Ser. Biol. Nauk, pp. 21-30 (1962).

Sidorenko, A. V., "Eolian differentiation of material in the desert," Izv. Akad. Nauk SSSR, Ser. Geogr., No. 3, pp. 3-22 (1956).

Vistelius, A. B., and Sarsadskii, N. N., "Nature of changes in mineral content of concentrates during successive washing of sands," Zap. Vses. Mineralog. Obshchestva, Pt. 80, No. 2, pp. 143-152 (1952).

Vistelius, A. B., and Yanovskaya, T. B., "Programming problems of geology and geochemistry for use with all-purpose electronic computers," Geol. Rudn. Mestorozhd., No. 3, pp. 34-48 (1963).

Williams, G., "The effect of eolian transport of sand at various heights on the grain-size distribution," Dokl. Akad. Nauk SSSR, Vol. 149, No. 2, pp. 410-412 (1963).

Williams, G., "Some aspects of eolian saltation load," (U.S. Geol. Surv., Wash.) Sedimentology, No. 3, pp. 275-287 (1964).

Znamenskii, A. I., "Movement of barchan sands," Tr. Turkmensk. Fil. Akad. Nauk SSSR, Vol. 2 (1942).

Znamenskii, A. I., "New principals of protection of industrial and domestic buildings against drifting sand," in: Objectives and Potentials of Prospecting-Exploratory Work on Oil and Gas in the Western Parts of Central Asia, Ashkhabad (1959).

VI. Various Geological Problems

HEAT CONDUCTION CALCULATIONS ON THE THERMAL HISTORY OF CONTACT AUREOLES

F. Hori

Institute of Earth Science
College of General Education
University of Tokyo
Tokyo, Japan

The problem of temperature distribution around a magmatic source and of the effect of these temperatures on the formation of contact aureoles is examined. Three models of heat transfer, represented by differential equations, are proposed. Results of the model studies are interpreted in terms of metamorphic facies.

There is some agreement among geologists about quantitative correlation of the mineral facies with temperature and depth. Then, the mineralogy of rocks surrounding an intrusive body may be regarded as the key to the temperature conditions prevailing in the aureole. Certainly, the source of heat in contact metamorphism is the igneous intrusion; and within the aureole a gradient in maximum temperature of metamorphism is produced. However, it is not always certain whether rocks of individual facies found in a given aureole are products of metamorphism at maximum temperatures attained there. In reality, contact metamorphism is a matter of time as well as place. The principal difficulty met with in the study of contact metamorphism arises from the situation that the factors governing the development of temperature cannot easily be discriminated from observed mineral assemblages.

To overcome this difficulty, it may be profitable to investigate the reverse problem of elucidating the dependence of mineral assemblages upon known factors of thermal history. This paper is a mathematical approach to the problem through the use of heat conduction calculations.

Calculations of Models

Temperature Patterns Around an Intrusive Body

Heat energy out of an intrusive rock is transferred principally by lattice vibrations and flow of fluid, and the radiative component is negligible in a relevant range of temperature [Clark, 1957; MacDonald, 1959; Hori, 1964]. In the following, it will be assumed that conduction is the principal element in heat transfer. Thus, the nature of an intrusion and the distribution of thermal conductivity and diffusivity in a country rock are viewed as having critical influence on the thermal history of the contact aureole.

In calculating the thermal history of a contact aureole, general assumptions will be made. These are: 1) the intrusion occurs instantaneously; 2) thermal conductivity and diffusivity are independent of both position and temperature; 3) the latent heat of fusion can be neglected; 4) there is no heat production in the country rock; 5) the temperature in the intrusive body and the country rock are the same at the contact surface, and the flux of heat is continuous over this surface; and, 6) the isothermal surfaces are planes parallel to the contact surface, and flow of heat is linear, the lines of flow being parallel to an axis of x.

From the assumptions made, the heat conduction equation which describes temperature patterns is

$$\frac{1}{k}\frac{\partial T}{\partial t} - \frac{\partial^2 T}{\partial x^2} = 0, \qquad (1)$$

where T is temperature, k is thermal diffusivity, t is the time measured from the instant of intrusion, and x is the distance measured from the igneous contact. Equation (1) was solved previously for the following three models of intrusion [Hori, 1963].

Model 1 assumes that an intrusive body is of semi-infinite extent bounded by a contact plane, and kept at a constant temperature of T' after its intrusion. The temperature of the country rock at the time of the intrusion is assumed to be a linear function of distance from the contact, and written as

$$(T)_{t=0} = f(x) = T_0 + ax \quad \text{for} \quad x > 0. \qquad (2)$$

Then the solution of equation (1) is

$$T = T_0 + (T' - T_0)\, \text{erfc}\, \frac{x}{2\sqrt{kt}} + ax \quad \text{for} \quad x > 0. \qquad (3)$$

In model 2, owing to heat loss from the contact surface, a semi-infinite intrusive body begins to cool immediately after the intrusion. Let the thermal conductivity and diffusivity across the contact plane be constant, so that we have

$$T = T_0 + \frac{T' - T_0}{2}\left(1 + \text{erf}\, \frac{|x|}{2\sqrt{kt}}\right) \quad \text{for} \quad x < 0, \qquad (4a)$$

$$T = T_0 + \frac{T' - T_0}{2}\, \text{erf}\, \frac{x}{2\sqrt{kt}} + ax \quad \text{for} \quad x > 0. \qquad (4b)$$

Model 3 differs from model 2 in that the intrusive body has a finite thickness, and its shape is approximated by an infinite slab r thick. Then we obtain

$$T = T_0 + \frac{T' - T_0}{2}\left(\text{erf}\, \frac{x+r}{2\sqrt{kt}} + \text{erf}\, \frac{|x|}{2\sqrt{kt}}\right) \quad \text{for} \quad -r < x < 0, \qquad (5a)$$

$$T = T_0 + \frac{T' - T_0}{2}\left(\text{erf}\, \frac{x+r}{2\sqrt{kt}} - \text{erf}\, \frac{x}{2\sqrt{kt}}\right) + ax \quad \text{for} \quad x > 0. \qquad (5b)$$

Although the initial temperatures of deep-seated igneous rocks are not precisely known, they probably fall within a range of 600-1000°C. The temperature of intrusions near the outer surface of the earth probably ranges from 800-1200°C.

Table 1. Temperature Distributions (T, °C) for Models 1 and 3 with T' = 800°C, T_0 = 0°C, k = 0.010 cm^2 · sec^{-1}, and r = 2 km

x, km	t, years					x, km	t, years				
	$1 \cdot 10^2$	$1 \cdot 10^3$	$1 \cdot 10^4$	$1 \cdot 10^5$	$1 \cdot 10^6$		$1 \cdot 10^2$	$1 \cdot 10^3$	$1 \cdot 10^4$	$1 \cdot 10^5$	$1 \cdot 10^6$
	Model 1						Model 3				
0.05	420	680	770	790	800	0.05	210	340	380	230	80
0.1	170	550	720	780	800	0.1	84	280	360	230	80
0.2	8	350	640	750	790	0.2	4	170	320	220	80
0.5	0	37	420	680	770	0.5	0	20	210	210	80
1		0	170	550	720	1		0	84	180	80
2			10	340	620	2			4	120	76
3			0	180	550	3			0	72	72
4				100	490	4				36	64
5				40	450	5				16	60

The initial temperature in a country rock depends primarily upon depth of burial in the earth's crust. The temperature distribution in the crust has probably reached the steady state before an intrusion occurs. In the steady state, an average gradient of temperature decreases by about 0.9 times with every kilometer of depth, and the temperature at a depth of 15 km comes within a factor of 1.5 of 300°C [Hori, 1962]. The geothermal gradient in the outer few tens of kilometers can roughly be averaged by a value of $2.0 \cdot 10^{-4}$ °C cm^{-1}.

In the present one-dimensional models, T_0 in equation (2) can be regarded as the temperature at a given depth, and α is the average geothermal gradient. It is considered then that the initial temperature in a country rock is constant at T_0 in a case where the contact plane is vertical and the lines of heat flow are horizontal; and it equals $(2.0 \cdot 10^{-4})$ x (in cm) in a case where the contact plane is horizontal and the lines of heat flow are upward.

Equation (1) has been derived by assuming that conductivity and diffusivity are independent of both position and temperature. Birch and Clark [1940] found by experimental work that there was no significant variation in thermal conductivity with temperatures up to 400°C. Lubimova [1958] suggested, on the basis of lattice-conduction theory, that conductivity increases with pressure. The theory of thermal conductivity applicable to silicate minerals is difficult, and experimental data relating to conductivity at high pressures are scarce. At present, therefore, we are not certain how much the effect of pressure would be. Within the limited range of temperature and pressure relevant to contact metamorphism, however, it is likely that the conductivity of crustal materials remains unchanged during the metamorphism.

Birch [1942] reviewed measurements of conductivities for a wide range of crustal materials. These values show pronounced variations with physical properties rather than with mineral composition. Sedimentary layers of which a thick column of a geosynclinal prism consists may have various conductivities depending on age, depth of burial, and degree of tectonic disturbance. Although there is a rather wide range in individual values, conductivities and diffusivities of most of unconsolidated and consolidated sediments come within a few tenths of a percent of 0.015 joule · cm^{-1} · sec^{-1} · °C^{-1}, and 0.010 cm^2 · sec^{-1}, respectively.

Taking T' as 800°C, T_0 as 0°C, α as 0°C cm^{-1}, k as 0.010 cm^2 · sec^{-1}, and r as 2000 m, numerical calculations of equations (3) and (5) have been carried out for values of t up to $1 \cdot 10^6$ years. The calculated temperatures are presented in terms of x and t in Table 1. Effects of variations in the parameters on the calculated values have been discussed elsewhere [Hori, 1963].

Table 2. Temperature Gradients ($\partial T/\partial x$, °C·cm^{-1}) for Models 1 and 3 with $T' - T_0 = 800$°C and $k = 0.010$ cm^2·sec^{-1}

r, km	x, km	\$t\$, years				
		$1 \cdot 10^2$	$1 \cdot 10^3$	$1 \cdot 10^4$	$1 \cdot 10^5$	$1 \cdot 10^6$
		Model 1				
	0	$8.0 \cdot 10^{-2}$	$2.6 \cdot 10^{-2}$	$8.0 \cdot 10^{-3}$	$2.6 \cdot 10^{-3}$	$8.0 \cdot 10^{-4}$
	0.05	$6.6 \cdot 10^{-2}$	$2.4 \cdot 10^{-2}$	$8.1 \cdot 10^{-3}$	$2.6 \cdot 10^{-3}$	
	0.1	$3.0 \cdot 10^{-2}$	$2.3 \cdot 10^{-2}$	$8.0 \cdot 10^{-3}$	$2.6 \cdot 10^{-3}$	$8.0 \cdot 10^{-4}$
	0.2	$1.7 \cdot 10^{-3}$	$1.8 \cdot 10^{-2}$			$8.0 \cdot 10^{-4}$
	0.5		$3.6 \cdot 10^{-3}$	$6.6 \cdot 10^{-3}$		
	1			$3.0 \cdot 10^{-3}$	$2.3 \cdot 10^{-3}$	$8.0 \cdot 10^{-4}$
	2			$3.3 \cdot 10^{-4}$	$1.8 \cdot 10^{-3}$	$8.0 \cdot 10^{-4}$
	3				$1.3 \cdot 10^{-3}$	
	5				$3.5 \cdot 10^{-4}$	$6.6 \cdot 10^{-4}$
	10					$3.0 \cdot 10^{-4}$
		Model 3				
0.1	0	$2.2 \cdot 10^{-2}$	$9.8 \cdot 10^{-4}$	$3.2 \cdot 10^{-5}$	$1.3 \cdot 10^{-6}$	$4.0 \cdot 10^{-8}$
	0.05	$2.6 \cdot 10^{-2}$	$1.9 \cdot 10^{-3}$	$6.4 \cdot 10^{-5}$		
	0.1	$1.6 \cdot 10^{-2}$	$2.5 \cdot 10^{-3}$	$9.4 \cdot 10^{-5}$	$2.6 \cdot 10^{-6}$	$9.7 \cdot 10^{-8}$
	0.2	$1.6 \cdot 10^{-3}$	$3.1 \cdot 10^{-3}$	$1.5 \cdot 10^{-4}$	$5.2 \cdot 10^{-6}$	$1.6 \cdot 10^{-7}$
	0.5	0	$1.0 \cdot 10^{-3}$	$2.8 \cdot 10^{-4}$	$1.3 \cdot 10^{-5}$	$4.0 \cdot 10^{-7}$
	1			$2.9 \cdot 10^{-4}$	$1.9 \cdot 10^{-5}$	$6.4 \cdot 10^{-7}$
	2			$4.7 \cdot 10^{-5}$	$3.0 \cdot 10^{-5}$	$1.3 \cdot 10^{-6}$
	3			$1.2 \cdot 10^{-6}$	$3.0 \cdot 10^{-5}$	$1.8 \cdot 10^{-6}$
	5				$1.4 \cdot 10^{-5}$	$2.6 \cdot 10^{-6}$
	10					$2.8 \cdot 10^{-6}$
2	0	$4.0 \cdot 10^{-2}$	$1.3 \cdot 10^{-2}$	$3.8 \cdot 10^{-3}$	$3.5 \cdot 10^{-4}$	$1.3 \cdot 10^{-5}$
	0.05	$3.3 \cdot 10^{-2}$	$1.3 \cdot 10^{-2}$	$3.9 \cdot 10^{-3}$		
	0.1	$1.8 \cdot 10^{-2}$	$1.2 \cdot 10^{-2}$	$3.9 \cdot 10^{-3}$	$3.8 \cdot 10^{-4}$	$1.3 \cdot 10^{-5}$
	0.2	$1.7 \cdot 10^{-3}$	$9.3 \cdot 10^{-3}$	$3.8 \cdot 10^{-3}$	$4.0 \cdot 10^{-4}$	$1.5 \cdot 10^{-5}$
	0.5	0	$1.8 \cdot 10^{-3}$	$3.3 \cdot 10^{-3}$	$4.8 \cdot 10^{-4}$	$1.9 \cdot 10^{-5}$
	1		$5.1 \cdot 10^{-6}$	$1.8 \cdot 10^{-3}$	$5.5 \cdot 10^{-4}$	$2.4 \cdot 10^{-5}$
	2			$1.7 \cdot 10^{-4}$	$5.7 \cdot 10^{-4}$	$3.5 \cdot 10^{-5}$
	3				$4.5 \cdot 10^{-4}$	$4.5 \cdot 10^{-5}$
	5				$1.5 \cdot 10^{-4}$	$5.8 \cdot 10^{-5}$
	10					$5.3 \cdot 10^{-5}$

Instantaneous Gradients of Temperature

Temperature gradient within a contact aureole can be derived by differentiating equations (3), (4b), and (5b), respectively, with respect to x. Thus we obtain

$$\frac{\partial T}{\partial x} = -\frac{T' - T_0}{\sqrt{\pi k t}} \exp\left(-\frac{x^2}{4kt}\right) + \alpha \quad \text{for model 1,} \tag{6}$$

$$\frac{\partial T}{\partial x} = -\frac{T' - T_0}{2\sqrt{\pi k t}} \exp\left(-\frac{x^2}{4kt}\right) + \alpha \quad \text{for model 2,} \tag{7}$$

$$\frac{\partial T}{\partial x} = -\frac{T' - T_0}{2\sqrt{\pi k t}} \left[\exp\left(-\frac{x^2}{4kt}\right) - \exp\left\{-\frac{(x+r)^2}{4kt}\right\}\right] + \alpha \quad \text{for model 3.} \tag{8}$$

It is evident from the above equations that the temperature gradient at a given distance at a given time is maximum for model 1 and minimum for model 3, and that this maximum value is twice as large as the value for model 2. It is also seen that the instantaneous temperature gradient is proportional to the initial difference between the internal and external temperatures in an intrusive body. Effects of variations in diffusivity vary in relation to distance, time, and size of intrusive body.

Since every intrusive body has a finite thickness, the temperature gradient in a contact aureole has always a maximum value. At a given time, the temperature gradient reaches a maximum value at a certain distance; and this distance becomes larger as the thickness of an intrusive body decreases. At a given distance, the temperature gradient reaches a maximum value at a certain time; and this time increases with increasing thickness.

Table 2 shows the instantaneous temperature gradients calculated for models 1 and 3 with $(T' - T_0) = 800°C$, $k = 0.010$ cm$^2 \cdot$ sec^{-1}, $\alpha = 0°C \cdot$ cm^{-1}, and $r = 100$ or 2000 m.

Instantaneous Flow Rates of Heat

The outward flux of heat by conduction per unit area and per unit time is

$$\frac{dq}{dt} = -K \frac{\partial T}{\partial x}, \qquad (9)$$

where K is the thermal conductivity of a body. Hence, the instantaneous rate q_x of heat flow into a country rock through any unit area parallel to the contact plane at any distance is given by

$$q_x = \frac{K(T' - T_0)}{\sqrt{\pi k t}} \exp\left(-\frac{x^2}{4kt}\right) - K\alpha \quad \text{for model 1,} \qquad (10)$$

$$q_x = \frac{K(T' - T_0)}{2\sqrt{\pi k t}} \exp\left(-\frac{x^2}{4kt}\right) - K\alpha \quad \text{for model 2,} \qquad (11)$$

$$q_x = \frac{K(T' - T_0)}{2\sqrt{\pi k t}} \left[\exp\left(-\frac{x^2}{4kt}\right) - \exp\left\{\frac{(x+r)^2}{4kt}\right\}\right] - K\alpha \quad \text{for model 3.} \qquad (12)$$

The total heat flow Q_x per unit area parallel to the contact plane at any distance between t_1 and t_2 is

$$Q_x = \int_{t_1}^{t_2} q_x dt. \qquad (13)$$

Since the exponential function in equations (10) and (11) equals 1 for the contact plane, the total heat inflow Q_0 per unit area at the contact between times t_1 and t_2 becomes

$$Q_0 = \frac{2K(T' - T_0)}{\sqrt{\pi k}} (\sqrt{t_2} - \sqrt{t_1}) \quad \text{for model 1,} \qquad (14)$$

$$Q_0 = \frac{K(T' - T_0)}{\sqrt{\pi k}} (\sqrt{t_2} - \sqrt{t_1}) \quad \text{for model 2.} \qquad (15)$$

It is clear that the value for Q_x at a given distance in a given length of time in model 3 is no more than that at the same distance in the same length of time in model 2.

Effects of variations in thermal conductivity are variable, for an increase in thermal conductivity generally results in an increase in thermal diffusivity. If the product of density and specific heat for a country rock remains constant, an increase in conductivity leads to an increase in both q_x and Q_0; a variation in conductivity by a factor of n produces a variation in Q_0 by a factor of $\sqrt{n}$.

Instantaneous flow rates of heat and total heat inflows calculated with $T' - T_0 = 800°C$, $K = 0.015$ joule $\cdot$ cm$^{-1} \cdot$ sec$^{-1} \cdot$ °C^{-1}, $k = 0.010$ cm$^2 \cdot$ sec^{-1}, $\alpha = 0$, and $r = 100$ or 2000 m are given in Tables 3 and 4, respectively.

Table 3. Instantaneous Flow Rates of Heat ($q_x \cdot 10^6$ cal $\cdot$ cm^{-2} $\cdot$ sec^{-1}) for Models 1 and 3 with $T' - T_0 = 800°C$, $k = 0.010$ cm^2 $\cdot$ sec^{-1}, and $K = 0.015$ joule $\cdot$ cm^{-1} $\cdot$ sec^{-1} $\cdot$ °C^{-1}

r, km	x, km	t, years				
		$1 \cdot 10^2$	$1 \cdot 10^3$	$1 \cdot 10^4$	$1 \cdot 10^5$	$1 \cdot 10^6$
		Model 1				
	0	290	93	29	9.3	2.9
	0.05	240	86	29	9.3	
	0.1	110	83	29	9.3	2.9
	0.2	6.0	65			
	0.5	0	13	24		2.9
	1		0.04	11	8.3	
	2			1.2	6.5	
	3				4.7	
	5				1.2	2.4
	10					1.1
		Model 3				
0.1	0	80	3.5	0.12	0.0047	0.00014
	0.05	95	6.6	0.23		
	0.1	59	8.9	0.35	0.0092	0.00035
	0.2	5.9	11	0.56	0.018	0.00057
	0.5	0	3.8	1.0	0.045	0.0014
	1		0	1.0	0.068	0.0023
	2			0.17	0.11	0.0047
	3			0.0044	0.11	0.0066
	5				0.048	0.0095
	10					0.010
2	0	140	47	14	1.2	0.045
	0.05	120	45	14		
	0.1	66	42	14	1.4	0.048
	0.2	6.0	33	14	1.5	0.054
	0.5	0	6.3	12	1.7	0.068
	1		0.018	6.6	2.0	0.089
	2			0.60	2.1	0.13
	3				1.7	0.17
	5				0.54	0.21
	10					0.20

Table 4. Total Heat Inflow in the First t_1 Years for Model 1 with $T' - T_0 = 800°C$, $k = 0.010$ cm^2 $\cdot$ sec^{-1}, and $K = 0.015$ joule $\cdot$ cm^{-1} $\cdot$ sec^{-1} $\cdot$ °C^{-1}

t_1, years	$1 \cdot 10^2$	$1 \cdot 10^3$	$1 \cdot 10^4$	$1 \cdot 10^5$	$1 \cdot 10^6$
Q_0, kcal $\cdot$ cm^{-2}	$1.8 \cdot 10^3$	$6.2 \cdot 10^3$	$1.8 \cdot 10^4$	$6.2 \cdot 10^4$	$1.8 \cdot 10^5$

Instantaneous Variation Rates of Temperature

If both an intrusive body and a country rock are of semi-infinite scale, the country rock will continuously be heated until the whole system reaches a final thermal equilibrium. If, however, the intrusive body has a finite dimension, temperatures at any distance in the country rock will first rise to a certain maximum value, and then drop to reach a final thermal equilibrium. The temperature at which the whole system reaches the thermal equilibrium is T', $(T' + T_0)/2$, and T_0, respectively, in models 1, 2, and 3. Before these thermal equilibriums are reached, temperatures at any distance rise or drop at various rates.

The instantaneous variation rates in temperature at any distance are given by differentiating equations (3), (4), and (5), respectively, with respect to t. Thus we have

$$\frac{\partial T}{\partial x} = \frac{(T' - T_0) x}{2 \sqrt{\pi k t^3}} \exp\left(-\frac{x^2}{4kt}\right) \quad \text{for model 1,} \tag{16}$$

$$\frac{\partial T}{\partial x} = \frac{(T' - T_0) x}{4 \sqrt{\pi k t^3}} \exp\left(-\frac{x^2}{4kt}\right) \quad \text{for model 2,} \tag{17}$$

$$\frac{\partial T}{\partial x} = \frac{(T' - T_0) x}{4 \sqrt{\pi k t^3}} \left[\exp\left(-\frac{x^2}{4kt}\right) - \left(1 + \frac{r}{x}\right) \exp\left\{-\frac{(x+r)^2}{4kt}\right\}\right] \quad \text{for model 3.} \tag{18}$$

Again, the rates of temperature variations are proportional to $(T' - T_0)$. In model 3 the rate at a given distance at a given time increases with thickness.

Variation rates in metamorphic temperature in cases where $T' - T_0 = 800°C$, $k = 0.010$ cm$^2 \cdot$ sec^{-1}, and $r = 100$ or 2000 m are listed in Table 5.

Petrological Implications

The equations in the preceding section illustrate the dependence of the thermal history of an intrusion on the thermal conditions and size of the intrusive rock, the depth of the intrusion, and the distribution of thermal conductivity and diffusivity in the country rock. This implies that the mineralogy and texture of rocks formed in individual contact aureoles show some variations depending on differences in these controlling factors. The numerical estimates listed in Tables 1-5 can be combined to gain an insight into main features of contact aureoles. However, any details of a geological setting can never be reproduced in the mathematical models. It is therefore necessary to consider how intrusions in nature may deviate from the assumed models.

Table 5. Variation Rates of the Temperature of Metamorphism ($\partial T/\partial t$ in °C $\cdot$ yr^{-1}) for Models 1 and 3 with $T' - T_0 = 800°C$ and $k = 0.010$ cm$^2 \cdot$ sec^{-1}

r, km	x, km	t, years				
		$1 \cdot 10^2$	$1 \cdot 10^3$	$1 \cdot 10^4$	$1 \cdot 10^5$	$1 \cdot 10^6$
			Model 1			
	0	0	0	0	0	
	0.05	1.7	0.060	0.0020		
	0.1	1.5	0.12	0.0040	$1.3 \cdot 10^{-4}$	
	0.2	0.16	1.8			
	0.5		0.090	0.017		$2.0 \cdot 10^{-5}$
	1			0.015	0.0012	$4.0 \cdot 10^{-5}$
	2			0.0033	0.0018	
	3				0.0020	
	5				$8.8 \cdot 10^{-4}$	$1.7 \cdot 10^{-4}$
	10					$1.5 \cdot 10^{-4}$
			Model 3			
0.1	0	−1.8	−0.12	−0.0040	$-1.3 \cdot 10^{-4}$	$-4.0 \cdot 10^{-6}$
	0.05	0.32	−0.049	−0.0020		
	0.1	0.74	−0.035	−0.0019	$-6.4 \cdot 10^{-5}$	$-2.0 \cdot 10^{-6}$
	0.2	0.16	−0.0068	−0.0017	$-6.3 \cdot 10^{-5}$	$-2.0 \cdot 10^{-6}$
	0.5	0	0.022	$-8.0 \cdot 10^{-4}$	$-5.9 \cdot 10^{-5}$	$-2.0 \cdot 10^{-6}$
2	0	0	0	−0.0034	−0.0019	$-7.8 \cdot 10^{-5}$
	0.1	0.91	0.059	$7.2 \cdot 10^{-4}$	$-8.8 \cdot 10^{-4}$	$-3.9 \cdot 10^{-5}$
	0.2	0.17	0.093	0.037	$-8.3 \cdot 10^{-4}$	$-3.8 \cdot 10^{-5}$
	0.5	0	0.044	0.0079	$-6.6 \cdot 10^{-4}$	$-3.8 \cdot 10^{-5}$
	1		$2.6 \cdot 10^{-4}$	0.0090	$-3.6 \cdot 10^{-4}$	$-3.6 \cdot 10^{-5}$
	2		0	0.0017	$2.1 \cdot 10^{-4}$	$-3.2 \cdot 10^{-5}$
	5			0	$3.5 \cdot 10^{-4}$	$-1.3 \cdot 10^{-5}$

Model 2 assumes the semi-infinite extent of an intrusive body. Certainly, this assumption does not hold for naturally occurring intrusions. Every intrusive body has a finite thickness, and thus must be approximated in model 3. Since

$$\text{erf} \frac{x+r}{2\sqrt{kt}} - \text{erf} \frac{x}{2\sqrt{kt}} \approx \text{erf} \frac{x}{2\sqrt{kt}} \tag{19}$$

for sufficiently large values of $(x + r)/\sqrt{t}$, however, the development of temperature either for short periods after the intrusion, or at distances far removed from the contact can fairly well be approximated by calculating model 2.

Model 1 differs from model 2 in that the temperature at a contact surface is constant for some time interval after intrusion. If, in any instance, heat is transported to the contact plane through some much more effective mode than conduction, the contact may be regarded as kept at constant temperatures. Calculating equation (4a), we see that the temperature of a large intrusive body at some 100 m inside the contact is still as high as 90% of its initial value in the first 10^3 years, even if heat is transferred only by conduction. These figures suggest that an intrusive body of regional scale may remain mobile except in a marginal part for a considerable length of time. In this event, owing to rapid and effective heat supply through mass movement of mobile material, further cooling of the marginal part may be markedly retarded. If there is heat supply at a rate compensating for heat loss from the contact, the temperature at the contact will be constant nearly at its initial value for certain lengths of time. Model 1 is applicable to such a situation.

Unfortunately, it is uncertain how long an intrusive body of a given size retains effective mobility. However, it seems natural to consider that the larger the intrusive body, the longer the time required before it consolidates as a whole. Thus, model 1 may be regarded as approximating batholithic intrusions, as far as their first 10^4 years or so are concerned. Even for a rather small intrusive body some 100 m thick, however, the thermal development in a region immediately adjoining the contact may be approximated by calculating model 1.

In short, it is expected that temperature patterns in a narrow region adjacent to an igneous contact lie between those calculated in models 1 and 2, and temperature patterns in a region far removed from the contact lie between those calculated in models 2 and 3. The former approximation applies to larger distances from the contact with increasing size of the intrusive body.

The present numerical experiments apply to a system in which heat is transferred exclusively by conduction. It is uncertain how far such an assumption may be relied upon. If, in any instance, it is evident that the country rock is open with respect to some components, heat transported by the flow of the mobile materials must be taken into consideration. Above all, the effect of water may be noticeable.

If vapor pressure in a cooling magma becomes higher than that in a surrounding rock, water will flow out of the magma. At depths where free openings are sealed, water flows mainly through an interconnected pore system. The equation of linear flow of water in the horizontal direction is taken in the form [Hori, 1964]

$$\frac{\partial P}{\partial t} = \frac{h}{fc\mu} \frac{\partial^2 P}{\partial x^2}, \tag{20}$$

where P is water pressure, h and f are the permeability and effective porosity, respectively, of a country rock; c and μ are the compressibility and viscosity, respectively, of water; t is the time from the instant at which water begins to flow out of an intrusive body, and x is the distance from the contact.

In a case where an initial difference in water pressure at an igneous contact remains constant at the order of 10^2 bars, a total volume of 10^3 cm^3 of water flows through any unit area at the contact plane into a country rock with a porosity of 10^{-3} and a permeability of 10^{-4} darcy in the first 10^3 years. This total volume of inflow results in water flow at volume-flow rates of 10^{-7}–10^{-8} cm · sec^{-1} at a distance of 1 km from the contact throughout the first 10^3 years. It will be seen from Tables 3 and 4 that the amount of heat transported by this water flow is rather comparable to that by conduction.

A volume-flow rate of 10^{-7} cm · sec^{-1} is average rather than extreme. Ability of fluids to carry heat depends on geological environments. Under physical conditions prevailing in the outer several kilometers of the earth, any fluid, if it flows at all out of a magma, may be an effective and long-distance heat carrier. In cases where fluid flows may be regarded as dominant over conduction, the detailed numerical estimates obtained here may be regarded, to the extent that this is true, as restrictive.

Break in Temperature Across an Igneous Contact

The temperature of contact metamorphism depends critically upon the length of time required before the temperature of an intrusive rock at the contact begins to drop appreciably. As will be seen from equation (3), temperatures similar to T' can be reached in an aureole, so long as the contact plane does not cool at all. Even if, however, the contact is constantly kept at T', it takes some 10^7 years for the country rock to reach temperatures regionally similar to T'. Probably, the initial temperature at any igneous contact does not remain constant for such lengths of time. If the temperature of an intrusive rock at the contact begins to drop immediately after the intrusion, temperatures in excess of $(T' + T_0)/2$ cannot be attained in the aureole. It follows that there commonly is a drastic drop in maximum temperature of metamorphism within a narrow range of distances from the contact.

Across the contact plane kept at T' for a length of time of t_0, there is a difference in temperature of

$$\Delta T_1 = (T' - T_0) \operatorname{erf} \frac{x}{2\sqrt{kt_0}}. \tag{21}$$

In view of the nature of the error function, we see that ΔT_1 is infinitesimal for small values of x, and increases rapidly with distance; and rates of this increase become larger as t_0 decreases. For example, a temperature of 90% of T' is reached at distances of about 10 and 10^2 m for t_0 of 10^2 and 10^4 years, respectively. Then, it turns out that temperatures of metamorphism similar to the initial temperature of an intrusion can be reached only in a region immediately adjoining the contact surface.

This may be the reason why rocks of the sanidinite facies, which corresponds to maximum temperatures and minimum pressures of metamorphism, occur exclusively as xenoliths immersed in basalt (consistent with a high-temperature magma), and as narrow aureoles adjoining necks or dikes of basic volcanic rocks.

In a case where an igneous contact begins to cool immediately after the intrusion, a difference of

$$\Delta T_2 = (T' - T_0) \left\{ 1 - \frac{1}{2}\left(\operatorname{erf} \frac{x+r}{2\sqrt{kt!}} - \operatorname{erf} \frac{x}{2\sqrt{kt}} \right) \right\}, \tag{22}$$

arises between the initial temperature of the intrusion and the temperature of metamorphism. At shorter distances ΔT_2 decreases rapidly, approaching

$$(T' - T_0)\left(1 - \frac{1}{2} \operatorname{erf} \frac{r}{2\sqrt{kt}}\right), \qquad (23)$$

as the contact plane is approached. At a given distance, ΔT_2 varies inversely with increasing thickness. If with a given value of $x/\sqrt{kt}$, r is sufficiently large, equation (22) is reduced to

$$\Delta T_2 = \frac{T' - T_0}{2}\left(1 + \operatorname{erf} \frac{x}{2\sqrt{kt}}\right). \qquad (24)$$

In this case we obtain for a time of t_0

$$\Delta T_2 - \Delta T_1 = \frac{T' - T_0}{2} \operatorname{erf} c \frac{x}{2\sqrt{kt}}. \qquad (25)$$

It is clear then that the value of $(\Delta T_2 - \Delta T_1)$ becomes larger as the value of t_0 increases.

From these it can be inferred that there is a marked break between the initial temperature of an intrusive rock and the maximum temperature of contact metamorphism except, perhaps, in a region immediately adjoining the contact surface. Equations (21), (22), and (25) illustrate the dependence of the break in temperature upon $(T' - T_0)$, r, and t_0; the influence of the last factor is most dominant. Thus, a break that is much in excess of 100-200°C appears unreasonable at the contact of deep-seated, low-temperature, large granite batholiths. On the other hand, a break in excess of 400-500°C appears possible at the contact of necks or dikes of basic volcanic rocks. Naturally, intrusions at higher temperatures favor the formation of rocks of higher grades. However, it is also reasonable to expect that high-grade rocks happen to form as aureoles around intrusive rocks of lower temperatures rather than higher temperatures.

The temperature of magmas probably ranges between 600-1200°C. Hence, temperatures of metamorphism of 600-800°C are possibly reached in a somewhat broad region if geologic environments are such that no marked break in temperature is developed across the contact plane. Within the aureole there is a gradient of the maximum temperature of metamorphism at more or less constant pressures. Thus, in many aureoles the pyroxene hornfels facies is developed immediately adjacent to the igneous contact and passes into an outer zone characterized by assemblages of the hornblende hornfels (amphibolite) facies (for example, the Oslo aureole [Goldschmidt, 1911]). In many other aureoles, however, rocks of the innermost zone are characteristic of the hornblende hornfels facies, and development of higher grade facies is hardly noticeable (for example, the Orijärvi aureole [Eskola, 1914]).

The diverse temperature conditions in these two types of aureole must be explained in terms not only of the temperature of the intrusion, but also of the break in temperature across the contact surface. High-temperature magmas, large intrusive bodies, intrusions at depths, initial rocks having great thermal diffusivity, and especially the retardation of cooling of contact surfaces, favor regional formation of high-grade rocks.

Eskola [1915] argued that pressure was higher and temperature lower in the Orijärvi than in the Oslo region. If the Orijärvi intrusion took place at deeper levels than the Oslo intrusion, then the formation of rocks of the pyroxene hornfels facies not in the Orijärvi aureole but in the Oslo may be explained in four ways: 1) the Oslo intrusion occurred at significantly higher temperatures than the Orijärvi intrusion; 2) the Oslo intrusive was of larger scale than the Orijärvi intrusive; 3) the thermal diffusivity of the country rock is greater in the Oslo than in the Orijärvi region; or, 4) the Oslo intrusive retained its initial temperature at the contact for a longer time after the intrusion. The retardation of the cooling might be due to the larger size of the intrusion, or to other factors unknown. The alternative is: because of the difference in the depth of intrusion, water pressure was lower in the Oslo than in the Orijärvi region. As a result, rocks of the pyroxene hornfels facies could form at significantly lower temperature in the Oslo region.

Patterns of Metamorphic Zones

Temperature and pressure have been viewed as the principal conditions governing mineral facies. It should be noted, however, that partially independent variables are involved in the general category of pressure conditions. In reality, metamorphic systems are not strictly invariant, unless chemical potentials of mobile components are dependent on the temperature and pressure distribution in the earth's crust [Thompson, 1955]. Furthermore, mineral phases characteristic of individual mineral facies are far from isochemical because of the prevalence of isomorphous substitution of cations in natural silicates. Accordingly, each facies must be regarded as corresponding to a rather broad range of physical conditions. Even at a given pressure, an assemblage characteristic of any facies will be stable over a broad range of temperature.

For the same reasons, boundaries between any two facies cannot be represented by single lines on a pressure (depth)—temperature plane. A rock of a given facies will transform into an assemblage of different facies over a broad range of temperatures even at constant pressures. Thus, it is expected that rocks of transitional facies form between any two assemblages belonging to different facies.

At more or less constant pressures, the patterns of metamorphic zones in a given aureole depend upon both the range of the stability temperatures for individual facies and the temperature gradient prevailing when the zones formed. Similarly, the width of a transitional zone depends upon a range of the transitional temperature and a temperature gradient in the aureole.

Equation (8) illustrates the exponential dependence of the temperature gradient upon r, x, k, and t. The gradient also varies inversely with increasing value for $\sqrt{kt}$. Hence, the temperature gradients that would be produced at any distance from the contact of a given intrusive body are variable, depending on t. It is the temperature gradient prevailing when maximum temperatures of metamorphism were reached that is concerned here. The maximum temperature of metamorphism is highest at the igneous contact and becomes lower as distance from the contact increases; and the longer the distance, the longer the time required before the maximum temperature is reached. In any given aureole, therefore, rocks of higher grades form under greater temperature gradients in earlier stages of metamorphism than rocks of lower grades. Thus, it may be said, in general, that rocks of a low-temperature facies occur as a broader aureole than rocks of a higher-temperature facies, around a given intrusive body, provided that the ranges of stability temperatures for the individual facies are not significantly different from each other.

Just how broad the stability field may be is not precisely known. For most contact metamorphic facies, it may be on the order of 200°C [Fyfe, Turner, and Verhoogen, 1958]. Then, incorporating the figures in Tables 1, 2, and 5, it is possible to derive information on the patterns of metamorphic zones.

Commonly, rocks of the pyroxene hornfels facies form under a temperature gradient of the order of 10^{-2} °C cm^{-1}, whereas rocks of the albite—epidote hornfels (epidote—amphibolite) facies form under a temperature gradient of the order of 10^{-3}-10^{-4} °C cm^{-1}. Therefore, it is expected that a zone of rocks of the albite—epidote hornfels facies is a factor of 10-10^2 wider than that of rocks of the pyroxene hornfels facies in any one aureole; and the width of a zone of rocks of the hornblende hornfels facies is intermediate between these two. Rocks of the pyroxene hornfels facies probably occur as narrow aureoles less than few hundred meters in width around intrusive bodies. On the other hand, aureoles composed of hornfelses of lower grades are not rarely several kilometers in extent around large granite batholiths.

Clue to Reaction Kinetics

It has been considered that the common mineral assemblages to which the facies concept has been applied approximate thermodynamic equilibrium under conditions prevailing when they formed. Thus, occurrence of zones of progressive metamorphism in an aureole has been viewed as representing a gradient in the maximum temperature of metamorphism. Clearly, this presupposes that every assemblage characteristic of individual facies has remained unchanged in the declining stage of contact metamorphism.

It depends upon both reaction rates and variation rates in temperature whether the mineralogy of rocks belonging to a given facies is subjected to alteration under conditions corresponding to different facies. Reaction rates are generally accelerated by increasing temperature, and retarded drastically with falling temperature. If a low-grade rock suffers from heating, it will transform successively into assemblages of progressively higher grades, unless heating rates are very rapid. If, however, a high-grade assemblage once forms, it will persist against retrogressive metamorphism, unless cooling rates are extremely slow.

The numerical estimates listed in Table 5 indicate that heating rates during progressive metamorphism are possibly but rarely in excess of $1°C \cdot yr^{-1}$. The heating rates decrease with increasing distance, and thus with decreasing maximum temperature of metamorphism. They are probably on the order of 10^{-2}-10^{-3} $°C \cdot yr^{-1}$ except in a region immediately adjoining an igneous contact. These heating rates might be sufficiently slow for most metamorphic reactions to reach thermodynamic equilibrium.

It should be noted that the thermal history of contact aureoles is characterized by extremely slow cooling rates rather than slow heating rates. Indeed, cooling rates soon after a maximum temperature is reached are commonly less than 10^{-3} $°C \cdot yr^{-1}$ even in a region neighboring an igneous contact. Cooling rates at distances more than some 100 m from the contact are not in excess of 10^{-3} $°C \cdot yr^{-1}$. Especially in later stages of the metamorphism, they are reduced to 10^{-5} $°C \cdot yr^{-1}$. Consequently, maximum temperatures attained in a region adjoining the contact remain nearly constant for 10-10^3 years. At distances of more than 10^2 m from the contact, there is no marked drop in maximum temperature of metamorphism in 10^4-10^6 years. From these statements it may be inferred that rocks of the pyroxene hornfels facies form during time intervals of the order of 10^4-10^5 years in earlier stages of metamorphism; rocks of lower grades form during much longer intervals of time, on the order of 10^5-10^6 years during later stages of metamorphism.

As seen from equations (5b) and (18), both the maximum temperature of metamorphism and variation rates in temperature depend upon $(T' - T_0)$ and r. Thus, it can be said that rates of cooling from a given maximum temperature of metamorphism are slower around deep-seated granitic rocks of regional scale than around near-surface basaltic rocks of small scale. This implies that mineral reactions during retrogressive metamorphism are apt to take place more extensively in aureoles around granite batholiths.

Because of difficulties in kinetic theory applicable to silicate minerals, reaction rates pertaining to metamorphism are unknown. As a result, we are not certain as to how mineral assemblages characteristic of individual facies respond to the temperature conditions varying at such rates as estimated here. Nevertheless, the following remarks appear reasonable: 1) the current assumption that the whole association of mineral assemblages in almost all facies approximates internal equilibrium may well be justified in view of the slow variation rates in temperature in contact aureoles. The only exception perhaps is the sanidinite facies, which corresponds to temperature conditions prevailing in a region immediately adjoining the igneous contact and thus undergoing rather rapid variations; 2) mineral assemblages found in aureoles around granite batholiths cannot necessarily be regarded as products of metamorphism at maximum temperature attained. Owing to extremely slow cooling, they might form in equilibrium with temperatures substantially lower than the maximum temperature during the declining stage of metamorphism.

Thermally Active Areas

The contact aureole may be several kilometers at most in extent even around large granite stocks, batholiths, and intrusive bodies of similar size. However, the figures listed in Table 5 suggest that the heat flow out of an intrusive body reaches not only the site of recrystallization, but greater distances beyond. Thus, at the surface of the site of magmatic disturbances, the effect of the intrusion may be felt over considerable lengths of time.

Essentially, equation (1) is based on the assumption of no heat production within the country rock. Actually, long-lived radioactive isotopes are contained to more or less extent in most crustal materials. As a result, there is heat production at rates on the order of 10^{-5} ergs $\cdot$ cm^{-3} $\cdot$ sec^{-1} in the outer layers of the earth's crust. The only effect of the heat production that is allowed for in the calculations is the existence of an initial temperature gradient in the country rock. Coupled with a certain amount of heat coming from below the crust, the heat produced within the crust sets up an equilibrium geothermal gradient in the crust. In Table 3, the calculations have been made by assuming that initially there is no temperature gradient in the country rock. Accordingly, the estimated values of heat flow must be regarded as representing the outward flux of heat in the horizontal direction. So long as the heat flow toward the outer surface of the crust is concerned, therefore, it is necessary to add the flux of heat due to the geothermal gradient to the calculated values.

A surface flux of heat of about 50 ergs $\cdot$ cm^{-2} $\cdot$ sec^{-1} ($1.2 \cdot 10^{-6}$ cal $\cdot$ cm^{-2} $\cdot$ sec^{-1}) is observed in areas far removed from the scene of magmatic disturbances [Lee and MacDonald, 1963; MacDonald, 1963]. The surface heat flow at this rate is thought of as being due to the geothermal gradient in the steady state. In thermally active areas, surface flux a few times to ten times as high as 50 ergs $\cdot$ cm^{-2} $\cdot$ sec^{-1} is often observed. In some areas, as pointed out by Donaldson [1962], a major proportion of the high heat flux values may be due to a convective circulation of water in the upper crustal layers. However, it appears that a significant proportion of unusually high heat flux values observed in many other areas can be explained by direct connection with igneous activity. In this case, the estimated values in Table 3 are considered to represent values of the surface heat flux additional to an average of $1.2 \cdot 10^{-6}$ cal $\cdot$ cm^{-2} $\cdot$ sec^{-1}.

It will be seen from equation (12) that the additional values depend on the initial temperature, depth, age, and size of an intrusive body, and on the distribution of conductivity and diffusivity in overlying layers. If, in any way, some of these parameters can be suitably assigned values, then the additional surface flux may be estimated in terms of the other parameters.

For example, with $K = 0.015$ joule $\cdot$ cm^{-1} $\cdot$ sec^{-1} $\cdot$ °C^{-1}, $k = 0.010$ cm^2 $\cdot$ sec^{-1}, $T' - T_0 = 800$°C, and $x = 5$ km, the surface flux at 10^5 and 10^6 years after the intrusion of a sheetlike body 2 km thick will be 50 and 20% higher, respectively, than the average value of 1.2 cal $\cdot$ cm^{-2} $\cdot$ sec^{-1}. It is likely that surface flux several times as high as the average value is observed during some 10^5 years in an area where an intrusive body of regional scale intrudes at few kilometers below the surface. In a case where a sheet some 100 m thick intrudes at depths of 2-3 km, on the other hand, the influence of the intrusion is hardly felt at the surface, unless thermal conductivity is drastically increased. A practical application of these relations would be an approach to the problem of heat source in thermally active areas.

Acknowledgments. I wish to thank Professors Nobuo Katayama and Toshio Kimura, of the University of Tokyo, whose interest in the thermal problems has been most helpful.

REFERENCES

Birch, F., "Thermal conductivity and diffusivity," in: Handbook of Physical Constants (F. Birch, J. F. Schairer, and H. C. Spicer, eds.), Geol. Soc. Am. Spec. Paper 36, pp. 243-266 (1942).

Birch, F., and Clark, H., "The thermal conductivity of rocks and its dependence upon temperature and composition," Am. J. Sci., Vol. 238, pp. 529-558 (1940).

Clark, S. P., "Radiative transfer in the earth's mantle," Trans. Am. Geophys. Union, Vol. 38, pp. 931-938 (1957).

Donaldson, I. G., "Temperature gradients in the upper layers of the earth's crust due to convective water flows," J. Geophys. Res., Vol. 67, pp. 3449-3459 (1962).

Eskola, P., "On the petrology of the Orijärvi region in southwestern Finland," Comm. Geol. Finlande. Bull., Vol. 40, pp. 1-277 (1914).

Eskola, P., "On the relations between the chemical and mineralogical composition in the metamorphic rocks of the Orijärvi region," Comm. Geol. Finlande. Bull., Vol. 44, pp. 1-107 [in Finnish], pp. 109-145 [in English] (1915).

Fyfe, W. S., Turner, F. J., and Verhoogen, J., Metamorphic Reactions and Metamorphic Facies, Geol. Soc. Am. Mem., Vol. 73 (1958), p. 258.

Goldschmidt, V. M., Die Kontaktmetamorphose in Kristianiagebiet, Oslo Vidensk. Skr., I. Mat.-Naturv. Kl., II (1911), 405 p.

Hori, F., "On the load metamorphic formation of rhodonite, tephroite and manganosite," Sci. Papers Coll. Gen. Educ., Univ. Tokyo, Vol. 12, pp. 117-142 (1962).

Hori, F., "On the contact metamorphism of a manganiferous carbonate rock," Sci. Papers Coll. Gen. Educ., Univ. Tokyo, Vol. 13, pp. 83-94 (1963).

Hori, F., " On the role of water in heat transfer form a cooling magma," Sci. Papers Coll. Gen. Educ., Univ. Tokyo, Vol. 14, pp. 121-127 (1964).

Lee, W. H. K., and MacDonald, G. J. F., "Global variation of terrestrial heat flow," J. Geophys. Res., Vol. 68, pp. 6481-6492 (1963).

Lubimova, H. A., "Thermal history of the earth with consideration of the variable thermal conductivity of the mantle," Geophys. J. Roy. Astron. Soc., Vol. 1, pp. 115-134 (1958).

MacDonald, G. J. F., "Calculations on the thermal history of the earth," J. Geophys. Res., Vol. 64, pp. 1967-2000 (1959).

MacDonald, G. J. F., "Deep structure of continents," Rev. Geophys., Vol. 1, pp. 587-665 (1963).

Thompson, J. B., Jr., "The thermodynamic basis for the mineral facies concept," Am. J. Sci., Vol. 253, pp. 65-103 (1955).

VARIANCE OF SOME SELECTED ATTRIBUTES IN GRANITIC ROCKS

E. H. T. Whitten

Department of Geology
Northwestern University
Evanston, Illinois

The use of a single statistical model for different geologic situations is commonly complicated by considerable variance of the investigated attributes. Two granitic massifs, similar in chemical composition and specific gravity, have been compared. Variance of the investigated attributes is found to differ for the two granitic rocks. The attributes determining the larger part of the variance also differ. It becomes necessary to obtain more specialized models of the geologic processes, but these may be developed by petrographers only from proper investigations.

In the past decade considerable attention has been focused on the quantitative nature of granitic rock massifs. Classical petrography emphasized differences between rock species, and there was a tendency to base estimates of the nature of whole plutons (or even batholiths) on single chemical analyses or on isolated petrographic descriptions. However, it is now widely recognized that many (if not most) igneous rock bodies possess considerable compositional variability [Whitten, 1963a]. Similarly, current techniques permit a very large number of attributes (200-300) to be measured for a single rock sample. For specimens of a stated size, each attribute may have a dissimilar variance, and, in general, a particular attribute has a dissimilar variance in samples of different size [cf. Baird and others, 1964, 1967; Hahn-Weinheimer and Ackermann, 1963; Hahn-Weinheimer and Johanning, 1963].

Many of the common concepts about petrology and petrography have been based on the assumption that the quantitative nature of a rock unit is known or can be readily determined. For example, that the modal and chemical composition of a granite stock or a dolerite flow are known correctly. However, the continuing problems associated with determining good modes and good chemical analyses emphasize that, although technological advances have materially assisted in developing accurate data, many attributes cannot be measured accurately and precisely by every practicing geologist who needs such data.

Experimental petrology has shed valuable light on the petrogenesis of common igneous rock assemblages. Sufficient knowledge is now available to permit erection of simple conceptual process—response models on the basis of experimental and theoretical geochemical data.

However, because the significance and magnitude of many variables remain uncertain, or unknown, it is commonly possible to erect several dissimilar process models which lead to dissimilar response models for a particular rock unit [Whitten, 1964; Whitten and Boyer, 1964]. The accuracy of a predicted conceptual response model should be susceptible to combined quantitative and qualitative testing on the basis of observed data collected from the actual rock unit concerned. For example, geochemical principles governing the crystallization of a magma could be used to predict the three-dimensional mineralogical and chemical nature and variability of a granite stock. If geology is to be included among the physical sciences, it is necessary to remove petrogenetic concepts from the domain of subjective and intuitive judgments, and to use rigorous objective tests of quantitative models. Unfortunately, at the present time, it remains difficult to assess accurately the quantitative nature and variability of many attributes needed to test the conceptual models.

Because many attributes can be measured much more precisely, accurately, and cheaply than others, it might be anticipated that the more difficult attributes could be predicted with sufficient accuracy on the basis of the more easily assayed attributes. For example, sequential multivariate regression methods might be applied to a suite of easily measured attributes, in order to predict the three-dimensional variation of an attribute that can only be measured with difficulty. Vistelius [1962] demonstrated that the P_2O_5 percentage of some granitic rocks of central Tien Shan can be predicted with the aid of a general linear equation, where the independent variables are modal percentages of quartz, potash feldspar, plagioclase, and mafic minerals.

In addition to testing petrogenetic models, and the general desirability of knowing the variance of rock attributes at different levels of sampling, there are other important reasons for investigating the extent to which observed data can be used to predict other attributes. Two obvious cases relate to (a) prospecting and evaluation in economic geology, and (b) the rapidly developing field of remote sensing of environment on earth and in outer space. In many types of economic geology, the economic "prize" (the sought-after commodity) occurs in small amounts that are difficult to detect during field exploration and economic development. However, it is commonly realistic to construct a quantitative statistical model to relate other geological variables to the "prize," so that, on the basis of the readily observed attributes, the quantitative three-dimensional distribution of the prize is predicted [cf. Howd, 1964]. If sufficient can be learned about the nature and interrelationships of the common attributes of rocks, such predictions should become relatively simple.

Although certain facets of remote sensing (e.g., seismic and gravity exploration work, electric logging of oil wells, etc.) have been used extensively for some years, new powerful tools are constantly being developed that hold tremendous promise in many phases of geology. While x-ray fluorescence and infrared methods for quantitative bulk chemical analysis are well known, current work with the whole electromagnetic spectrum augurs well for the future of remote sensing of many geological attributes. As additional electromagnetic sensors are used to observe the characteristics of rock — in the laboratory, in the field, down deep drill holes, or on the surface of neighboring planets observed from orbiting satellites — a whole plethora of new attributes will be measured [Colwell, 1963; Brewer, 1964; Badgley and others, 1965]. Such attributes will be interesting in their own right, but a major use of the information will be for making predictions about those attributes of rocks with which geologists are currently familiar. For example, when electric logs are made for oil wells, the remotely sensed data can be used to predict lithological characteristics of the penetrated section. The logical use of data collected by a remote sensor would be to erect statistical models to predict the nature and variability of the rocks observed (in terms of the well-known traditional attributes). As with methods based on the results of experimental petrology, it is an initial requirement that, whenever possible, such models be tested (and corrected as necessary) on the basis of "field" studies of the actual rocks.

Unhappily, surprisingly little is known about the quantitative behavior of, and the interrelationships between, those traditional attributes that have engaged the interest of geologists for decades. Baird and others [1964, 1967] showed that much remains to be learned about the necessary sampling methods required to achieve an adequate assessment of the composition and variability of a simple igneous complex. Hence, many uncertainties are involved in making any prediction about the three-dimensional behavior of one traditional attribute on the basis of numerous other traditional attributes. For example, with what degree of confidence can the specific gravity at different points within a granite be predicted from a knowledge of the chemistry and mineralogy of a suite of samples?

Some Representative Data for Granitic Rocks

In an initial attempt to assess the interrelationships between attributes for, say, granitic rocks, it might be assumed that a miscellaneous set of samples and analyses could be compared without specific regard to their geographical origins. For example, in arriving at generalizations about granitic rocks, Tuttle [Tuttle and Bowen, 1958] and Chayes [1951] culled from the literature large numbers of published analyses for different granitic masses in the USA and prepared composite ternary diagrams. Even when a single rock unit is involved, plotting modal or chemical information on a ternary diagram divorces that data from its essential geographical location. Similarly, Schmidt equal-area projections and Wulff stereographic projections, although essential tools in the quantitative study of folded rocks, also divorce the observations from their geographical sites.

Such techniques can divert attention from the sampling problems inherent in the collection of all geological data. It is not necessary to emphasize that no amount of mathematical manipulation can rectify inadequate sampling. Special care must be given to defining the target and sampled populations of interest before the samples are selected and analyzed [Whitten, 1961]. Recently, Baird and others [1964, 1967] described their extensive sampling studies of the granitic rocks in Rattlesnake Mountain Pluton, California. On the basis of 863 x-ray spectrographic analyses of 465 rocks specifically collected according to a stratified sampling plan, Baird and others [1964] used analysis of variance techniques to evaluate the components of the standard deviation associated with each of four levels of sampling. Their results confirmed that "... without estimates of smaller scale variability it is not possible to judge the significance of large-scale variations. The collection of at least two specimens at each locality permits an estimate of the small-scale chemical variability... The figures presented suggest that the variability at the scale of the outcrop (even with cores 30 cm in length) may require the collection of more than two specimens per locality." Being unique, these data cannot be compared with results for other rock units, but, intuitively, it might be anticipated that a large number of granitic massifs will prove to possess variance at the outcrop level that is as large, or larger, than that for the Rattlesnake Mountain pluton.

Hence, it becomes even more important to discriminate between the sampled and the target populations when evaluating a series of chemical analyses for a granite mass. In what follows some relationships shown by small groups of analyses for two granites are discussed. The limitations of such small sets of analyses must be borne in mind clearly.

The Malsburg Granite and the Aulanko Granodiorite

Two granitic complexes that have received extensive recent petrologic study were chosen for examination during 1962. First, the Aulanko Granodiorite (Fig. 1), a small Svecofennidic

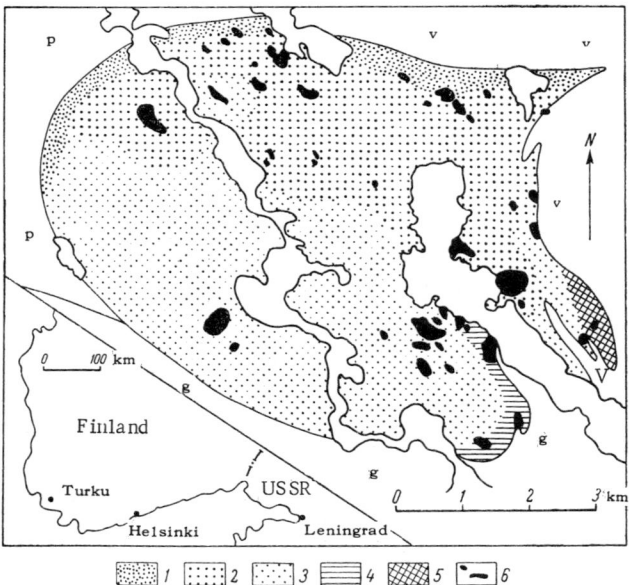

Fig. 1. The Aulanko Granodiorite, Finland, showing the petrographic variations recognized and mapped by Simonen [1948]. Inset is location map. 1) Granitized granodiorite without hornblende; 2) granitized granodiorite with hornblende; 3) granodiorite; 4) gneissose granodiorite; 5) marginal facies of granodiorite; 6) actual outcrops of Aulanko Granodiorite mass seen by Simonen. The envelope rocks are intermediate and basic metavolcanic rocks (v), uralite porphyrite (p), and microcline granite (g).

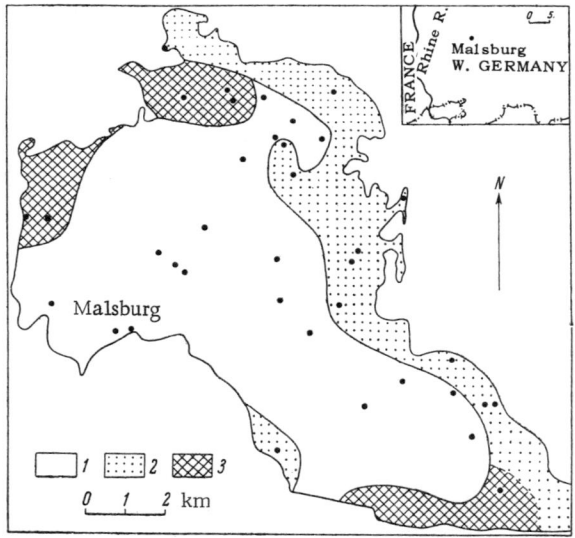

Fig. 2. The Malsburg Granite, southern Black Forest, Germany, showing the petrographic variations recognized by Mehnert [1960] and Mehnert and Willgallis [1961]. 1) "Central" zone; 2) "marginal" zone; 3) "marginal" zone rich in feldspar megacrysts. Dots are locations of samples used in present study.

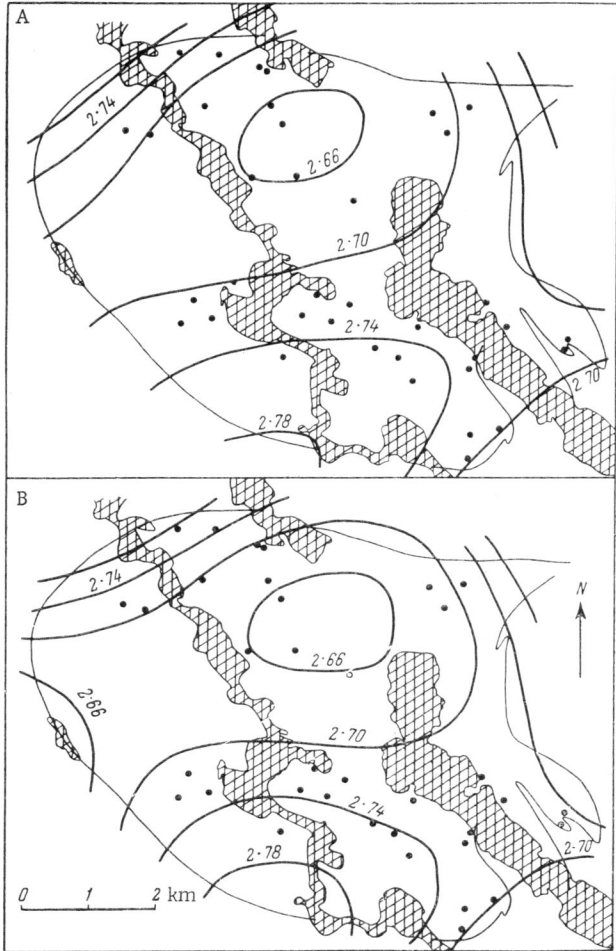

Fig. 3. Degree 3 trend surfaces for specific gravity of samples collected by Whitten from the Aulanko Granodiorite, Finland. A and B relate to sampled populations A and B, respectively. Dots show sample locations.

post-tectonic unit in central southern Finland, which was described and mapped by Simonen [1948]. Second, the Malsburg Granite, which is rather poorly exposed in the southern Black Forest, Germany (Fig. 2); this granite has been a subject of several petrographical and chemical studies [Mehnert and Willgallis, 1961; Rein, 1961; Whitten, 1962; Hahn-Weinheimer and Ackermann, 1963].

More detailed results will be incorporated in a subsequent paper. At this time it is intended to expose certain differences between these two masses revealed when sequential multivariate analysis is used in an attempt to predict P_2O_5 percentage, TiO_2 percentage, or specific gravity on the basis of the ten major oxides analyzed in samples collected from well-scattered outcrops throughout the granites.

Specific Gravity

The Aulanko Granodiorite is poorly exposed and outcrops are few. Two samples (1-2 m apart) were collected from each available outcrop (Fig. 1); the two samples at each outcrop

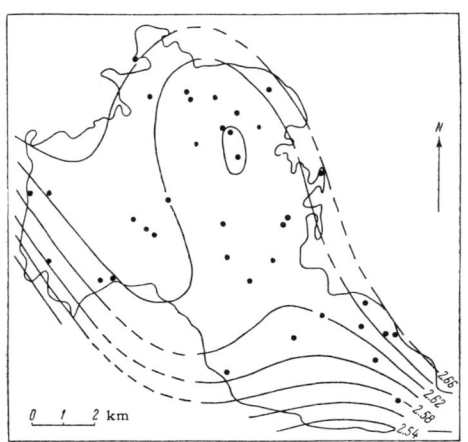

Fig. 4. Degree 3 trend surface for specific gravity of samples collected by Whitten from the Malsburg Granite, Germany. Dots show sample locations. Despite the relief of the area, W was assumed constant and the surface computed accounts for only 25.85% of the total variability.

were arbitrarily assigned to sampled populations A and B. The degree 3 trend surfaces (Fig. 3) for specific gravity of the sampled populations A and B account for 68.03 and 64.42% of the sum of squares, respectively. The 36 samples constituting population A were chemically analyzed. Then sequential multivariate analysis was used to determine whether a general linear equation could provide a reasonable basis for predicting specific gravity when SiO_2, Al_2O_3, Fe_2O_3, FeO, MgO, CaO, Na_2O, K_2O, P_2O_5, and TiO_2 are considered independent variables.

The general linear equation may be written as:

$$Y = a_0 + a_1 X_1 + \ldots + a_{10} X_{10}, \qquad (1)$$

where Y is specific gravity and $X_1, \ldots, X_{10}$ are the ten analyzed oxides. Solving the equation by the method of least squares, using all ten independent variables, permits 94.83% of the total variability of specific gravity to be predicted.* How-

Table 1. Best Predictions of Specific Gravity Based on 10 Oxides and Spatial Coordinates for the Aulanko and Malsburg Granitic Masses

No. of indep. variables taken at a time	Aulanko Granodiorite, Finland		Malsburg Granite, Germany	
	independent variables	sum of squares reduction, %*	independent variables	sum of squares reduction, %**
1	CaO	90.12	CaO	37.76
	SiO_2	83.66	FeO	31.89
	Al_2O_3	58.32	SiO_2	18.64
2	CaO, Fe_2O_3	92.50	CaO, FeO	43.11
	CaO, SiO_2	92.43	CaO, SiO_2	40.24
	CaO, MgO	92.19	CaO, Fe_2O_3	40.13
	CaO, P_2O_5	91.79	CaO, Al_2O_3	39.90
	CaO, TiO_2	91.77	CaO, MgO	39.28
3	CaO, Fe_2O_3, P_2O_5	94.17	CaO, Fe_2O_3, SiO_2	45.26
	CaO, Fe_2O_3, MgO	94.03	CaO, FeO, Na_2O	44.28
	CaO, SiO_2, Al_2O_3	93.76	CaO, FeO, U	44.13
	CaO, Fe_2O_3, U	93.38	CaO, Fe_2O_3, TiO_2	44.11
4	CaO, Fe_2O_3, P_2O_5, U	94.46	CaO, Fe_2O_3, SiO_2, U	47.64
	CaO, Fe_2O_3, P_2O_5, MgO	94.46	CaO, Fe_2O_3, SiO_2, P_2O_5	47.56
	CaO, Fe_2O_3, P_2O_5, SiO_2	94.45	CaO, SiO_2, P_2O_5, Al_2O_3	46.94

* Maximum with all ten oxides as independent variables, 94.83%. Maximum with all ten oxides and also U and V, 95.12%.

** Maximum with all ten oxides as independent variables, 49.77%. Maximum with all ten oxides and also U, V, and W, 53.47%.

---

* These and similar calculations were made with the aid of the FORTRAN computer program prepared by Krumbein and others [1964].

Table 2. Percentage of Total Sums of Squares Accounted for by Trend Surfaces for Specific Gravity

Degree	Aulanko Granodiorite		Malsburg Granite
	population A	population B	
1	23.63	13.65	2.95
2	24.54	17.44	8.41
3	68.03	64.42	25.85

ever, for this hornblende granodiorite, it is instructive to examine the strongest equations for prediction when only a few independent attributes are included. Table 1 shows that

$$Y = a_0 + a_1(CaO) \qquad (2)$$

accounts for 90.12% of the total sum of squares. Hence, for many purposes, inclusion of the other nine oxides is redundant, i.e., the inclusion of one or two oxides as independent variables additional to CaO adds little to the predictive power of the equation. Notice, too, that the degree 3 trend surface based on geographic coordinates only accounts for 68.03% of the variability.

Figure 4 shows the degree 3 trend surface for specific gravity based on samples collected from sites in the Malsburg Granite shown in Fig. 2. These samples were analyzed for the same ten oxides as the Aulanko samples, but now, using all ten oxides as independent variables in equation (1), only 49.77% of the total sum of squares is accounted for. Table 1 shows that CaO again provides the strongest equation when only one independent variable is used, but, for this granite, an equation like (2) accounts for only 37.76% of the total variability; use of both FeO and CaO as independent variables permits 43.11% of the total variability to be predicted.

The reason for this dissimilarity between the Aulanko and Malsburg complexes is not immediately obvious. Although the Aulanko granodiorite is considerably more mafic than the Malsburg Granite, the variances of both sets of specific gravity data are comparable (Table 3). The means and variances for all of the analyzed oxides are shown in Table 3. For the Malsburg complex the data are considered together, but they are also divided into two groups: (a) the central and (b) the porphyritic marginal zones mapped by Mehnert [1960] and Mehnert and Willgallis [1961]. On the basis of a study of some 25 suites of volcanic rocks, Chayes [1962a, 1962b, 1964] suggested that the variance of SiO_2 is commonly an order of magnitude greater than that for Al_2O_3, MgO, CaO, or (FeO + Fe_2O_3), and that the variances of Na_2O, K_2O, and TiO_2 are a magnitude smaller. Table 3 shows that the Malsburg analyses conform to this pattern, but that for the Aulanko rocks Al_2O_3 and CaO have larger variances (1.18 and 1.11, respectively).

Table 3. Data for the Samples Collected from the Aulanko Granodiorite and the Malsburg Granite

Attribute	Aulanko Granodiorite		Malsburg Granite			
			mean % of all analyses	variances		
	mean, %	variance		all analyses	central area analyses only	marginal area analyses only
SiO_2	67.41	3.47	66.09	2.31	1.44	2.39
Al_2O_3	14.76	1.18	15.61	0.83	0.79	0.79
Fe_2O_3	1.98	0.60	1.72	0.57	0.54	0.60
FeO	2.07	0.86	1.54	0.53	0.33	0.62
MgO	1.73	0.56	2.06	0.60	0.35	0.71
CaO	3.53	1.11	1.96	0.66	0.66	0.64
Na_2O	3.93	0.45	3.83	0.34	0.35	0.34
K_2O	2.88	0.44	4.48	0.36	0.36	0.33
TiO_2	0.39	0.12	0.41	0.12	0.06	0.14
P_2O_5	0.16	0.05	0.23	0.06	0.03	0.06
Sp gr	2.71	0.03	2.63	0.03	0.02	0.04

Table 4. Orders of Magnitude of Variances

Aulanko Granodiorite			Malsburg Granite (whole mass)*		
>1.0	>0.1	>0.01	>1.0	>0.1	>0.01
SiO_2 Al_2O_3 CaO	Fe_2O_3 FeO MgO Na_2O K_2O TiO_2	P_2O_5	SiO_2	Al_2O_3 CaO Fe_2O_3 FeO MgO Na_2O K_2O TiO_2	P_2O_5

* The same ranking applies to both the central and the marginal areas considered separately, except that TiO_2 should be transferred to the third column for the central area data.

Table 5. Best Predictions of TiO_2 Percentage Based on 9 Oxides, Specific Gravity, and Spatial Coordinates for the Aulanko and Malsburg Granitic Masses

No. of indep. var. taken at a time	Aulanko Granodiorite, Finland		Malsburg Granite, Germany	
	independent variables	sum of squares reduc., %*	independent variables	sum of squares reduc., %**
1	MgO	80.73	SiO_2	75.53
	SiO_2	66.19	MgO	47.31
	FeO	61.16	P_2O_5	47.23
2	MgO, SiO_2	84.98	SiO_2, V	81.69
	MgO, CaO	83.98	SiO_2, W	79.28
	MgO, Sp. gr.	83.79	SiO_2, U	77.27
	MgO, Na_2O	82.13	SiO_2, Fe_2O_3	76.90
	MgO, Fe_2O_3	81.79	SiO_2, CaO	76.87
3	MgO, SiO_2, Al_2O_3	88.29	SiO_2, U, V	83.43
	MgO, SiO_2, P_2O_5	86.63	SiO_2, V, Na_2O	82.48
	MgO, P_2O_5, Sp. gr.	86.45	SiO_2, V, Fe_2O_3	82.26
	MgO, P_2O_5, CaO	86.41	SiO_2, V, W	82.24
4	MgO, SiO_2, Al_2O_3, P_2O_5	89.88	SiO_2, U, V, CaO	84.27
	MgO, SiO_2, Al_2O_3, FeO	89.03	SiO_2, U, V, Na_2O	84.11
	MgO, SiO_2, Al_2O_3, Fe_2O_3	88.86	SiO_2, U, V, Al_2O_3	83.63

* Maximum with all nine oxides and specific gravity as independent variables, 91.25%.
Maximum with all nine oxides, specific gravity, U, and V, 91.57%.
** Maximum with all nine oxides and specific gravity as independent variables, 80.89%.
Maximum with all nine oxides, specific gravity, U, V, and W, 86.97%.

TiO_2

Multiple linear regression was used to study the contribution of specific gravity and the other nine oxides for predicting the variability of TiO_2 percentage. Using all ten independent variables, 91.25% and 80.89% of the total sum of squares are accounted for at Aulanko and Malsburg, respectively (Table 5). For both sets of samples, a large proportion of the total variability is accounted for, but one independent variable can account for a very large proportion of the total variability; the single variable is MgO for Aulanko and SiO_2 for Malsburg. While SiO_2 and MgO are the strongest pair of variables for Aulanko, there are eight pairs of attributes that are slightly stronger at Malsburg.

Table 6. Percentage of Total Sums of Squares Accounted for by Trend Surfaces for TiO$_2$ Percentage

Degree	Independent variables	Aulanko Granodior.	Malsburg Granite
1	U, V	21.87	13.16
2	U, V	22.75	21.24
	U, V, W	n.d.	28.01
3	U, V	47.17	29.26
	U, V, W	n.d.	55.39

n.d.: not determined.

At Aulanko all sample sites are at approximately the same elevation above sea level so that three-dimensional sampling was impracticable. Trend surfaces account for only a small proportion of the total variability of TiO$_2$; there is a strong linear correlation between TiO$_2$ and MgO at Aulanko, and their degree 3 trend surfaces account for similar proportions of the total variability (Table 7). While lack of relief precludes evaluation of the vertical variability of these oxides, the Malsburg samples have a vertical range of 620 m and the percentage of the total sum of squares for TiO$_2$ is almost doubled when the vertical dimension is included in the degree 3 trend component* (Table 4).

P$_2$O$_5$

When equation (1) is used, the dependent variable is divorced from its essential spatial coordinates. The regression equation can be extended easily to include U, V, and W; thus:

$$Y = a_0 + a_1 X_1 + \ldots + a_{10} X_{10} + a_{11} U + a_{12} V + a_{13} W, \tag{3}$$

where U, V, and W are orthogonal spatial coordinates with U and V being horizontal map coordinates and W the vertical direction. Only linear regression is considered in this paper, but equation (3) could be extended readily to include independent terms that are powers of U, V, W, or X$_n$. Table 7 shows that, for the Aulanko samples, the P$_2$O$_5$ percentage can be predicted on the basis of U and V almost as efficiently as the MgO or TiO$_2$ percentages. By contrast, for the Malsburg rocks, only 19.14% of the total sum of squares of P$_2$O$_5$ is predicted from U and V with the degree 3 polynomial; inclusion of W as an independent variable increases the percentage to 48.52.

At Aulanko, 80.65% of the total variability of the P$_2$O$_5$ percentage can be predicted on the basis of the nine other oxides, U, and V; the strongest single variable is MgO (60.53%) while SiO$_2$ is third (46.54%). Table 8 shows that over 70% of the variability can be accounted for by four combinations of three variables, all of which include MgO and two of which include the spatial coordinate V.

At Malsburg, use of the nine other oxides, U, V, and W together only permits 75.09% of the variability of P$_2$O$_5$ to be predicted. Unlike the Aulanko samples, SiO$_2$ is the strongest single attribute (only 56.34%) and MgO is the fourth in rank order and accounts for only 27.51% of the P$_2$O$_5$ variability. Hence, there is a marked difference between the P$_2$O$_5$: MgO correlations in the two suites of granitic samples. Inclusion of linear powers of U, V, and W is not useful for prediction of P$_2$O$_5$, although Table 7 suggests that inclusion of the degree 3 powers of U, V, and W (including the cross-product terms) would significantly increase the success of predictions.

Conclusion

In a subsequent paper, it is proposed to evaluate the significance of the full chemical analysis in terms of the detailed petrography. In this preliminary account, it has been intended to demonstrate that predictive models developed for one rock unit do not necessarily apply to

---

* These trend surface components were calculated by the use of computer programs made available by Whitten [1963b], Whitten and others [1965], and Peikert [1963].

Table 7. Percentages of Total Sums of Squares Accounted for by Some Trend Surfaces for the Aulanko and Malsburg Granitic Masses

Depend. variables	Independent variables	Aulanko Granodiorite (36 samples)			Malsburg Granite (37 samples)		
		degree			degree		
		1	2	3	1	2	3
TiO_2	U, V	21.87	22.75	47.17	13.15	21.24	29.26
	U, V, W	—	—	—	—	28.01	55.39
P_2O_5	U, V	27.85	31.01	42.25	10.59	11.73	19.14
	U, V, W	—	—	—	10.67	17.51	48.52
MgO *	U, V	25.06	31.80	49.10	10.86	23.04	28.18
FeO	U, V	60.98	66.45	76.32†	18.35	19.11	25.13

*MgO is strongest single independent variable for predicting TiO_2 at Aulanko.
†Strongest degree 3 trend component among the ten oxides and specific gravity at Aulanko.

Table 8. Best Predictions of P_2O_5 Percentage Based on 9 Oxides, Specific Gravity, and Spatial Coordinates for the Aulanko and Malsburg Granitic Masses

No. of indep. var. taken at a time	Aulanko Granodiorite, Finland		Malsburg Granite, Germany	
	Independent variables	sum of squares reduc., %*	independent variables	sum of squares reduc., %**
1	MgO	60.53	SiO_2	56.34
	Sp. gr.	52.91	TiO_2	47.23
	SiO_2	46.54	Al_2O_3	42.32
	FeO	44.29	MgO	27.51
2	MgO, Sp. gr.	67.09	SiO_2, Al_2O_3	62.57
	MgO, V	66.32	Al_2O_3, TiO_2	59.90
	MgO, CaO	65.99	SiO_2, Na_2O	59.37
	SiO_2, Na_2O	65.06	SiO_2, Sp. gr.	58.18
	MgO, U	63.14	SiO_2, V	58.13
3	MgO, Sp. gr., TiO_2	72.48	Al_2O_3, Fe_2O_3, FeO	65.54
	MgO, Sp. gr., V	71.15	SiO_2, Al_2O_3, Na_2O	65.42
	MgO, CaO, TiO_2	71.14	SiO_2, Al_2O_3, Fe_2O_3	64.60
	MgO, CaO, V	70.73	Al_2O_3, Na_2O, TiO_2	64.24
	MgO, Sp. gr., Na_2O	69.22	SiO_2, Al_2O_3,	63.99

*Maximum with all nine oxides, specific gravity, U and V as independent variables, 80.65%.
**Maximum with all nine oxides, specific gravity, U, V, and W as independent variables, 75.09%.

another, broadly analogous, rock unit. Clearly, one aim of petrography must be to erect generalized conceptual models. Petrogenetic concepts lead to the erection of process—response models and it would be very significant if petrographers could discover what general rules are involved in the behavior and interrelationship of common attributes of granitic complexes. It is regrettable that, at the present time, extraordinarily little is known about the variance of chemical attributes in individual granitic complexes, or about the interrelationships of common attributes.

Clearly, there is an urgent need to acquire more information on these subjects, and this paper is offered in the hope of stimulating more geologists to apply themselves to obtaining and publishing this much-needed data.

REFERENCES

Badgley, P. C., Fischer, W., and Lyon, R. J. P., "Geologic exploration from orbital altitudes," Geotimes, Vol. 10, No. 2, pp. 11-14 (1965).

Baird, A. K., McIntyre, D. B., and Welday, E. E., "Geochemical and structural studies in batholithic rocks of the southern California batholith, Pt. II: Sampling of the Rattlesnake Mountain pluton for chemical composition, variability, and trend analysis," Geol. Soc. Am., Bull. Vol. 78, pp. 191-222 (1967).

Baird, A. K., McIntyre, D. B., Welday, E. E., and Madlen, K. W., "Chemical variations in a granitic pluton and its surrounding rocks," Science, Vol. 146, pp. 258-259 (1964).

Brewer, W. A., "Exploration potential of remote sensing," Bull. Am. Assoc. Petroleum Geologists, Vol. 48, pp. 518-519 (1964).

Chayes, F., "Modal composition of granites," Carnegie Inst. Wash., Vol. 50, pp. 41-42 (1951).

Chayes, F., "Numerical correlation and petrographic variation," J. Geol., Vol. 70, pp. 440-452 (1962a).

Chayes, F., "Variance relations in some published Harker diagrams," Carnegie Inst. Wash., Vol. 69, pp. 118-119 (1962b).

Colwell, R. N. (Chairman), "Basic matter and energy relationships involved in remote reconnaissance," Photogramm. Eng., Vol. 29, pp. 761-799 (1963).

Hahn-Weinheimer, P., and Ackerman, H., "Quantitative röntgen—spektral-analytische Bestimmung von Kalium, Rubidium, Strontium, Barium, Titan, Zirkonium und Phosphor," Z. Analyt. Chem., Vol. 194, pp. 81-101 (1963).

Hahn-Weinheimer, P., and Johanning, H., "Geochemische Untersuchungen an magmatisch differenzierten Gesteinskörpern des Südschwarzwaldes I: Die Verteilung von Kalium, Rubidium, Calcium, Strontium, Titan, Zirkonium, Niob und Phosphor in Hinblick auf ihre Verwendung als Leitelemente zur Untergleiderung des Bärhalde-Granits," Beitr. Mineral. Petrog., Vol. 9, pp. 175-197 (1963).

Howd, F. H., "The taxonomy program — a computer technique for classifying geologic data," Colorado School Mines Quart., Vol. 59, No. 4, pp. 207-222 (1964).

Krumbein, W. C., Benson, B. T., and Hempkins, W. B., Whirlpool, a Computer Program for "Sorting Out" Independent Variables by Sequential Multiple Linear Regression, Office Naval Res. Geogr. Branch Techn. Rept., No. 14 (1964).

Mehnert, K. R., "Zur Geochemie der Alkalien im tiefen Grundgebirge," Beitr. Mineral. Petrog., Vol. 7, pp. 318-339 (1960).

Mehnert, K. R., and Willgallis, A., "Die Alkaliverteilung im Malsburger Granit (Südschwarzwald), jb. geol. Landesamt Baden-Württ., Vol. 5, pp. 117-139 (1961).

Peikert, E. K., IBM-709 Program for Least-Square Analysis of Three-Dimensional Geological and Geophysical Observations, Office of Naval Res. Geogr. Branch. Techn. Rept., No. 4 (1963).

Rein, G., "Die quantitativ-mineralogische Analyse des Malsburger Granitplutons und ihre Anwendung auf Intrusionsform and Differentiations-verlauf," Jb. geol. Landesmat Baden-Württ., Vol. 5, pp. 53-115 (1961).

Simonen, A., "On the petrology of the Aulanko area in southwestern Finland," Bull. Comm. Geol. Finland, Vol. 143 (1948).

Tuttle, O. F., and Bowen, N. L., Origin of Granite in the Light of Experimental Studies in the System $NaAlSi_3O_8—KAlSi_3O_8—SiO_2—H_2O$, Geol. Soc. Am. Mem. 74 (1958).

Vistelius, A. B., "Phosphorus in granitic rocks of central Tien Shan," Geochemistry, No. 2, pp. 116-132 (1962).

Whitten, E. H. T., "Quantitative areal model analysis of granitic complexes," Geol. Soc. Am. Bull., Vol. 72, pp. 1331-1360 (1961).

Whitten, E. H. T., "Areal variability of alkalies in the Malsburg granite, Germany," Neues jb. Mineral. Mh., Vol. 9, pp. 193-200 (1962).

Whitten, E. H. T., "Application of quantitative methods in the geochemical study of granitic massifs," Spec. Publ. Roy. Soc. Canada, No. 6, pp. 76-123 (1963a).

Whitten, E. H. T., A Surface-Fitting Program Suitable for Testing Geological Models which Involve Areally Distributed Data, Office Naval Res. Geogr. Branch Techn. Rept., No. 2 (1963b).

Whitten, E. H. T., "Process—response models in geology," Geol. Soc. Am. Bull., Vol. 75, pp. 422-464 (1964).

Whitten, E. H. T., and Boyer, R. E., "Process—response models based on heavy mineral content of the San Isabel Granite, Colorado," Geol. Soc. Am. Bull., Vol. 75, pp. 841-862 (1964).

Whitten, E. H. T., Krumbein, W. C., Waye, I., and Beckman, W. A., Jr., A Surface-Fitting Program for Areally Distributed Data from the Earth Sciences and Remote Sensing, NASA Contr. Rep., CR-318 (1965).

THE ROLE OF MATHEMATICAL STATISTICS IN IMPROVED ORE VALUATION TECHNIQUES IN SOUTH AFRICAN GOLD MINES

D. G. Krige

University of Johannesburg
Johannesburg, Transvaal
Union of South Africa

Mathematical statistical techniques developed for use in ore valuation on the South African gold mines are reviewed briefly in the following categories: 1) Small random sampling theory for the three-parameter lognormal, mainly for borehole valuations; 2) Quality control techniques: a) for underground sampling based on a specially developed distribution model, b) for routine assaying based on the lognormal with variance negatively correlated with gold values; 3) Correlation, regression, and weighted moving average trend-surface techniques for routine ore reserve determinations.

The first statistical analysis of gold values in South African mines was published by Watermeyer in 1919, and was followed some 10 years later by Truscott [1929]. These attempts were not successful because of the lack of a proper distribution model and the fact that both authors discarded the straight arithmetic mean as biassed because of difficulties in reconciling underground sampling values with mill head values [Krige, 1964].

Subsequent research initiated by Sichel [1947] and followed up by Ross [1950] and Krige [1951, 1952, 1960] established the three-parameter lognormal frequency distribution as the appropriate model for gold values and showed that the bias between underground and mill head values was due partly to gold lost in mining and partly to sampling and assay biases.

All of the more recent developments in statistical techniques in South African gold mines have therefore been concentrated on using the three-parameter lognormal model to the best advantage in both one and two dimensions, and on developing efficient quality control measures to reduce sampling and assay errors to a minimum.

Table 1

Ratio: $(t + a)$ estimate/geometric mean of n (observations $+ a$)

No. of observations per sample $= n \rightarrow$	3	5	7	10	15	20	25	30	50	70	100	∞
0.1	1.051	1.051	1.051	1.051	1.051	1.051	1.051	1.051	1.051	1.051	1.051	1.051
0.2	1.103	1.103	1.104	1.104	1.104	1.105	1.105	1.105	1.105	1.105	1.105	1.105
0.3	1.156	1.158	1.159	1.159	1.160	1.161	1.161	1.161	1.161	1.161	1.162	1.162
0.4	1.210	1.214	1.216	1.217	1.218	1.219	1.220	1.220	1.220	1.221	1.221	1.221
0.5	1.266	1.272	1.275	1.277	1.279	1.280	1.281	1.282	1.282	1.283	1.283	1.284
0.6	1.323	1.332	1.336	1.339	1.343	1.344	1.345	1.346	1.347	1.348	1.349	1.350
0.7	1.382	1.393	1.399	1.404	1.409	1.411	1.412	1.414	1.415	1.416	1.417	1.419
0.8	1.442	1.457	1.465	1.472	1.478	1.481	1.483	1.484	1.487	1.489	1.490	1.492
0.9	1.503	1.523	1.533	1.542	1.550	1.554	1.557	1.558	1.562	1.564	1.565	1.568
1.0	1.566	1.591	1.604	1.615	1.625	1.630	1.634	1.636	1.641	1.643	1.645	1.649
1.1	1.630	1.661	1.677	1.691	1.703	1.710	1.714	1.717	1.724	1.726	1.728	1.733
1.2	1.696	1.733	1.753	1.770	1.785	1.793	1.798	1.802	1.810	1.813	1.816	1.822
1.3	1.764	1.807	1.831	1.851	1.870	1.880	1.886	1.891	1.899	1.905	1.908	1.916
1.4	1.832	1.884	1.912	1.937	1.958	1.971	1.978	1.984	1.992	1.997	2.004	2.014
1.5	1.903	1.963	1.996	2.025	2.051	2.065	2.075	2.081	2.091	2.098	2.106	2.117
1.6	1.975	2.044	2.082	2.117	2.147	2.164	2.175	2.183	2.195	2.204	2.212	2.226
1.7	2.049	2.128	2.172	2.212	2.247	2.267	2.280	2.289	2.304	2.315	2.323	2.340
1.8	2.124	2.214	2.265	2.310	2.352	2.375	2.390	2.400	2.418	2.231	2.440	2.460
1.9	2.201	2.303	2.361	2.413	2.460	2.487	2.504	2.517	2.538	2.552	2.563	2.586
2.0	2.280	2.395	2.460	2.519	2.574	2.604	2.624	2.638	2.664	2.679	2.692	2.718
2.1	2.360	2.489	2.563	2.630	2.691	2.726	2.749	2.765	2.796	2.813	2.827	2.858
2.2	—	2.586	2.669	2.744	2.814	2.854	2.880	2.898	2.935	2.954	2.969	3.004
2.3	—	2.686	2.778	2.863	2.942	2.987	3.016	3.037	3.081	3.102	3.118	3.158
2.4	—	2.788	2.891	2.986	3.074	3.125	3.159	3.182	3.234	3.257	3.274	3.320
2.5	—	2.894	3.008	3.113	3.212	3.270	3.307	3.334	3.395	3.420	3.438	3.490
2.6	—	3.003	3.129	3.245	3.356	3.420	3.462	3.492	3.565	3.590	3.610	3.669

$$V = 5.3019 \times \left[\begin{array}{l} \text{mean of squares of common logs of } n \text{ individual (observations } + a) \\ - \{\text{Mean of common logs of } n \text{ (observations } + a)\}^2 \end{array} \right]$$

The Three-Parameter Lognormal Frequency Distribution

This distribution, in its normalized form, can be expressed as

$$f(x)\,dx = [\sigma \sqrt{2\pi}]^{-1} \exp\left[\frac{-1}{2\sigma^2}(x-\xi)^2\right] dx, \qquad (1)$$

which is the Gaussian normal frequency distribution with mean ξ and standard deviation σ, normalization being effected on the basic untransformed variable z by

$$x = \log_e(z+a). \qquad (2)$$

The three parameters for the distribution are, therefore, ξ, σ, and an additive constant a, and the distribution function for z becomes:

$$\psi(z)\,dz = [\sigma \sqrt{2\pi}]^{-1} \exp\left[\frac{\sigma^2}{2} - \xi - \frac{1}{2\sigma^2}\{\ln(z+a) - \xi + \sigma^2\}^2\right] dz, \qquad (3)$$

where ξ = mean of $\log_e (z + a)$.

The shape of this distribution and the small sampling estimates based thereon are not very sensitive to changes in a [Krige, 1960] and in practice are accepted as known *a priori* at a level based on regional data for the mine or goldfield concerned, or are estimated from the observations themselves if the sample is large. The simplest effective method for estimating a has been found to be the graphical trial and error solution which yields a straight line fit on log-probability paper.

The mean value for the z population

$$\theta = \exp\left(\xi + \frac{\sigma^2}{2}\right) - a. \qquad (4)$$

In a small sample the maximum likelihood estimator for θ is given by [Sichel, 1952]

$$t = e^{\bar{x}}\left[1 + \frac{1}{2}V + \frac{(n-1)}{2^2 \cdot 2!\,(n+1)}V^2 + \frac{(n-1)^2}{2^3 \cdot 3!\,(n+1)(n+3)}V^3 + \ldots\right] - a, \qquad (5)$$

which for large n reduces to the same form as (4), i.e.,

$$t' = \exp\left(\bar{X} + \frac{V}{2}\right) - a, \qquad (6)$$

where $\bar{X}$ is the mean of natural logarithms of (observations + a), V is the unadjusted variance of the natural logarithms of the (observations + a), and n is the number of observations.

Formula (5) can be expressed as

$$t = \left[\begin{array}{c}\text{geometric mean of}\\ \text{(observations } +a)\end{array}\right] \times [f(V, n)] - a. \qquad (7)$$

Table 1 gives the values for $f(V, n)$ over the practical ranges of V and n, and is based on the original tables published by Sichel [1952].

Another useful form of maximum likelihood estimator for θ, when σ^2 is known *a priori*, is given by Krige [1952, 1962]:

$$t'' = \left[\begin{array}{c}\text{geometric mean of}\\ \text{(observations + } a)\end{array}\right] \times \exp\left[\left(\frac{n-1}{2n}\right)\sigma^2\right] - a. \qquad (8)$$

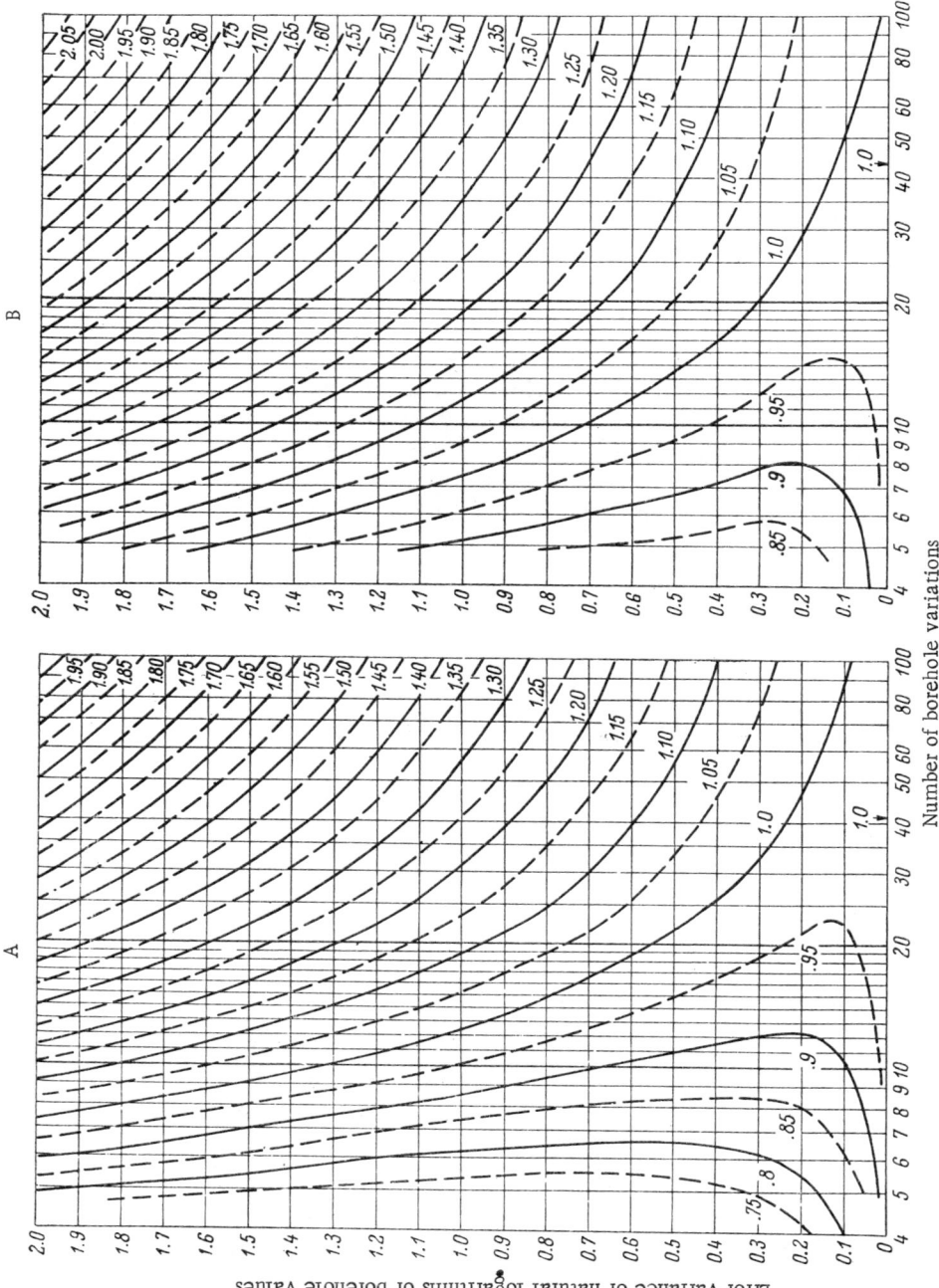

Fig. 1. Graphical representation of approximations of the complex distribution (3) for the lower 5 and 10% limits of error. A) Relation of the lower 5% limit to the geometric mean of borehole evaluations; B) relation of 10% limit to the geometric mean of borehole evaluations.

This estimator, t", can in practice serve as a useful limiting value for t where V as calculated on the observations falls outside the range dictated by the *a priori* known practical limits of σ^2.

The estimator (t" + *a*) is lognormally distributed with the approximate error variance

$$(t'' + a)^2 \exp\left[\frac{\sigma^2}{n} - 1\right], \tag{9}$$

and limits of error can therefore be conveniently estimated on normal theory [Krige, 1962].

In the case of the t estimator, the error variance was given by Sichel [1952] but the form of the error distribution is very complicated, being the lognormal equivalent of Student's distribution for the normal case. In an unpublished paper read to the South African Statistical Association in 1966, Sichel has given a close approximate solution to this problem, and he has prepared tables based on computer solutions of the complicated functions. A graphical representation of these solutions for the lower 5 and 10% limits of error is shown in Fig. 1.

The superior efficiency of the t estimator relative to the straight arithmetic mean has been clearly demonstrated by Sichel [1952].

Borehole Valuations

Outcrop mining has long become a feature mainly of historical interest in South Africa, and for several decades the only method of finding a new large gold mine and estimating its likely ore potential has been by deep drilling from the surface. For a whole mine the data would normally be obtained from between, say, 10 and 30 boreholes, usually spaced on a fairly regular grid, and the sampling fraction could be as low as 1 in 3000 million. Sampling is therefore not random, but experience has shown only a slight correlation between adjacent borehole values, and these may be accepted in practice as random.

A representative set of 10 borehole values from an Orange Free State mine with an average value of 1000 inch-dwt* can cover a range from 50 to 5000 inch-dwt, but in practice the lowest value could well be a trace and the highest 50,000. The orthodox arithmetic mean of such a set of borehole values, therefore, is clearly very unreliable and extremely sensitive to the level of the highest value in the set. The statistical t estimator, on the other hand, being based on the far more stable geometric mean of the observations, is not very sensitive to extreme values and provides a more efficient estimate of the average grade, with limits of error.

After the average gold grade for the whole mine has been estimated, allowance must be made for the effect of selective mining of ore blocks with grades only above the pay limit or cutoff grade (below which the ore would be mined at a loss). Again, the lognormal model provides a suitable basis for estimating the percentage and average grade of the ore blocks to be mined selectively, since analysis has shown [Krige, 1959] that the grades of ore block units are also lognormally distributed. The logarithmic variance (σ^2) of such ore block distributions is naturally much lower than that for borehole value distributions and normally lies in the range from 0.1-0.2. It can be estimated rather closely by *a priori* evaluation for a specific reef in a mine from known data in adjoining mines.

---

* Measure used in South Africa for gold concentration per unit area of reef (i.e., ore-horizon), 2880 inch-dwt being equivalent to 1 oz. of gold per square foot.

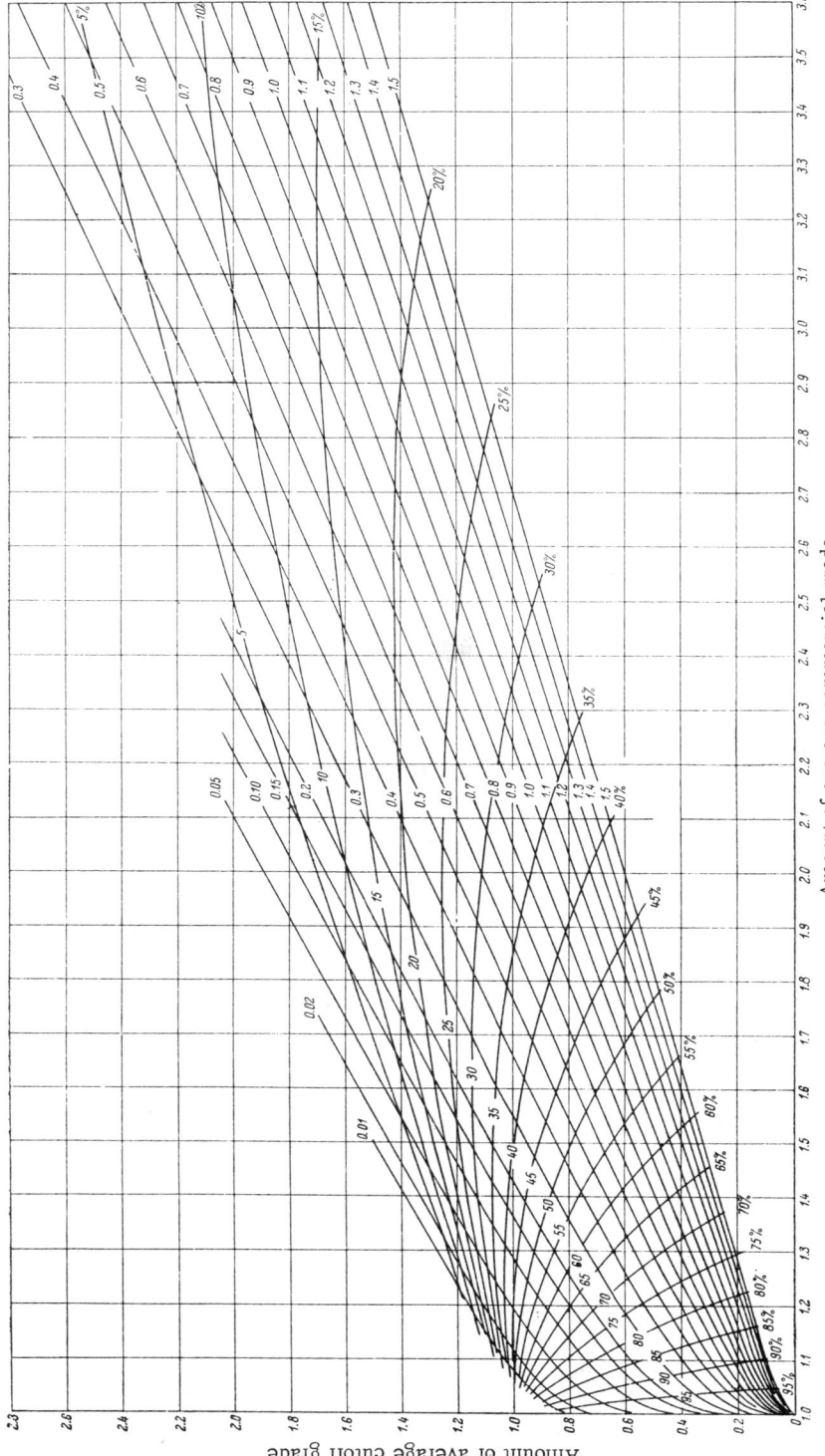

Fig. 2. Relation of the combination of average cutoff and commercial grades to the percentage of commercial ore. Lognormal distribution given in relative values.

The frequency of z values above the pay limit value of z_1 is given for distribution (3) by Krige [1952, 1953]:

$$\frac{1}{\sqrt{2\pi}} \int_{W_i}^{\infty} \exp\left(-\frac{W^2}{z}\right) dW, \tag{10}$$

where

$$W_i = \frac{1}{\sigma}[\log_e(z_i + a) - \xi] \tag{11}$$

or

$$W_i = \frac{1}{\sigma}\left[\ln(z_i + a) - \ln(\theta + a) + \frac{\sigma^2}{2}\right]; \tag{12}$$

and where the integral is determined from tables of the normal distribution.

Similarly the average of all z values above the value z_i is given by

$$(\theta + a) \frac{\int_{W_i - \sigma}^{\infty} \exp\left(-\frac{W^2}{2}\right) dW}{\int_{W_i}^{\infty} \exp\left(-\frac{W^2}{2}\right) dW} - a, \tag{13}$$

with w_i defined as above.

A convenient graphical solution of equations (10)-(13) is provided in Fig. 2.

Further practical adjustments in estimating a mine's recovery grade from borehole values, covering factors such as milling width, percentage of waste sorted, and gold losses were dealt with by the author in 1961.

Quality Control of Routine
Underground Mine Sampling

This subject was covered exhaustively by Rowland and Sichel [1960]. The basic control data comprises pairs of original and check samples taken underground in or adjacent to the same groove by the sampler himself or by the sampler and a senior experienced sampler, respectively. Such duplicate samples are taken in about 10% of all the sample grooves cut underground.

Conventional statistical control techniques for differences between duplicate measurements cannot be applied because the standard deviation of these differences is a function of the gold content of the ore samples.

A study of duplicate values has shown that the proper control unit is the logarithmic ratio of the original value to the check value. Furthermore, if the pairs of values are first grouped in categories corresponding to their average gold contents, the logarithmic variances of these ratio distributions show a negligible variation from category to category.

The distribution of the logarithms of these ratios is symmetrical but distinctly nonnormal. The distribution is more peaked than the normal and has longer tails.

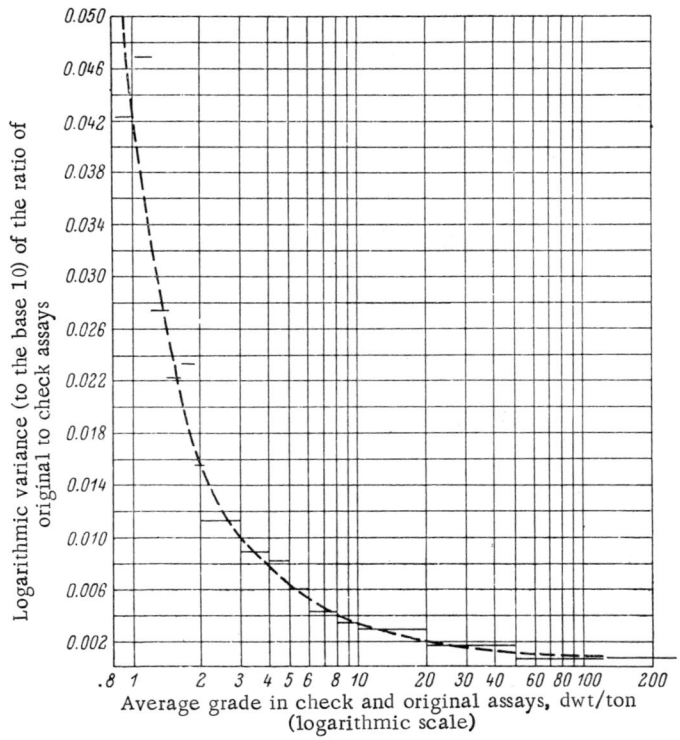

Fig. 3. Negative correlation between variance and gold content (for a typical gold mine).

The equation suggested by Rowland and Sichel is:

$$f(R) = \frac{\left(\frac{a}{\pi}\right)^{\frac{1}{2}}}{2^{V-\frac{1}{2}} \Gamma\left(V + \frac{1}{2}\right)} \frac{1}{R} K_V\left(\sqrt{2a}\, |\log_e R|\right) \quad \text{for} \quad 0 \leq R \leq \infty, \tag{14}$$

where a is a scale constant, V is a shape parameter, and $K_V(\sqrt{2a}|\log_e R|)$ is the auxiliary modified Bessel function of the second kind, of order V and argument $(\sqrt{2a}|\log_e R|)$.

On substitution of $z = \log_e R$, equation (14) reduces to

$$\gamma(z) = \frac{\left(\frac{a}{\pi}\right)^{\frac{1}{2}}}{2^{V-\frac{1}{2}} \Gamma\left(V + \frac{1}{2}\right)} K_V\left(\sqrt{2a}\, |z|\right) \quad \text{for} \quad -\infty \leq z \leq \infty. \tag{15}$$

This distribution is symmetrical with zero mean, and is nonnormal, but tends to normality when $V \to \infty$. The parameters a and V are estimated from the data using tables of the required Bessel function, and the control limits are then determined.

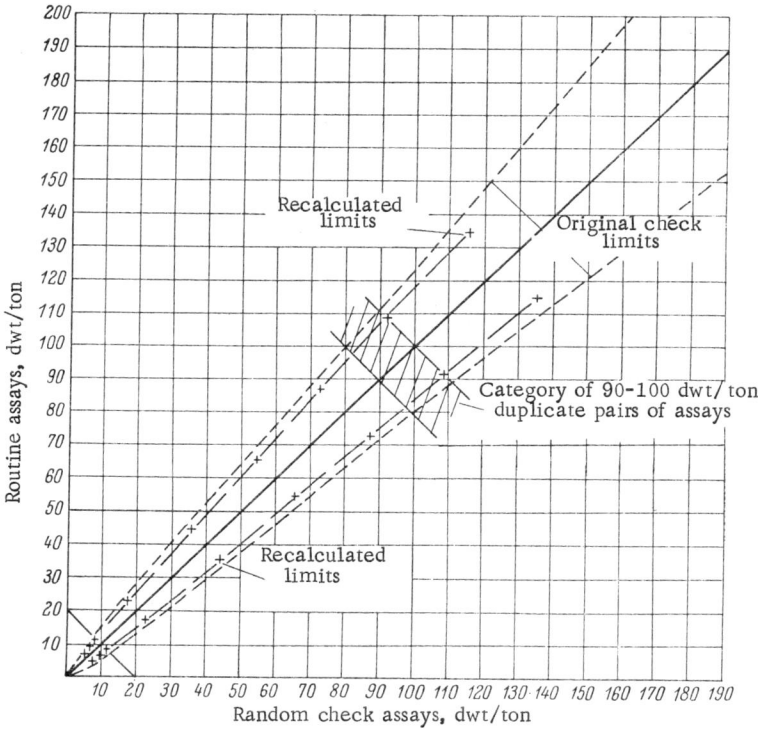

Fig. 4. Control chart for testing assays (with 99% probability limits for repeated pairs of assays).

In practice, the author has found that where an adequate number of ratios are available, say more than 1000, these may be plotted on logarithmic-probability paper and a smooth S curve fitted by eye to provide the required 1 in 1000 "action" and $2\frac{1}{2}$ in 100 "warning" control levels. These are drawn on the customary control charts and check ratios for individual samplers are plotted on the charts for quality control purposes.

These control charts have provided a relative standard of measurement of the efficiency of individual samplers and of the sampling staff as a whole, in regard to both variability and bias. They also serve to disclose gross sampling blunders and the likelihood of assay blunders.

Quality Control of Routine Mine Assays

Routine assays of mine samples are performed in the South African gold mining industry at the rate of about 15,000 per day on a total of some 50 large gold mines or, say, 300 per day per mine. These assays are performed in bulk and an estimated correction is made for the silver contained in the button weighed after cupellation. From 5-10% of these assays are selected at random for checking purposes, a reassay is performed and the button is parted before weighing to give a closer estimate of the fine gold content.

These check assays provide, as in the case of check underground samples, a control measure for disclosing any overall bias in the routine assaying as well as an estimate of the overall relative efficiency of assaying. As for check mine samples, the appropriate control unit is again the logarithm of the ratio of original to check assays.

The distribution in this case can be accepted as normal, but its variance increases in the lower grade categories, i.e., the variance is negatively correlated with gold grade. This is clearly illustrated graphically on Fig. 3 for a typical gold mine.

The control chart developed by Coxon and Sichel [1959] takes the form of a "fan chart" as shown on Fig. 4. The 45° line represents the center of the "fan" and corresponds to perfect agreement between routine and check assays. Spaced symmetrically on either side of this line are the 99% probability limits which serve as the required control limits. The diagram shows the original limits and the improvement achieved as reflected by the recalculated limits after a period of use of the original fan chart.

These limits are determined from normal theory for the full range of grade categories using the variance—grade relationship shown in Fig. 3.

Example: For the category of 90-100 dwt/ton shown hatched in Fig. 4, the logarithm variance (to the base 10) from Fig. 3 is 0.0008, and the log standard deviation = 0.0283. The 99% limits are: $0.0283 \cdot 2.576 = 0.0729$; $\text{antilog}_{10}(.0729) = 1.183$; and if original = 100 dwt/ton, check = $100/1.183 = 84.5$ dwt/ton, which corresponds to a point on the upper control line (recalculated) in Fig. 4.

In practice, these fan charts are used on the same basis as the customary quality control charts.

Correlation and Regression Techniques Based on the Lognormal

These techniques have been applied successfully in underground ore-reserve valuations based on available sample values obtained in development headings and along stope faces.

Sampling of the ore is generally done manually at 5-foot intervals, in drives and raises and at 10-20 foot intervals along advancing stope faces. Stope faces are usually resampled after having advanced some 20 ft. Valuations of blocks of ore are based on whatever development and/or stope sampling information is obtainable from the exposed part of the periphery of a block, ranging in extent from one exposed side to the optimum case where all four sides are exposed.

Such valuations, therefore, incorporate two basic sources of error, i.e., the inaccuracy of the valuation of the (partial) periphery and the error in assuming the value of the ore along such (partial) periphery to be identical with that of the ore inside the block. The first of these two errors is usually insignificant compared to the second, i.e., the skin of actual ore as exposed on the periphery is reasonably valued but the extrapolation into the unknown ore behind the exposed periphery is inaccurate.

Gold values have a very high variability, and this accounts for the fact that after normalizing the distribution of the block inch-dwt value estimates on the basis of $\log_e$ (value + constant), where the constant varies between 20 and 60 inch-dwt, the error variance for a single block value on a reef of below average variability is of the order of 0.05, corresponding to 90% confidence intervals of −29% to +48%.

After valuation of an ore block and as it is being mined out, further samples are taken along stope faces inside the ore block, thus allowing a direct comparison between the peripheral block value and the internal value subsequently obtained. These observations, if collected for a large number of ore blocks, can be analyzed as a straightforward correlation problem.

In a typical case, normalization of 200 sets of block and corresponding internal inch-dwt values from a large gold mine was effected on the basis of $\log_e$ (value + 50), and the following results were obtained:

	Block values	Internal values
Marginal variances:	0.18 (σ_x^2)	0.16 (σ_y^2)
Mean of logarithms:	5.815 ($\bar{x}$)	5.853 ($\bar{y}$)

The correlation coefficient of block to internal values = 0.80 and the variance of differences (y-x) = 0.068.

The regression line for internal on block values can then be estimated as follows: slope of line equals

$$r\sqrt{\frac{\sigma_y^2}{\sigma_x^2}} = 0.754.$$

To pass through $\bar{x}$ and $\bar{y}$, the formula for the line is then

$$y = 1.468 + 0.754x.$$

Conditional variance $S_y^2 = \sigma_y^2(1 - r^2) = 0.058$.

The correction [Krige, 1962a] for difference between geometric and arithmetic means of lognormal distribution of observations in y arrays equals $S_y^2/2 = 0.029$.

The corrected regression formula is then

$$y = 1.497 + 0.754x,$$

and in the untranslated form

$$\log_e (\text{expected internal value} + 50) = 1.497 + .754 \log_e (\text{block value} + 50)$$

or

$$\log_{10} (\text{expected internal value} + 50) = 0.650 + .754 \log_{10} (\text{block value} + 50) \quad (16)$$

For a low block value of 200, expected internal value is 237; for a high value of 600, expected internal value is 540.

This example clearly illustrates the phenomenon which was already observed in the earlier days of the Witwatersrand goldfield, but only explained statistically as recently as 1952 [Krige], i.e., that ore block values in the lower grade categories underestimate the true gold content of the blocks, and ore block values in the higher grade categories overestimate the true gold contents. Where the pay limit, as is usually the case, falls in the lower grade categories, the effect of relying on the block values can therefore introduce a serious bias by classifying ore blocks, which in fact are payable on average, as unpayable. Adjustment of all ore block values on the basis of a regression formula such as equation (16) above, has therefore become standard practice at almost all South African gold mines. The improved efficiency of ore valuation due to this practice is illustrated in the above example by the improvement in the log difference variance from 0.068 to 0.058. As the "internal" block values are themselves subject to error, the actual error variances of the orthodox and "regressed" block values are somewhat smaller than these two figures respectively. In practice it is necessary to determine regression formulas independently for different high- and low-grade sections of a mine to avoid certain other bias errors as explained in detail by Krige and Ueckermann [1963].

Multiple Correlation Techniques Using Regional Value Contours

The need to split the mine into sections and to determine a regression formula for each section led logically to the further development of avoiding arbitrary divisions between mine sections and determining for each ore block its own sectional or regional value to be used in the regression formula. This was achieved by using a straightforward two-dimensional moving average on a regional scale to obtain a value contour (or trend) surface.

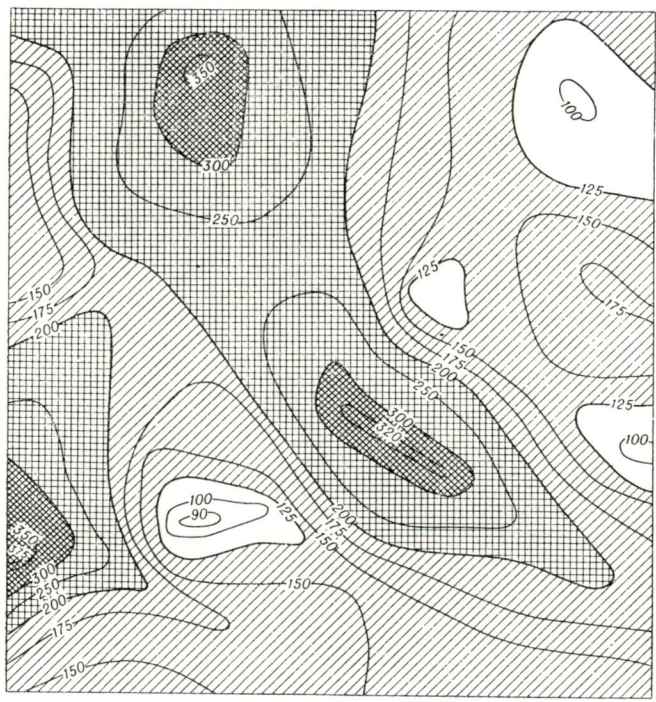

Fig. 5. Contour plan showing regional value trends of gold (inch-dwt).

The procedure is to accept only development values as these are normally available on a reasonably regular and unbiased grid (whereas stope values would tend to be concentrated in payable areas only), and to average all available values within overlapping circular areas of about 500 ft. radius. The averages are plotted at the centers of the circles and the overlap is such that averages are obtained on a grid of, say, 500 × 500 ft. This grid of values is then used, as in the case of topographic data, to prepare a smooth contour plan showing clearly all regional value trends. An example of such a contour map for a section of a gold mine is shown in Fig. 5.

Regression of ore block values is then effected as in the straightforward case dealt with above, but in addition to the block and internal values for each block, a regional contour value is now also available. A multiple lognormal-correlation model is therefore set up with contour and block values as the two independent variables (x_1 and x_2 after normalization) and the internal value as the dependent variable (y after normalization). The data for the same 200 blocks used in the above example were as follows [Krige and Ueckermann, 1963]:

	Log variances	Mean of logs
Block values	$x_1^2 = 0.18$	$\bar{x}_1 = 5.815$
Contour values	$x_2^2 = 0.13$	$\bar{x}_2 = 5.874$
Internal values	$y^2 = 0.16$	$\bar{y} = 5.853$

The correlation coefficients are $r_{x_1 y} = 0.80$, $r_{x_2 y} = 0.70$, $r_{x_1 x_2} = 0.56$.

The multiple regression solution for these data takes the form

$$y = k + b_1 x_1 + b_2 x_2,$$

with $b_1 = 0.40$, $b_2 = 0.55$, $S_y^2 = 0.043$.

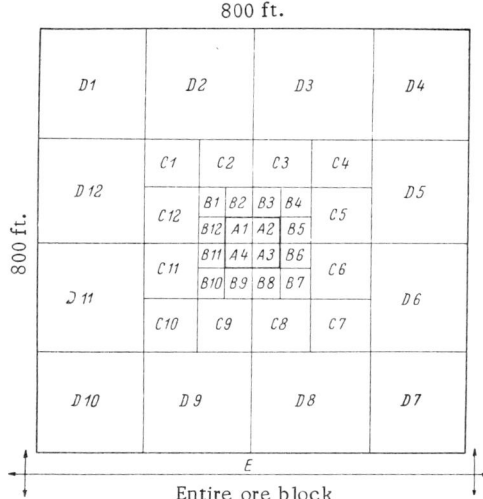

Fig. 6. Distribution of data blocks for constructing weighted moving average trend-surfaces. Explanation is in the text.

The unadjusted formula is

$$y = 0.296 + 0.40x_1 + 0.55x_2, \text{ adjustment } \tfrac{1}{2} S_y^2$$
$$= 0.022, \text{ giving}$$
$$y = 0.318 + 0.40x_1 + 0.55x_2,$$

i.e.,

$\log_e$ (expected internal value + 50)
= $0.318 + .40 \log_e$ (block value + 50) +
+ $0.55 \log_e$ (contour value + 50)

or

$\log_{10}$ (internal value + 50) =
= $0.138 + 0.40 \log_{10}$ (block value + 50) +
+ $0.55 \log_{10}$ (contour value + 50).

The improved efficiency of valuation obtained is clear from the new S_y^2 of 0.043, compared with 0.068 and 0.058 in the previous two cases.

This multiple regression approach to ore reserve valuations has been used successfully on four of the Anglovaal group of mines during the years 1963-64.

Weighted Moving Average Trend-Surfaces

The most recent development in the study progress toward more efficient ore reserve valuations has been the introduction of weighted two-dimensional moving averages. Viewed objectively, the multiple regression approach just explained gives an estimate for each ore block which, in essence, is a weighted average of (1) the peripheral value of the block, (2) a regional value for an area surrounding the ore block, and (3) the overall average value for the whole mine or mine section concerned.

It seems anomalous that the peripheral values should carry a certain weight and that all other values up to, say, 500 ft. away from the periphery should all together (through the regional contour value) carry a weight of similar order, plus some weight attaching to the overall grade of the whole mine section. The anomaly is highlighted by the fact disclosed by many correlation analyses [Krige, 1962a, 1962b] that gold values are serially correlated relative to their positions in space, starting with a high correlation (say 0.8) for values close together and decreasing to a negligible correlation for values 400 ft. apart. Any weighted-average, or regression, approach should therefore be even more efficient if the data around an ore block could be segregated into concentric "rings" with progressively less weight attached to the outer "rings."

For practical reasons the circular approach, although theoretically more correct, was abandoned in favor of square data blocks. The basic data unit was accepted as a 25 × 25 ft. block, and the basic valuation unit (or minimum size of ore block to be valued) was 100 × 100 ft. The basic data units were combined and averaged to form 50 × 50 ft. units, these in turn to yield 100 × 100 ft. units, and these again to form 200 × 200 ft. units. The configuration of data units for calculation of the weighted moving-average estimate for one 100 × 100 ft. ore block (i.e., for one point on the eventual weighted moving average contour-surface) is shown in Fig. 6. The central four 50 × 50 ft. squares represent the 100 × 100 ft. area to be valued, and data are potentially acceptable for such a valuation up to a distance of some 400 ft. away. The data are used in 12 segments (3 per quadrant) radiating from the center; e.g., blocks A_1, B_1, C_1, and D_1 form 1 of the 12 segments.

The weighted system to be used is based on an expansion of the type of analysis explained above for arriving at a regression solution of the form defined by (17). Data are collected from mined-out areas of the mine for at least 100-200 areas such as depicted in Fig. 6. Unless the section dealt with is very large, the 100-200 areas will be partly overlapping. A series of multiple regression solutions as for (17) is then obtained but with not only two but several independent variables assuming for each analysis a particular configuration of available data units.

In addition, the error variance for the valuation of the central 100×100 ft. square when values are available for all four of its component squares, is also determined on the basis of correlating the average values for two interpenetrating sets of basic 25×25 ft. squares (8 squares to each average value).

A typical analysis on this approach would be as follows (not for the same mine covered in the previous example):

Error variance for value of 100×100 ft. square based on data within square: normalization effected on $\log_e$ (value + 20); variance of 140 differences between logs of means of two interpenetrating sets of 25×25 ft. squares = 0.07; log error variance for 100×100 ft. square if valued on one such set only = 0.035; log error variance for 100×100 ft. square if valued on all data within square (i.e., both sets) = 0.02.

Referring now to Fig. 6, 140 sets of data were first analyzed on the following basis:

$$\log_e \text{ (average for squares } A_1, \ldots, A_4 + 20) = x_1,$$
$$\log_e \text{ (average for squares } B_1, \ldots, B_{12} + 20) = x_2,$$
$$\log_e \text{ (average for squares } C_1, \ldots, C_{12} + 20) = x_3,$$
$$\log_e \text{ (average for squares } D_1, \ldots, D_{12} + 20) = x_4,$$

i.e., the four independent variables are x_1, x_2, x_3, and x_4. $\log_e$ (average actual value $A_1, \ldots, A_4 + 20$) = y, i.e., the dependent variable to be estimated.

$$\sigma^2_{x_1} = 0.18; \quad \sigma^2_{x_2} = 0.10; \quad \sigma^2_{x_3} = 0.04; \quad \sigma^2_{x_4} = 0.02;$$
$$\sigma^2_y = \sigma^2_{x_1} - \text{ error variance } = 0.18 - 0.02 = 0.16;$$
$$r_{x_1 x_2} = 0.67; \quad r_{x_1 x_3} = 0.34; \quad r_{x_1 x_4} = 0.01; \quad r_{x_2 x_3} = 0.57;$$
$$r_{y x_1} = 1 : \sqrt{1 + \frac{\text{error variance}}{\sigma^2_y}} = \frac{1}{1.0606} = 0.94.$$

Similarly:

$$r_{y x_2} = r_{x_1 x_2} \cdot 1.0606 = 0.71;$$
$$r_{y x_3} = 0.36; \quad r_{y x_4} = 0.01.$$

Ignoring x_4 because of negligible correlation, the multiple regression solutions on the above evidence are as follows.

Let b_1, b_2, and b_3 be the coefficients for the independent variables x_1, x_2, and x_3, respectively.

With x_1, x_2, and x_3 known, $b_1 = 0.79$, $b_2 = 0.20$, $b_3 = -0.03$, $S_y^2 = 0.017$.

With x_1 and x_2 only, $b_1 = 0.79$, $b_2 = 0.19$, $S_y^2 = 0.017$.

With x_1 only, $b_1 = 0.89$, $S_y^2 = 0.019$.

With x_2 and x_3, $b_2 = 0.94$, $b_3 = -0.13$, $S_y^2 = 0.08$.

With x_2 only, $b_2 = 0.90$, $S_y^2 = 0.08$.

With x_3 only, $b_3 = 0.72$, $S_y^2 = 0.14$.

This evidence suggests that: 1) when x_1 and x_2 are available, both should be used but x_3 and x_4 should be disregarded; 2) when x_1 is not known, x_2 should be used and x_3 and x_4 disregarded; 3) when both x_1 and x_2 are unknown, x_3 should be used and x_4 disregarded; and, 4) when x_1, x_2, and x_3 are unknown, x_4 should carry only a nominal weight.

If a simple weighting system is permissible for the untransformed values of blocks A-D on Fig. 6, with weights proportional to the coefficients in the multiple regression solutions given above for the logarithmically transformed variables, the weights could be calculated as follows. The overall population mean values for blocks A-D are all equal to the average value E for the mine section as a whole and the coefficient (b_5) applicable to the section value (E in Fig. 6) may be therefore accepted as equal to unity less the sum of the coefficients $b_1, \ldots, b_4$.

Let blocks $A_1, \ldots, A_4$ carry a weight of, say, 60 each, i.e., total weight of 240. If blocks A and B are used, weight for $B_1, \ldots, B_{12} = (0.19/0.79) \cdot 240 = 58$ (5 each); and weight for mean $(E) = (0.02/0.79) \cdot 240 = 5$ (or $\frac{1}{2}$ for each of 12 segments).

If $A_1, \ldots, A_4$ only are known, weights $A_1, \ldots, A_4 = 60$ each, and weight $E = 30$ (2 per segment. If $B_1, \ldots, B_{12}$ only are known, weight $B_1, \ldots, B_{12} = k$ each, and weight $E = 0.11k$ (per segment). If $C_1, \ldots, C_{12}$ are known, weight $C_1, \ldots, C_{12} = l$ each, and weight $E = 0.39l$ per segment).

This whole approach was repeated for three further configurations of data, e.g., only the squares on the left-hand side of Fig. 6 known, or only the squares in the top left quadrant of Fig. 6 known, etc. The conclusions (1)-(4) above were confirmed for these configurations as well. The final results obtained can be summarized as follows: weight for each A square accepted as 60; weight for each B square relative to 60 for A: 5, 6, 9, 11 (average 8); weight for mean (per segment) relative to B at 8: 1, 2, 2, 2 (average 2); weight for each C square relative to mean at 2: 5, 7, 3, 3 (average $4\frac{1}{2}$); weight for each D square accepted nominally at 1.

As a check on the loss in efficiency of estimation in adopting the above somewhat arbitrary system of straight weighting of the inch-dwt values, rather than the correct theoretical multiple regression solution for each case, 50 of the original 140 sets of observations were used to calculate both the weighted moving average and the multiple regression solution for two cases: (a) with all the A and B squares known, and (b) with two A squares and the six corresponding B squares known.

Correlation coefficients were calculated for the weighted moving average vs. the multiple regression solution after normalizing and were found to be 0.996 and 0.998, respectively, with lower 1% limits in excess of 0.99 in both cases.

In routine practice the use of the much simpler weighted moving average is therefore justified and will yield estimates the error variances of which will correspond very closely with those of the more laborious regression approach. The system is now in use on four large gold mines and is being tested on others. Data are collected on all mined-out and developed areas in the mine, and weighted moving averages are calculated on a 100 × 100 ft. grid over all

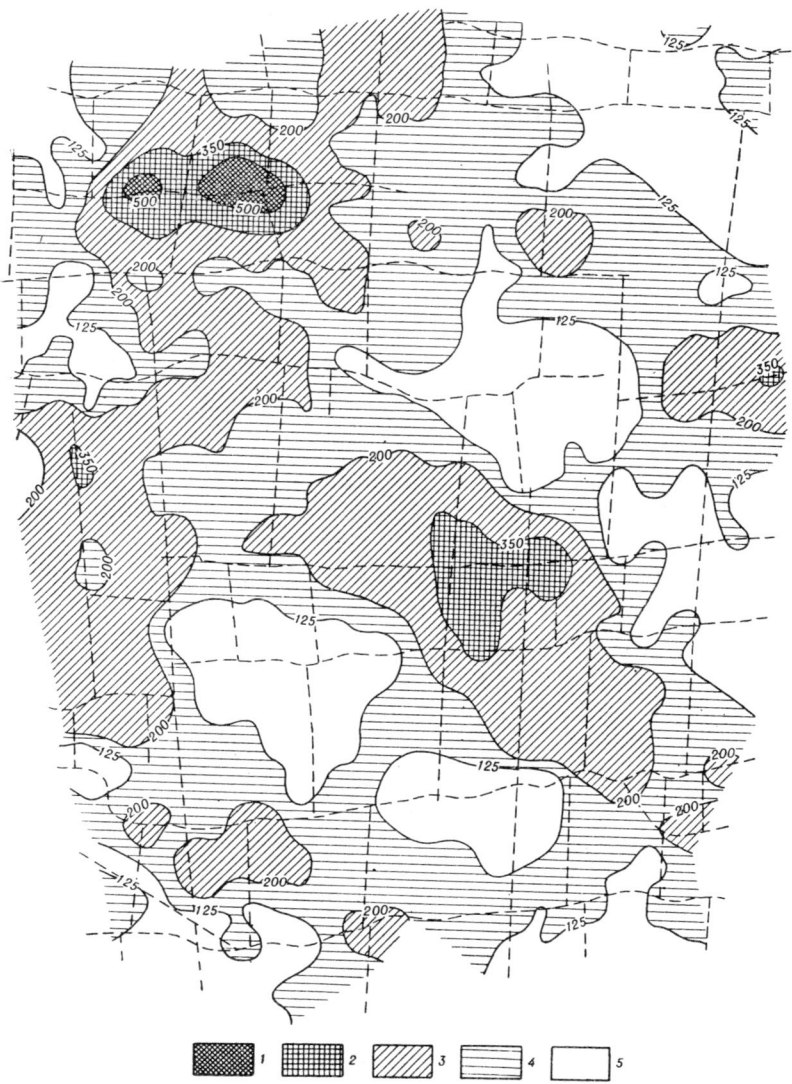

Fig. 7. Contour plan of gold values (inch-dwt) obtained by weighted moving averages for the same mine section covered by Fig. 5. 1) >500; 2) $350 \leq 500$; 3) $200 \leq 350$; 4) $125 \leq 200$; 5) <125.

these areas and then used to prepare a value contour plan. This plan provides a direct visual and quantitative representation of value trends as well as grade estimates for blocks of ore still to be mined. Calculations were initially done manually but, in view of the time and effort involved, these have now been programmed for a large computer which enables updated versions to be obtained at short notice.

Figure 7 shows the weighted moving-average contours for the same mine section covered by Fig. 5 and based on the same development data used for the latter figure. The increased detail as compared with the regional contours for Fig. 5 is clearly evident. In cases such as for this area, where a distinctive grade trend is evident, the weighting system would obviously allow for higher than average weights along the trend direction and below average weights across the trend.

Table 2

Valuation procedure:	Orthodox blocking	Straight Regression	Regression on regional contours	Weighted moving average
Mine / No. of blocks valued		σ^2 of ratios		
A — 400	0.21	0.14	0.12	0.10
B { reef 1 — 28	0.42	0.29	-	0.23
B { reef 2 — 38	0.38	0.27	-	0.17
C — 33	0.05	-	0.05	0.04
D { gold values — 77	0.10	0.09	0.09	0.08
D { uranium values — 77	0.11	0.11	0.11	0.09

Table 3

Data squares available				Error variance S_y^2	Sum of wts.	Data squares available				Error variance S_y^2	Sum of wts.
A	B	C	D			A	B	C	D		
4	12	—	—	.017	384*	2	—	—	—	.06	144
4	—	—	—	.019	264	—	6	—	—	.10	84
—	12	—	—	.08	144	—	—	6	—	.12	54
—	—	12	—	.14	84	1	3	—	—	.08	114
3	9	—	—	.03	294	1	—	—	—	.09	84
3	—	—	—	.031	204	—	3	—	—	.10	54
—	9	—	—	.10	114	—	—	3	—	.13	39
—	—	9	—	.14	69	—	—	—	—	.16**	24**
2	6	—	—	.05	204						

* 4 A squares at 60 each, 12 B squares at 10 each plus regional value at 12 × 2, totals 384.
** Regional value only; error variance then equals variance (σ_y^2) of dependent variable.

Comparison of Various Valuation Methods

Proof of the progressive improvements in efficiency of ore reserve valuations associated with the developments outlined above is provided by the evidence from four large gold mines where the different methods were used on the same basic data for a number of ore blocks and these valuations are compared with the "internal" block values based on subsequent stope sampling results. The criterion for comparison is the natural logarithmic variance of the ratios of these block valuations (inch-dwt + constant) and their corresponding internal block values (+ constant).

Similar tests done on more limited data using up to eighth-order polynomial fits of sampling data in terms of their north-south and east-west coordinates have so far not provided valuations as efficient as the weighted moving average.

Limits of Error for Weighted Moving Average

Due to a recent change in sampling procedures of the mine covered above in the detailed example on the weighted moving-average technique, the weighting system finally used for this mine was somewhat amended to make the weights for the B and C squares 10 and 5, respectively instead of 8 and 4½ as indicated above. The sums of weights used in arriving at the weighted moving averages in this example were then compared with the error variances (S_y^2) estimated on the corresponding multiple-regression solutions as shown in Table 3.

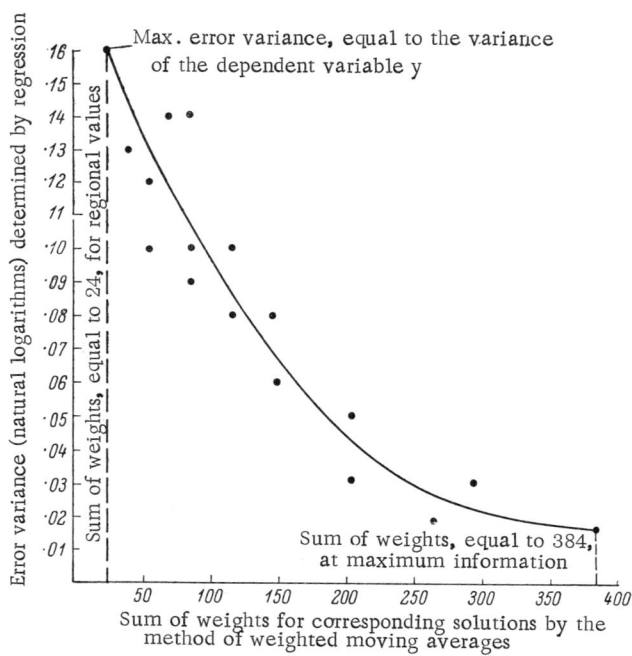

Fig. 8. Trend line and observed values, reflecting variance of weighted moving average.

Figure 8 shows a graphical plot of these observations with a trend line fitted by eye. The correlation indicated is such as to suggest that in practice limits of error can be set off for the weighted moving average using the sum of weights as a criterion and a graph calculated on the basis of Fig. 8. The calculation of limits of error are based on normal theory applied to the transformed values, i.e., $\log_e$ (value + constant), on the same basis as the 99% assay control limits were determined in the example dealt with in the relevant paragraph above.

Acknowledgment. The author gratefully acknowledges the permission by Anglovaal Consolidated Investment Company Ltd. to publish this paper as well as the efforts of the various workers who have, either directly through their own efforts, or indirectly through assistance and discussion, contributed to the development during the last two decades of all the above techniques.

REFERENCES

Coxon, C. H., and Sichel, H. S., "Quality control of routine mine assaying and its influence on underground valuation," J. S. Africa Inst. Mining Met., Vol. 59, No. 10, p. 489 (1959).

Krige, D. G., "A statistical approach to some basic mine valuation problems on the Witwatersrand," J. Chem. Met. Min. Soc. S. Africa, Vol. 52 (1951/2), p. 119, discussion pp. 201, 264; and Vol. 53, pp. 25, 255 (1952).

Krige, D. G., "A statistical analysis of some of the borehole values in the Orange Free State goldfields," J. Chem. Met. Min. Soc. S. Africa, Vol. 53, 1952/3, p. 47 (1953).

Krige, D. G., "A study of the relationship between development values and recovery grades on the South African goldfields," J. S. African Inst. Mining Met., Vol. 59, p. 317; discussion, Vol. 59, p. 469; Vol. 60, p. 126 (1959).

Krige, D. G., "On the departure of ore value distributions from the lognormal model in South African gold mines," J. S. African Inst. Mining Met., Vol. 61, p. 231; discussion, Vol. 61, p. 333, Vol. 62, p. 63 (1961).

Krige, D. G., "Effective pay limits for selective mining," J. S. African Inst. Mining Met., Vol. 62, p. 435 (1962a).

Krige, D. G., "Economic aspects of stoping through unpayable ore," J. S. African Inst. Mining Met., Vol. 62, p. 364 (1962b).

Krige, D. G., "Statistical applications in mine valuation. Parts I and II," J. Inst. Min. Surveyors S. Africa, Vol. 12, p. 45 (1962c).

Krige, D. G., "A review of the impact of statistics on mine valuation in the gold mining industry," J. S. African Inst. Mining Met., Vol. 64, pp. 373-380 (1964).

Krige, D. G., and Ueckermann, H. J., "Value contours and improved regression techniques for ore reserve valuations," J. S. African Inst. Mining Met., Vol. 63, p. 429 (1963).

Pirow, P. C., and Krige, D. G., "The use of mathematical statistics and computers for planning and control of the gold mines of the Transvaal and Orange Free State," Eighth Com. Min. Met. Congress, Australia and New Zealand (Proceedings in press, 1965).

Ross, F. W. J., The Development and Some Practical Applications of Statistical Value Distribution Theory for the Witwatersrand Auriferous Deposits, Master's thesis, Witwatersrand Univ. (unpublished, 1950).

Rowland, R. St. J., and Sichel, H. S., "Statistical quality control of routine underground sampling," J. S. African Inst. Mining Met., Vol. 60, p. 251; discussion, Vol. 60, pp. 384, 556 (1960).

Sichel, H. S., "An experimental and theoretical investigation of bias error in mine sampling with special reference to narrow gold reefs," Trans. Inst. Min. Met., London, Vol. 56, p. 403 (1947).

Sichel, H. S., "New methods in the statistical evaluation of mine sampling data," Trans. Inst. Min. Met., London, Vol. 61, p. 261; discussion, pp. 391, 463, 501 (1951).

Truscott, S. J., "The computation of the probable value of ore reserves from assay results," Trans. Inst. Min. Met., London, Vol. 39 (1929).

Watermeyer, G. A., "Application of the theory of probability in the determination of ore reserves," J. Chem. Met. Min. Soc. S. Africa, Vol. 19 (1919).

FUNDAMENTAL PROBLEMS OF COMPUTING RESERVES OF MINERAL RESOURCES

V. Nemeč

Geological Exploration
Prague, Czechoslovakia

1. The computation of mineral reserves should be made with proper consideration of the geological history of formation of the deposit. It cannot be reduced to a single operation with contents of tabulated elements, but must be based on an understanding of the relationships among the minerals of the entire deposit.

2. Strict criteria are necessary for determining details when making computations of a given category. Attention to unnecessary details leads to futile waste of means, and it makes the work more expensive; but insufficient data lead to extensive mining operations without adequate basis for the work.

3. The necessity of determining the form of a deposit and the distribution of mineral concentrations within it requires various approaches: geometric approximations of the ore body, on the one hand, accomplished by geometrical methods of surveying, and investigation of concentrations, on the other, by studying random samples (in the mathematical sense of the word). At present we see an overemphasis on geometrical studies and a distinct inadequacy in the development of statistical work.

4. Mathematical geology, developed in great measure by the works of A. B. Vistelius, has tremendous value for working out a science of computing mineral reserves. It introduces genetic concepts of geologists into algorithms for computation, and it thus permits us to make all our computations in strict compliance with the geologic features of a deposit. It is necessary that investigations in mathematical geology be more widely undertaken, and that they be developed simultaneously with theoretical founding of the applied aspects, permitting us to formulate a theory of the simplest means of attaining our goal.

5. Formulation of the problem of computing mineral reserves as a problem of mathematical geology requires the combined efforts of people concerned with theoretical geology and those occupied in practical exploratory work. In this endeavor, the joint efforts of these individuals and others should be directed toward development of models of processes that may be expressed mathematically. This fundamental idea of mathematical geology may be firmly

established only by combining scientific ideas with the abundance of data that are accumulating in metal mines, industrial organizations, and coal mines. It is necessary to close the gap between theoretical geochemistry and prospecting—exploration practice, a gap that is a source of detriment to both sides.

6. The modeling of geological processes and the determination of reserves on the basis of quantitative representations of these models require the development of techniques. All these problems may be solved only by the wide use of electronic computers. These computers must not serve merely as aids in computation, but must be used to solve logical problems and, above all else, to check geological hypotheses.

7. In concluding this summary with a traditional academic greeting to A. B. Vistelius — Vivat, crescat, floreat — we hope for a wide response in the geological community to the idea of organic mathematical geology and to the establishment of this field as a branch of the exact sciences, a result awaited with impatience by exploration geologists.

VISUAL DISPLAY OF COMPUTER OUTPUT AIDS GEOLOGICAL INTERPRETATION

D. F. Merriam

Division of Basic Geology
Kansas Geological Survey
and
Research Associate
The University of Kansas
Lawrence, Kansas

Methods of automatic illustration of solutions of geologic problems by means of electronic computers are discussed. Line printers have been used in order to obtain the illustrations. Examples of computational illustrations of trend surfaces and time—trend sequences are given.

Machine plotting and contouring of data are important aspects of computer applications in geology. Data can now be manipulated and displayed in almost any desired form for visual inspection by the investigator. Geologists, long accustomed to using pattern recognition methods, find graphic display of output very useful in helping to interpret results. By using automated procedures, the geologist is released from tedious and time-consuming work to allow him more time to explore problems and their solutions. In addition to various uses in data processing, high-speed, large-memory computers permit application of analytical techniques in problem solving that previously were difficult, if not impossible, to use.

Looking for underlying trends is important, especially in complex data, and can be best analyzed in many instances by contour-type maps. Another readily understood frame of reference is the time—trend sequence where some variable is plotted against time, or relative time.

For the most part, instructions for visual output are contained as a subroutine within a main program. Generally, results are displayed by use of line printers or teleprinters, but on- or off-line plotters and cathode-ray tubes have found increasing use. Plotters are superior to line printers because they can produce almost smooth, continuous lines in any direction by moving in very small north-south and east-west increments [Batcha and Reese, 1964, p. 10; Forgotson, 1963, p. 107]. Line printers have limitations imposed by the size and number of individual alphanumeric and special type characters. Nevertheless, line printers can handle most output if the program is designed with imagination and ingenuity [Harbaugh, 1962]. The image produced by the cathode-ray tube is not perfect, but a series of such projections can be produced very rapidly [Tobler, 1965].

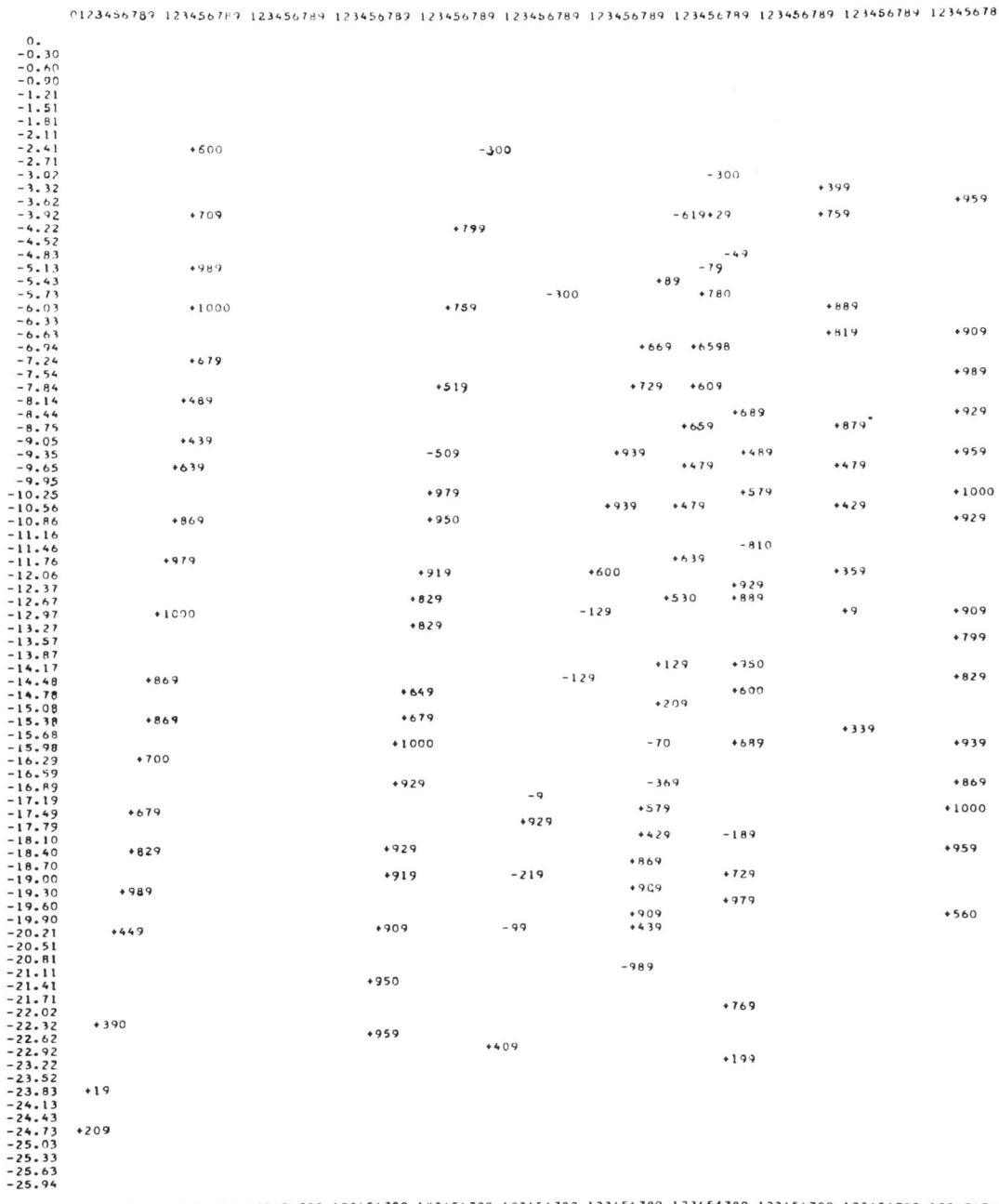

Fig. 1. Plot by the computer of sea magnetic data from the Antarctic. Data values are in their relative geographic positions ready for contouring by hand. North is to top of map and scale factors are noted (data courtesy of The Geophysics Section, Department of Geology, University of Birmingham).

Graphic Output for Data Processing

Data processing involves storage, processing, and retrieval of large amounts of information. The petroleum industry in the United States has assumed a position of leadership in geologic data processing through development of well-data systems. These systems, which have

```
            0123456789 123456789 123456789 123456789 123456789 123456789 123456789 123456789 123456789 123456789 123456789
7.00                   CCCCCCCCCCCCCCCC               DDDDDDDDDDDDDDDD               EEEEEEEEEEEEEEEE
6.83   B               CCCCCCCCCCCCCCCC               DDDDDDDDDDDDDDDD               EEEEEEEEEEEEEEEE
6.67   BBBBB            CCCCCCCCCCCCCCCCC              DDDCDDDDDDDDDDDDD              EELEEEEEEEEEEEEEE
6.50   BBBBBBBBBB         CCCCCCCCCCCCCCCC              DDDDDDDDDDDDDDDD                EEEEEEEEEEEEEEEE
6.33   BBBBBBBBBBBBBBB      CCCCCCCCCCCCCCCC              DCCDDDDDDDDDDDDD                EEEEEEEEEE
6.17        BBBBBBBBBBBBBBB      CCCCCCCCCCCCCCCC              DDDDDDDDDDDDDDDD                EEEEE
6.00          BBBBBBBBBBBBBBBB       CCCCCCCCCCCCCCCC            DDDDDDDDDDDDDDDD
5.83            BBBBBBBBBBBBBBBBBB     CCCCCCCCCCCCCCCC            DDDDDDDDDDDDDDDD
5.67   AAAAA       BBBBBBBBBBBBBBBB       CCCCCCCCCCCCCCCC           DDDDDDDDDDDDDDDD
5.50   AAAAAAAAA       BBBBBBBBBBBBBBBB      CCCCCCCCCCCCCCCC            DDDDDDDDDDDDDDDDD
5.33   AAAAAAAAAAAAAA     BBBBBBBBBBBBBBBBB     CCCCCCCCCCCCCCCCC              DDDDDDDDDDDDDDD
5.17    AAAAAAAAAAAAAAAA     BBBBBBBBBBBBBBBB       CCCCCCCCCCCCCCCC                DDDDDD
5.00         AAAAAAAAAAAAAAAA     BBBBBBBBBBBBBBBB      CCCCCCCCCCCCCCCC                  D
4.83           AAAAAAAAAAAAAAAA     BBBBBBBBBBBBBBBB       CCCCCCCCCCCCCCCC
4.67   ....      AAAAAAAAAAAAAAAA     BBBBBBBBBBBBBBBB         CCCCCCCCCCCCCCCC
4.50   ..........     AAAAAAAAAAAAAAAA    BBBBBBBBBBBBBBBB        CCCCCCCCCCCCCCCCC
4.33   ..............    AAAAAAAAAAAAAAAA    BBBBBBBBBBBBBBBB         CCCCCCCCCCC
4.17      ................    AAAAAAAAAAAAAAAA    BBBBBBBBBBBBBBBB         CCCCCCC
4.00         .................    AAAAAAAAAAAAAAAA    BBBBBBBBBBBBBBBBB           CC
3.83             ................    AAAAAAAAAAAAAAAA    BBBBBBBBBBBBBBBBB
3.67   000           .................    AAAAAAAAAAAAAAAA     BBBBBBBBBBBBBBBBB
3.50   COOOOOOO           ...............    AAAAAAAAAAAAAAAA      BBBBBBBBBBBBBBBB
3.33   000000000000000         ................    AAAAAAAAAAAAAAAA     BBBBBBBBBBBBBB
3.17   0000000000000000         ...............      AAAAAAAAAAAAAAAA      BBBBBBBB
3.00        0000000000000000        ................     AAAAAAAAAAAAAAAA        BBB
2.83          0000000000000000        ................      AAAAAAAAAAAAAAAA
2.67   11       0000000000000000        ................      AAAAAAAAAAAAAAAA
2.50   1111111     0000000000000C000        ................      AAAAAAAAAAAAAAAA
2.33   1111111111       000000000000000        ................        AAAAAAAAAAAAAA
2.17   1111111111111       0000000000000000        ...............         AAAAAAAA
2.00   1111111111111111       0000000000000000        ................           AAAA
1.83       1111111111111111       0000000000000000        ................
1.67   2        1111111111111111      0000000000000000        ................
1.50   222222       1111111111111111      0000000000000000          ................
1.33   22222222222      1111111111111111      0000000000000000          ...............
1.17   2222222222222222     1111111111111111      0000000000000000            ..........
1.00       2222222222222222     1111111111111111      0000000000000000                ....
0.83         2222222222222222     1111111111111111     00C000000000000000
0.67           2222222222222222     1111111111111111     0000000000000000
0.50   33333     222222222222222222     11111111111111111    0000000000C000
0.33   3333333333       2222222222222222     1111111111111111      0000C00000000000
0.17   333333333333333      2222222222222222     1111111111111111       00000C00000
0.00   333333333333333       2222222222222222        1111111111111111         000000
            0123456789 123456789 123456789 123456789 123456789 123456789 123456789 123456789 123456789 123456789 123456789
```

Fig. 2. Linear trend surface of absolute age determinations of detrital sand in Cretaceous rocks [Vistelius, 1964]. Reference contour (edge of dots next to "A") = 300 m/y; C. I. = 15 m/y. North is to top of map.

been described by Bonham [1963], Buller [1964], Dillon [1964], Dillon and Nichols [1965], Slack and others [1963], and many others, contain billions of items of information relating to geological characteristics, engineering properties, and production records for wells that have been drilled in the search for petroleum in many areas. These systems are providing the geologist with almost unlimited subsurface information that can be retrieved in any desired sequence or arrangement, and sorted and processed quickly for study of special problems.

An example of a simple use of the computer is shown in Fig. 1. A data-plot subroutine is used to display information at its appropriate geographic position. Here, magnetic data taken at sea in the Antarctic are plotted and ready for hand contouring.* Programs are also available for converting notational legal-land descriptions in different parts of the United States to Cartesian coordinates [Good, 1964b; Moser, 1963]. Many other similar applications of data processing could be described [see, for example, Assiter, 1960; Lovering and Davidson, 1964].

Graphic Output for Problem Solving

Graphic output for problem-solving applications differs in that each result presents a different and perhaps unique problem. Visual display is most effective if a pattern is simple and can be recognized by the geologist [see, for example, Merriam and Lippert, 1964].

---

*This plotting procedure is from a trend-surface program developed by Harbaugh [1963] and modified by Good [1964a].

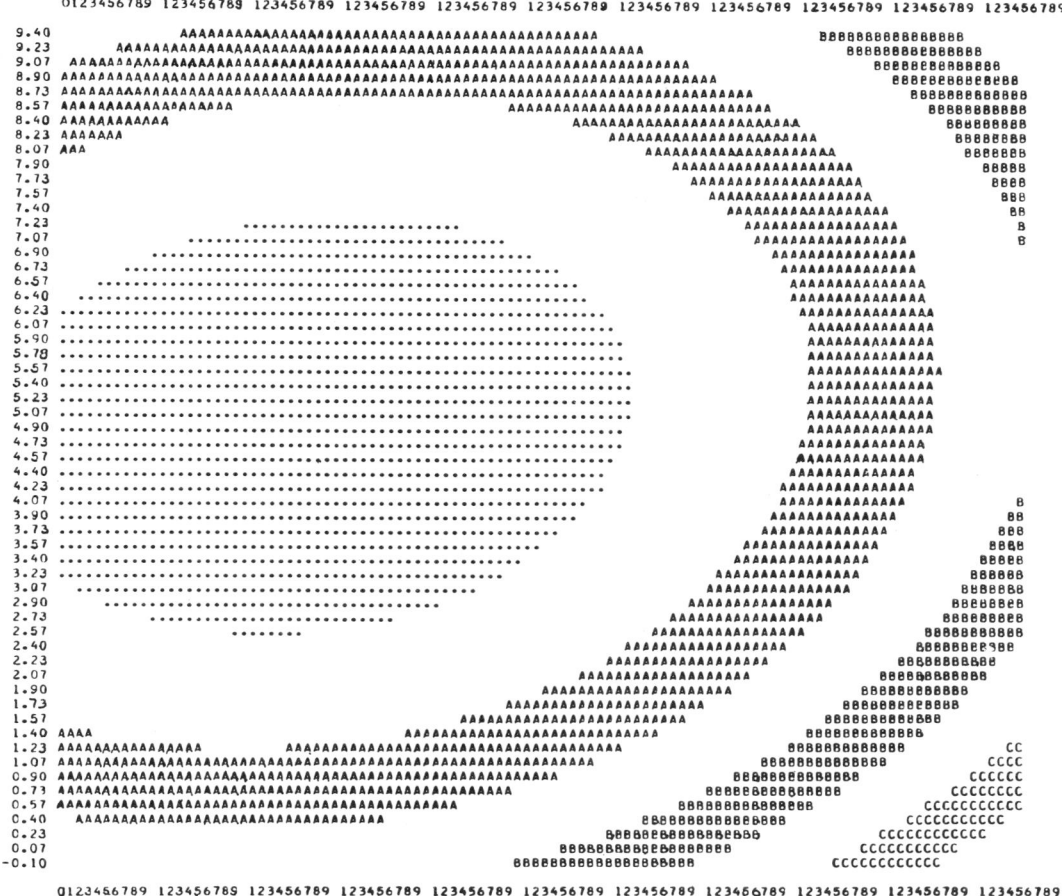

Fig. 3. Quadratic trend surface of percent disarticulation of brachiopod shells in *Cyrtina septosa* Band, Derbyshire, England [from Sadler and Merriam, 1967]. Reference contour = 85%; C. I. = 5%. North is to top of map.

Trend analysis is one important technique recently applied to geologic data. Many methods seeking trends are available and some applications are discussed in papers by Agterberg [1964], Krumbein [1959], Merriam [1964], Merriam and Harbaugh [1963, 1964], Miller [1956], Vistelius and Hurst [1964], and Whitten [1961]. At first, output was presented in tabular form, and it was necessary for the geologist to plot and contour the maps. Later, however, programs contained plotting and contouring capabilities.

Examples of machine-computed linear, quadratic, and cubic trend-surfaces output on a line printer are shown in Figs. 2, 3, and 4. Edges of the alternate bands of letters and numbers represent contour lines. The reference contour is the edge of the dots next to "A." The value of any contour can be determined easily if the contour interval is known.*

Other plotting and contouring programs are available and have been described in the literature [IBM, 1965; Kobetich, 1964]. Most of these programs are for three-dimensional problems (two geographic positions and one variable). Harbaugh [1964], however, has described a four-dimensional method of trend analysis in which low-ordered hypersurfaces are fitted to irregularly spaced data.

---

* These examples are from a program described by Good [1964a].

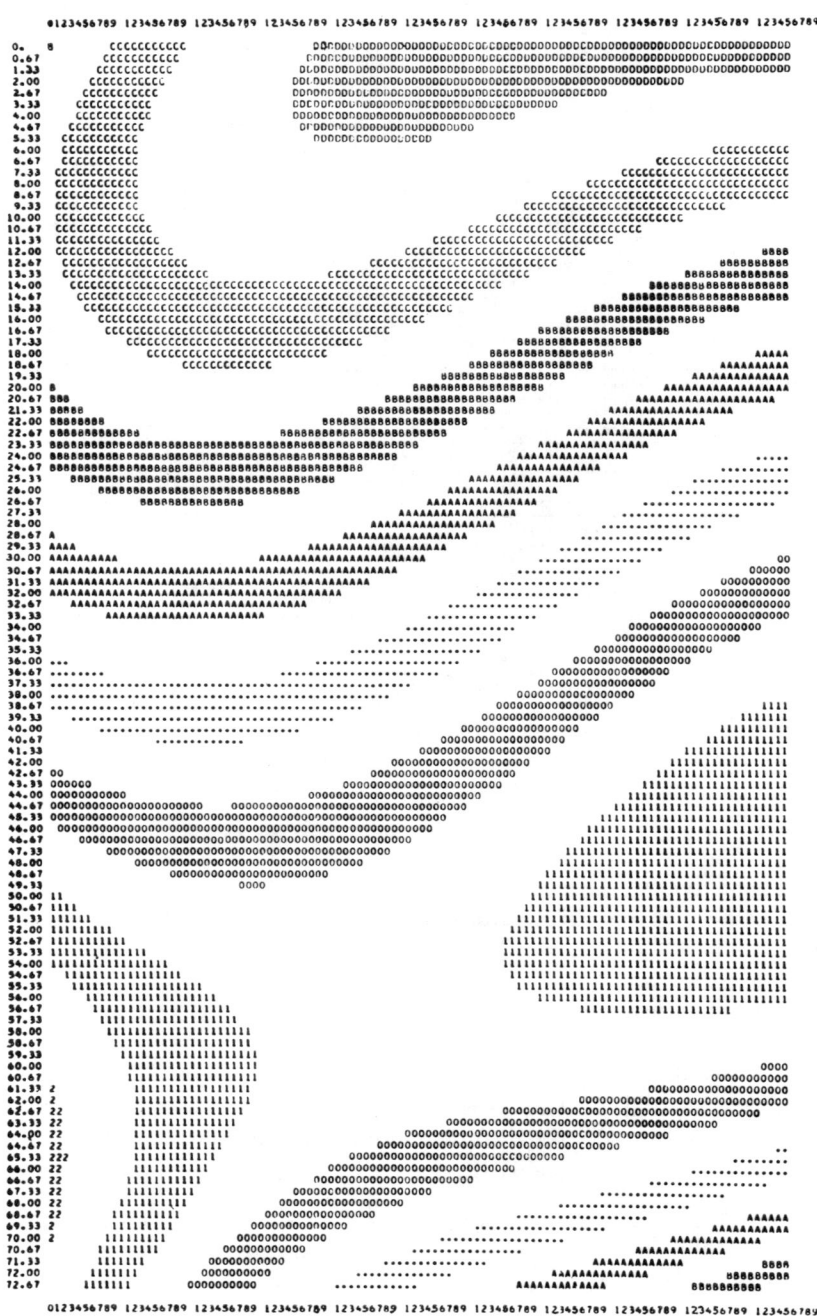

Fig. 4. Cubic trend surface of stratigraphic thickness data of Black Metals Group (Namurian) in Midland Valley, Scotland [from Read and Merriam, 1966]. Reference contour = 140 feet; C. I. = 15 feet. North is to top of map.

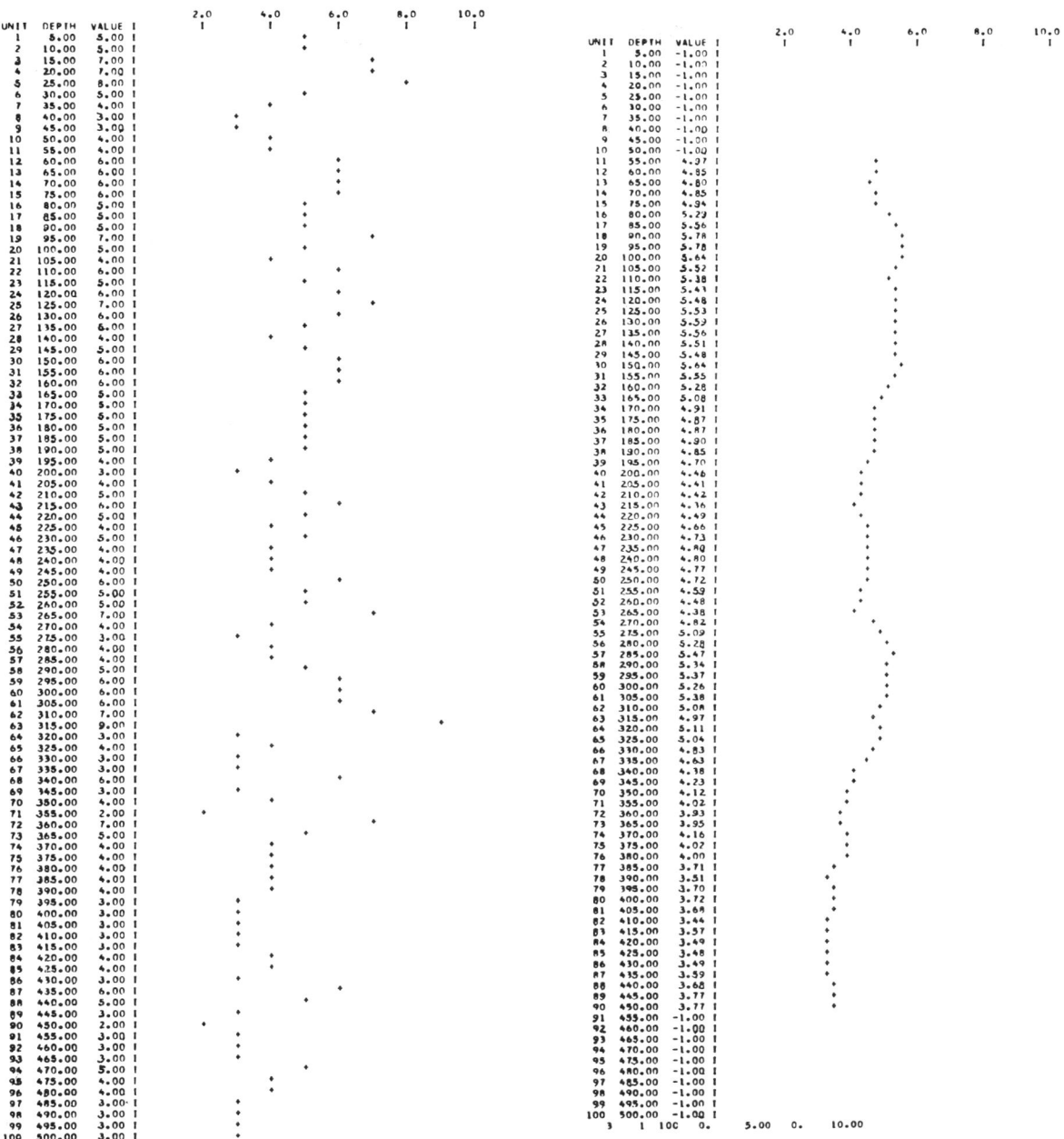

Fig. 5. Raw data of time—trend curve; time (horizontal axis) needed to drill successive 5-ft intervals of rock (shown vertically) is displayed. Harder units, such as limestone, take longer to drill and corresponding values are deflected to right.

Fig. 6. "Smoothed" time—trend curve for information shown in Fig. 5. Data taken from drilling-time log in Middle Pennsylvanian cyclic sequence — limestone, shale, sandstone — in eastern Kansas.

Fig. 7. Computers available to geologists of Kansas Geological Survey. a) Small relatively slow IBM 1401 computer used to control a high-speed line printer; b) an IBM 1620 now replaced by (c) a fast, medium-sized IBM 7040; a large, fast, IBM 7090/1401 computer system (d) is available at Stanford University for special problems. Bigger and faster machines, of course, are necessary for large, complex programs.

Time—trend information, data collected at successive intervals of time, can be automatically plotted [Fox, 1964; Merriam, 1965]. These plots show the raw data and smoothed curves (Figs. 5 and 6). Time—trend analysis has been successfully applied to geologic observations such as porosity, chemical and mineralogical composition, grain size, bed thickness, etc., taken along a traverse or stratigraphic section [Lee, 1963; Schwarzacher, 1964; Vistelius, 1961]. Automatic plots of Fourier analyses of geologic structure are shown in a paper by Harbaugh and Preston [1965].

Visual output is most important for any type of geologic modeling. As this aspect of computer application increases in use, various devices for best displaying the model will be explored [Krumbein, 1962; Raup and Michelson, 1965].

Availability of Equipment and Programs

For many installations in the United States, both academic and commercial, cost of computer operation is nominal. At many places, time is available to geologists working on research projects at no cost and can be obtained upon request. Geologists of the Kansas Geological Survey have available several computers on which to develop and use programs (Fig. 7).

At present, no clearing house exists for exchange and dissemination of computer programs in the earth sciences. Such an establishment, to be international in scope, has been proposed, however.

Summary

Much time and effort can be saved by showing the results of computer analyses in graphic form. Recognition of patterns displayed is eased by putting the results into a frame of reference familiar to geologists; contour-type maps and time—trend sequences are two such forms. Line printers, on- and off-line plotters, and cathode-ray tubes are used for displaying the output.

Because of the ready availability of computers, many operations in data processing and analyses in problem solving can be used routinely. The speed and versatility of present-day computers also allow more complex problems to be investigated.

Acknowledgments. I would like to thank Drs. W. W. Hambleton and F. W. Preston of The University of Kansas for critically reading the manuscript. All data were run at the Computation Center, the University of Kansas. Mrs. S. S. Brown, Mrs. N. C. Cocke, and Mr. R. H. Lippert, all of the Kansas Geological Survey, kindly helped prepare and process the data. Mr. Malcolm Barker, University of Leicester (U.K.), photographed the illustrations from the original computer printout.

REFERENCES

Agterberg, F. P., "Methods of trend surface analysis," Colorado School Mines Quart., Vol. 59, No. 4, pp. 111-130 (1964).

Assiter, E. J., "Electronic computers aid many exploration phases," World Oil, Vol. 150, No. 5, pp. 105-111 (1960).

Batcha, J. P., and Reese, J. R., "Surface determination and automatic contouring for mineral exploration, extraction, and processing," Colorado School Mines Quart, Vol. 59, No. 4, pp. 1-14 (1964).

Bonham, L. C., "Here's Sokal's experience with a mechanized well data system" World Oil, Vol. 156, No. 5, pp. 94-100 (1963).

Buller, J. V., "A computer oriented system for the storage and retrieval of well information," Dept. Min. Res. Saskatchewan (1964), 65 pp.

Dillon, E. L., "Electronic storage, retrieval, and processing of well data," Am. Assoc. of Petroleum Geologists Bull., Vol. 48, No. 11, pp. 1828-1836 (1964).

Dillon, E. L., and Nichols, C. W., "Handling of statistical well data by computer," Am. Assoc. of Petroleum Geologists Bull., Vol. 49, No. 9, pp. 1520-1531 (1965).

Forgotson, J. M., Jr., "How computers help find oil," Oil and Gas J., Vol. 61, No. 11, pp. 100-109 (1963).

Fox, W. T., FORTRAN and FAP Program for Calculating and Plotting Time—Trend Curves using an IBM 7090 or 7094/1401 Computer System, Kansas Geol. Survey Sp. Dist. Publ., 12 (1964), 24 pp.

Good, D. I., FORTRAN II Trend—Surface Program for the IBM 1620: Kansas Geol. Survey Sp. Dist. Publ., 14 (1964a), 54 pp.

Good, D. I., "Mathematical Conversion of Section, Township, and Range Notation to Cartesian Coordinates, Kansas Geol. Survey Bull. 170, Pt. 3 (1964b), 30 pp.

Harbaugh, J. W., "Direct printing of computer maps of facies data by computer (abstract)," Am. Assoc. of Petroleum Geologists Bull., Vol. 46, No. 2, p. 268 (1962).

Harbaugh, J. W., BALGOL Program for Trend—surface Mapping using an IBM 7090 Computer, Kansas Geol. Survey Sp. Dist. Publ. 3 (1963), 17 pp.

Harbaugh, J. W., A Computer Method for Four-Variable Trend Analysis Illustrated by a Study of Oil-Gravity Variations in Southeastern Kansas, Kansas Geol. Survey Bull. 171 (1964), 58 pp.

Harbaugh, J. W., and Preston, F. W., "Fourier series analysis in geology," College of Mines, Arizona Univ., Vol. 1, pp. R-1 to R-46 (1965).

IBM, Numerical Surface Techniques and Contour Map Plotting, IBM, Data Processing Application, White Plains, N. Y. (1965), 34 pp.

Kobetich, E. J., "General computer plotting subroutine," Kansas Acad. Sci. Trans., Vol. 67, No. 4, pp. 640-645 (1964).

Krumbein, W. C., "Trend surface analysis of contour-type maps with irregular control-point spacing," J. Geophy. Res., Vol. 64, No. 7, pp. 823-834 (1959).

Krumbein, W. C., "The computer in geology," Science, Vol. 136, No. 3522, pp. 1087-1092 (1962).

Lee, P. J., "Correlation of sediments by using an electronic digital computer," Petrol. Geol. Taiwan, No. 2, pp. 137-147 (1963).

Lovering, T. G., and Davidson, D. F., "Storage and retrieval of analytical data on geologic materials," Colorado School Mines Quart., Vol. 59, No. 4, pp. 247-257 (1964).

Merriam, D. F., "Use of trend-surface residuals in interpreting geologic structures," Stanford Univ. Publ., Geol. Sci., Vol. 9, No. 2, pp. 686-692 (1964).

Merriam, D. F., "Geology and the computer," New Scientist, Vol. 26, No. 444, pp. 513-516 (1965).

Merriam, D. F., and Harbaugh, J. W., "Computer helps map oil structures," Oil and Gas J., Vol. 61, No. 47, pp. 158-159, 161-163 (1963).

Merriam, D. F., and Harbaugh, J. W., Trend-Surface Analysis of Regional and Residual Components of Geologic Structure in Kansas, Kansas Geol. Survey Sp. Dist. Publ. 11 (1964), 27 pp.

Merriam, D. F., and Lippert, R. H., "Pattern recognition studies of geologic structure using trend-surface analysis," Colorado School Mines Quart., Vol. 59, No. 4, pp. 237-245 (1964).

Miller, R. L., "Trend surfaces: their application to analysis and description of environments of sedimentation," J. Geol., Vol. 64, No. 5, pp. 425-446 (1956).

Moser, F., A Computer Oriented System in Stratigraphic Analysis, Inst. Sci. and Tech., Michigan Univ. (1963), 32 pp.

Raup, D. M., and Michelson, A., "Theoretical morphology of the coiled shell," Science, Vol. 147, No. 3663, pp. 1294-1295 (1965).

Read, W. A., and Merriam, D. F., "Trend-surface analysis of stratigraphic thickness data from some Namurian rocks east of Sterling, Scotland," J. Scottish Geol., Vol. 2, No. 1, pp. 96-100 (1966).

Sadler, H. E., and Merriam, D. F., "Reinterpretation of the *Cyrtina septosa* data, Lower Carboniferous of Derbyshire, England, by computer analysis," Sedimentology, Vol. 8, pp. 55-61 (1967).

Schwarzacher, Walther, "An application of statistical time-series analysis of a limestone—shale sequence," J. Geol., Vol. 72, No. 2, pp. 195-213 (1964).

Slack, H. A., and others, "Now — map making made accurate, objective," Oil and Gas J., Vol. 61, No. 31, pp. 158-160 ff (1963).

Tobler, W. R., "Automation in the preparation of thematic maps," Cartogr. J., British Cartographic Soc., Vol. 2, No. 1, pp. 32-38 (1965).

Vistelius, A. B., "Sedimentation time—trend functions and their application for correlation of sedimentary deposits," J. Geol., Vol. 69, No. 6, pp. 703-728 (1961).

Vistelius, A. B., "Paleographic reconstructions of absolute age determinations of sand particles," J. Geol., Vol. 72, No. 4, pp. 483-486 (1964).

Vistelius, A. B., and Hurst, V. J., "Phosphorus in granitic rocks of North America," Geol. Soc. Am. Bull., Vol. 75, No. 11, pp. 1055-1091 (1964).

Whitten, E. H. T., "Modal variation and form of the Beinn an Dubhaich granite, Skye," Geol. Mag., Vol. 98, No. 6, pp. 467-472 (1961).

VII. Chronicle and Bibliography

TREND ANALYSIS OF GEOLOGIC DATA
(BASIC LITERATURE)

M. A. Romanova

Laboratory of Mathematical Geology
V. A. Steklov Mathematical Institute
Academy of Sciences of the USSR
Leningrad, USSR

In recent time the analytical method using a trend surface (trend analysis) has become more and more widely used in solving geologic problems. The method involves the approximation of observed values, belonging to an association homogeneous in their geologic relations, by a surface of two-dimensional regression on the geographic coordinates of the observation points. Each observed value Z_{ij} is considered the sum of a systematic component $\tilde{Z}_{ij}$, corresponding to the conditional mathematical expectation of a two-dimensional random process, and a random component ξ_{ij}.

In assuming that the systematic component is a function of the coordinates of the observation points, a two-dimensional regression surface is computed. Since the work of Oldham and Sutherland [1956], the method has been called trend-surface analysis.

The method was first used in solving geological problems by Krumbein [1956] and Miller [1956]. The introduction of electronic computers in practical geologic work has led to the wide use of trend analysis in solving various geological problems, and at the present time a large number of papers may be found in which the method is discussed.

In view of the fact that the method of trend analysis is very effective for solution of some problems, especially in association with discrimination of regional and local components of observed characteristics, we have given a bibliographic summary below of the principal works in which the method of trend analysis has been used in some form. The list cannot be considered exhaustive, but it embraces most types of geologic investigations in which approximation methods with computation of a trend surface of some order have been used.

LITERATURE

Agterberg, F. P., "Value contours and improved regression techniques for ore reserve valuations, contributions to discussion," J. S. African Inst. Mining Met., pp. 108-109 (1963).

Agterberg, F. P., "Methods of trend-surface analysis," International Symposium on Applications of Statistics, Operation Research and Computers in the Mineral Industry, Colorado School of Mines Quart., Vol. 59, No. 4, pp. 111-130 (1964).

Allen, P., and Krumbein, W. C., "Secondary trend components in the top Ashdown pebble bed: A case history," J. Geol., Vol. 70, No. 5, pp. 507-538 (1962).

Baird, A. K., McIntyre, D. B., and Welday, E. E., "Trend surfaces of high order (abstract)," Geol. Soc. Am., Spec. Paper 76, p. 1880 (1965).

Baird, A. K., McIntyre, D. B., and Welday, E. E., "Geochemical and structural studies in batholithic rocks of Southern California, Pt. II. Sampling of the Rattlesnake Mountain pluton for chemical composition, variability, and trend analysis," Bull. Geol. Soc. Am., Vol. 78, pp. 191-222 (1967).

Belonin, M. D., "Principal evolutionary features of the Sokolovy and Bagaevka structures in the Saratov part of the Volga region," Sov. Geol., No. 12, pp. 90-109 (1964).

Cain, J. A., and Dodd, J. R., "'Meaningful' patterns from random data," Geol. Soc. Am., annual meeting, Miami, Program, p. 24 (1964).

Chayes, F., and Suzuki, V., "Geological contours and trend surfaces (a discussion)," J. Petrol., Vol. 4, No. 2, pp. 307-312 (1963).

Chorley, R. J., "An analysis of the areal distribution of soil size facies on the Lower Greensand rocks of east-central England by the use of trend—surface analysis," Geol. Mag., Vol. 101, No. 4, pp. 314-321 (1964).

Connor, J. J., and Miesch, A. T., "Application of trend analysis to geochemical prospecting data from Beaver County, Utah," in: Computers in the Mineral Industries, Stanford Univ., Publ. Geol. Sci., Vol. 9, No. 1, pp. 110-125 (1964).

Dawson, K. R., and Whitten, E. H. T., "The quantitative mineralogical composition and variation of the Lacorne, La Motte, and Preissac granitic complex, Quebec, Canada," J. Petrol., Vol. 3, No. 1, pp. 1-37 (1962).

Dempsey, J. R., "A generalized two-dimensional regression procedure," Contribution 2, State Geol. Surv. Univ. Kansas, Lawrence, pp. 1-12 (1966).

Duff, P. M. D., and Walton, E. K., "Trend-surface analysis of coal fields," in: Developments in Sedimentology, Vol. 1, American Elsevier, New York (1963), pp. 114-122.

Duff, P. M. D., and Walton, E. K., "Trend-surface analysis of sedimentary features of the Modiolaris zone, East Pennine coal field, England," in: Deltaic and Shallow Marine Deposits: Developments in Sedimentology, Vol. 1, American Elsevier, New York (1964), pp. 114-122.

Fahrig, W. F., Geology of the Athabasca Formation, Geol. Soc. Canada Bull., Vol. 68 (1961).

Forgotson, J. M., Jr., "New computers help find oil," Oil and Gas J., Vol. 61, No. 11, pp. 100-109 (1963).

Good, D. J., FORTRAN II Trend-Surface Program for the IBM-1620, Kansas Geol. Surv., Spec. Distrib. Publ., No. 14 (1964).

Grant, F., "A problem in the analysis of geophysical data," Geophys., Vol. 22, pp. 309-344 (1961).

Harbaugh, J. W., "Direct printing of contour maps of facies data by computer," Am. Assoc. Petroleum Geologists Bull., Vol. 46, No. 2, p. 268 (1962).

Harbaugh, J. W., BALGOL Program for Trend-Surface Mapping using an IBM-7090 Computer, Kansas Geol. Surv., Spec. Distrib. Publ., No. 3 (1963), 17 pp.

Harbaugh, J. W., "A computer method four-variable trend analysis illustrated by a study of oil-gravity variations in southeastern Kansas," Kansas Geol. Surv. Bull., Vol. 171, p. 58 (1964).

Harbaugh, J. W., "Trend-surface mapping of hydrodynamic oil traps with the IBM 7090-94 computer," Colorado School of Mines Quart., Vol. 59, No. 4, Pt. B., pp. 557-578 (1964).

Harbaugh, J. W., "Application of four-variable trend hypersurfaces in oil exploration," Computers in the Mineral Industries (abstract), Stanford Univ. Publ., Geol. Sci., Vol. 9, No. 2, p. 693 (1964).

Harbaugh, J. W., and Preston, F. W., "Fourier series analysis in geology," Symposium on Computer Applications in Mining Exploration, Vol. 1, Univ. of Arizona (1965), pp. 1-46.

Hewlett, R. F., Computer Methods in Evaluation, Development and Operations of an Ore Deposit, Bureau of Mines Rep. Invest., U. S. Dept. of Interior (1963).

Hewlett, R. F., Polynomial Surface Fitting using Sample Data from an Underground Copper Deposit, Bureau of Mines Rep. Invest. 6522, U.S. Dept. of Interior (1964).

Krige, D. G., "A statistical approach to some basic mine valuation problems on the Witwatersrand," J. Chem. Met. Mining Soc. S. Africa, Vol. 52, p. 119; discussion, Vol. 52, pp. 210, 264; Vol. 53, pp. 25, 255 (1951).

Krige, D. G., "Statistical applications in mine valuations, Pts. I and II," J. Inst. Mining Surv. S. Africa, Vol. 12, pp. 45-95 (1962).

Krige, D. G., "The Use of mathematical statistics and computers for planning and control at the gold mines of the Transvaal and Orange Free State," Paper read at the 8th Commonwealth Mining and Metallurgical Congress, Australia (1965).

Krige, D. G., "A study of gold and uranium distribution patterns in the Klerksdorp gold field," Geoexploration, Vol. 4, pp. 43-53 (1965).

Krige, D. G., "A review of the impact of statistics on mine valuation in the gold mining industry," J. S. African Inst. Mining Metal., Vol. 64, Pt. II, pp. 373-380 (1966).

Krige, D. G., "Two-dimensional weighted moving average trend surfaces for ore valuation," J. S. African Inst. Mining Metal., Vol. 64, pp. 13-79 (1966).

Krige, D. G., and Ueckermann, H. J., "Value contours and improved regression techniques for ore reserve valuations," J. S. African Inst. Mining Met., Vol. 63, pp. 429-452 (1962).

Krumbein, W. C., "Regional and local components in facies maps," Am. Assoc. Petroleum Geologists Bull., Vol. 40, No. 8, pp. 2163-2194 (1956).

Krumbein, W. C., "Trend—surface analysis of contour-type maps with irregular control-point spacing," J. Geophys. Res., Vol. 64, No. 7, pp. 823-834 (1959).

Krumbein, W. C., "The sorting out of geological variables illustrated by regression analysis of factors controlling beach firmness," J. Sediment. Petrol., Vol. 29, No. 4, pp. 575-587 (1959).

Krumbein, W. C., "Computer analysis of stratigraphic maps," Am. Assoc. Petroleum Geologists Bull., Vol. 46, No. 2, pp. 270-271 (1962).

Krumbein, W. C., "Confidence intervals on low-order polynomial trend surfaces," J. Geophys. Res., Vol. 68, No. 20, pp. 5869-5879 (1963).

Krumbein, W. C., A Comparison of Polynomial and Fourier Models in Map Analysis, Techn. Rep., No. 2, ONR Task No. 388-078, Contract-Norn-1228(36), Dept. Geol., Northwestern Univ. (1966).

Krumbein, W. C., and Graybill, F. A., An Introduction to Statistical Models in Geology, McGraw-Hill (1965).

Lippit, L., "Statistical analysis of regional facies change in the Ordovician Cobourg limestone in northwestern New York and southern Ontario," Am. Assoc. Petroleum Geologists Bull., Vol. 43, No. 4, pp. 807-816 (1959).

Mandelbaum, H., "Statistical and geological implications of trend mapping with nonorthogonal polynomials," J. Geophys. Res., Vol. 68, No. 2, pp. 505-519 (1963).

McIntyre, D. B., "Program for computation of trend surfaces and residuals of degree 1 through 8," Depart. Geol. California, Vol. 4, pp. 1-24 (1963).

Merriam, D. F., "Use of trend—surface residuals in interpreting geologic structures," in: Computers in the Mineral Industries, 2, Stanford Univ. Publ., Geol. Sci., Vol. 9, No. 2, pp. 686-692 (1964).

Merriam, D. F., "Geology and the computer," New Sci., Vol. 26, No. 444, pp. 513-516 (1965).

Merriam, D. F., "Computer applications in the Earth sciences," Colloquium on Classification Procedures, Computer Contributions 7, Kansas Geol. Surv., Spec. Distrib. Publ. (1966).

Merriam, D. F., "Geologic use of the computer," Wyoming Geol. Assoc. 20th Annual Conf., Casper College, pp. 109-112 (1966).

Merriam, D. F., and Harbaugh, J. W., "Computer helps map oil structures," Oil and Gas J., Vol. 61, No. 47, pp. 158, 159, 161-163 (1963).

Merriam, D. F., and Harbaugh, J. W., Trend-Surface Analysis of Regional and Residual Components of Geologic Structure in Kansas, Kansas Geol. Surv., Spec. Distrib. Publ., No. 11 (1964), 27 pp.

Merriam, D. F., and Lippert, R. H., "Pattern recognition studies of geologic structure using trend-surface analysis," Colorado School of Mines Quart., Vol. 59, No. 4, pp. 237-245 (1964).

Merriam, D. F., and Lippert, R. H., "Geologic basin model studies using trend-surface analysis," J. Geol., Vol. 74, No. 3, pp. 345-357 (1966).

Miller, R. L., "Trend-Surface, their application to analysis and description of environments of sedimentation," J. Geol., Vol. 64, No. 5, pp. 425-446 (1956).

Miller, R. L., "Comparison analysis of trend maps," in: Computers in the Mineral Industries, Stanford Univ. Publ., Geol. Sci., Vol. 9, No. 2, pp. 669-685 (1964).

Miller, R. L., and Kahn, J. S., Statistical Analysis in the Geological Sciences, John Wiley and Sons (1962).

Nordeng, S. G., "Application of trend-surface analysis to semiquantitative geochemical data." Fifth Ann. International Symposium on Computers and Computer Applications in Mining Exploration, Univ. of Arizona, Vol. 1, Pt. I, pp. 1-36 (1965).

Nordeng, S. G., Ensigh, C. O., and Volin, M. E., "The application of trend-surface analysis to the White Pine copper district," in: Computers in the Mineral Industries, Stanford Univ. Publ. Geol. Sci., Vol. 9, No. 1, pp. 186-202 (1964).

O'Leary, M., Lippert, R. H., and Spitz, O. T., FORTRAN IV and Map Program for Computation and Plotting of Trend Surfaces for Degrees 1 through 6, Computer Contribution 3, State Geol. Surv. Univ. Kansas, Lawrence (1966), 48 pp.

Oldham, C. H. G., and Sutherland, D. B., "Orthogonal polynomials, their use in estimating the regional effect," Geophys., Vol. 20, No. 2, pp. 295-306 (1956).

Peikert, E. W., IBM-709 Program for Least-Squares Analysis of Three-Dimensional Geological and Geophysical Observations, Techn. Report No. 4 (1963).

Pirow, P. C., "Computer programs for moving average (weighted) trend surfaces," Symposium on Mathematical Statistics and Computer Applications in Ore Valuation, J. S. African Inst. Mining Met. (1966).

Pirow, P. C., Siebert, J. F., Knox, G. M., Claassen, P., and Lawton, J., "Computer programs for the estimation of a two-dimensional trend surface using a weighted moving average," J. S. African Inst. Mining Met., pp. 80-105 (1966).

Preston, F. W., and Harbaugh, J. W., BALGOL Programs and Geologic Application for Single and Double Fourier Series using IBM-7090-7094 Computers, State Geol. Surv. Univ. Kansas, Lawrence, No. 24 (1965), 72 pp.

Ragland, P. C., and Adams, J. A. S., Partial Trend Surfaces within Enchanted Rock Batholith, Llano Uplift, Texas, Geol. Soc. Am. Spec. Paper 73 (1963), p. 221.

Read, W. A., and Merriam, D. F., "Trend-surface analysis of stratigraphic thickness data from some Namurian rocks east of Sterling, Scotland," Scot. J. Geol., Vol. 2, No. 1 (1966).

Romanova, M. A., "Use of two-dimensional regressions for solving some geological problems," Summary Reports of the Seventh All-Union Conference on Mathematical Statistics and Probability Theory, Tbilisi (1963).

Romanova, M. A., "Recent deposits of central Kara Kum and the problem of studying subsurface structures," Sov. Geol., No. 12, pp. 70-89 (1964).

Saha, A. K., "Systematic quantitative areal variation in five granitic massifs (a discussion)," J. Geol., Vol. 70, pp. 116-118 (1962).

Saha, A. K., "A simple grid deviation technique of study of the areal composition variations in granitic bodies," Geol. Mag., Vol. 101, No. 2, pp. 145-150 (1964).

Shaw, D. M., and Kraft, E. F., "Homogeneity and random results in geochemistry," in: Chemistry of the Earth's Crust, II, Izd. Nauka, Moscow (1964).

Spitz, O. T., "Generation of orthogonal polynomials for trend-surfacing with a digital computer," in: Computers and Operation Research in Mineral Industries, Pennsylvania State Univ., 6th Annual Symposium (1966), pp. 1-6.

Vistelius, A. B., "Trend surfaces," Symposium on Mathematical Statistics and Computer Applications in Ore Valuation, Johannesburg, J. S. African Inst. Mining Met., pp. 66-72 (1966).

Vistelius, A. B., and Hurst, V. J., Phosphorus in granitic rocks of North America, Geol. Soc. Am. Bull., Vol. 75, pp. 1055-1092 (1964).

Vistelius, A. B., and Romanova, M. A., "Distribution of the heavy fraction in sands from deposits of the central Kara Kum," Dokl. Akad. Nauk SSSR, Vol. 158, No. 4, pp. 860-864 (1964).

Vistelius, A. B., and Yanovskaya, T. B., "Programming problems of geology and geochemistry for use with all-purpose electronic computers," Geol. Rudn. Mestorozhd., No. 3, pp. 34-48 (1963).

Vistelius, A. B., Guimaraes, Z., and Galibin, V., "Phosphorus and some trace elements in granitic rocks of Brazil," J. Geol., Vol. 75, No. 6 (1967).

Wadsworth, W. B., "Textural variation within a quartz diorite pluton (Twelvefoot Falls pluton), northeastern Wisconsin," Geol. Soc. Am. Bull., Vol. 74, pp. 243-250 (1963).

Whitten, E. H. T., "Composition trends in granite. Model variation and ghost-stratigraphy in part of the Donegal granite, Eire," J. Geophys. Res., Vol. 64, No. 7, pp. 835-948 (1959).

Whitten, E. H. T., "Data density necessary for quantitative modal analysis of granitic complex," Geol. Soc. Am. Bull., Vol. 70, p. 1697 (1959).

Whitten, E. H. T., "Quantitative evidence of palimpsestic ghost-stratigraphy from modal analysis of granitic complex," 21st International Geological Congress, Pt. 14, pp. 182-193 (1960).

Whitten, E. H. T., "Systematic quantitative areal variation of six granitic massifs (abstract)," Geol. Soc. Am. Bull., Vol. 71, No. 12, Pt. 2, pp. 2002-2003 (1960).

Whitten, E. H. T., "Quantitative areal modal analysis of granitic complexes," Geol. Soc. Am. Bull., Vol. 72, No. 9, pp. 1331-1360 (1961).

Whitten, E. H. T., "Systematic quantitative areal variation in five granitic massifs from India, Canada, and Great Britain," J. Geol., Vol. 69, pp. 619-646 (1961).

Whitten, E. H. T., "Modal variation and the form of the Beinn an Dubhaich granite, Skye," Geol. Mag., Vol. 98, pp. 467-472 (1961).

Whitten, E. H. T., "Quantitative distribution of major and trace components in rock masses," Am. Inst. Mining Met. Petroleum Engineers Trans., Vol. 220, pp. 239-246 (1961).

Whitten, E. H. T., "Sampling and trend surface analysis of granites, a reply," Geol. Soc. Am. Bull., Vol. 73, No. 3, pp. 415-418 (1962).

Whitten, E. H. T., "Areal variability of alkalies in the Malsburg granite, Germany," N. Jahrb. Miner. Monat., Vol. 9, pp. 193-200 (1962).

Whitten, E. H. T., "Systematic quantitative areal variation in five granitic massifs," J. Geol., Vol. 70, No. 1, pp. 119-121 (1962).

Whitten, E. H. T., "A reply to Chayes and Suzuki," J. Petrology, Vol. 4, No. 2, pp. 313-316 (1963).

Whitten, E. H. T., A Surface-Fitting Program Suitable for Testing Geological Models which Involve Areally Distributed Data, Techn. Rept., No. 2 (1963), 56 pp.

Whitten, E. H. T., "'Best' mathematical model for mapped variables; Gold variability at Virginia Mine, South Africa," Annual Meeting Geol. Soc. Am., Miami Beach, Florida, Program, p. 223 (1964).

Whitten, E. H. T., "Process—response models based on heavy mineral content of the San Isabel Granite, Colorado," Geol. Soc. Am. Bull., Vol. 75, pp. 841-862 (1964).

Wilks, W. W., "Statistical inferences in geology in the earth sciences," in: Problems and Progress in Current Research, William Marsh Rice Univ., Univ. of Chicago Press (1963).

MATHEMATICAL METHODS IN GEOLOGY (CHRONICLE FOR THE PERIOD FROM SEPTEMBER 1964 TO SEPTEMBER 1966)

M. E. Demina

Laboratory of Mathematical Geology
V. A. Steklov Mathematical Institute
Academy of Sciences of the USSR
Leningrad, USSR

Within the Soviet Union the Ministry of Geology of the USSR took a number of administrative actions in the indicated period, the purpose of which has been the introduction of mathematical methods and computers into geologic exploration practice. As a result, laboratories of appropriate design have been organized in the leading scientific research institutes of the Ministry. At the same time, a resolution of the Presidium of the Academy of Sciences, USSR has been carried out: to organize in the V. A. Steklov Mathematical Institute of the Academy of Sciences of the USSR, a Division of Mathematical Problems in Geology and Geophysics with two problem-solving laboratories (one for mathematical geology and another for mathematical methods of geophysics) with auxiliary laboratories for standard and automatic programming of problems in geology and geophysics. A seminar was organized in the Laboratory of Mathematical Geology, and at one of its expanded meetings the well-known French geologist B. A. Choubert gave a report on "the geochemical behavior of elements in the lithosphere in dependence on time." At this same meeting, A. B. Vistelius reported on the trends in mathematical geology. At the conclusion, brief communications were presented by D. N. Ivanov, Yu. V. Podol'skii, and V. V. Gruza on linear paragenetic associations and a report was given by M. A. Romanova and I. N. Golynko concerning the use of two-dimensional regressions in the solution of geological problems and the use of appropriate computer programs.

In addition, the Laboratory conducts consultations with various institutes both within the Soviet Union and in foreign countries. In the indicated period of time specialists from France, Czechoslovakia, East Germany, India, Japan, and the Union of South Africa visited the Laboratory. A study-method seminar was also conducted during this time at the Leningrad Mining Institute on mathematical methods in geology, and in the first half of 1965, under the auspices of the Mining Scientific, Engineering, and Technical Society, A. B. Vistelius presented a series of lectures for Leningrad geologists at the All-Union Scientific Research Institute of Methods and Exploration Techniques. In these lectures Vistelius examined questions associated with theories of evaluation and methods of testing statistical hypotheses. In the spring of 1966, Vistelius also presented a series of lectures in Sofia on problem solving in mathematical geology for the geologists of Bulgaria.

In our country, the following general meetings were held during the indicated time interval.

1. During December 10-16, 1965, a meeting was held at the Institute of Geology and Geophysics of the Siberian Branch of the Academy of Sciences of the USSR, in which about 300 individuals from various cities in the Union participated. From the author's viewpoint, the most significant reports for the development of geology as a science, at this meeting, were those on stochastic modeling of geologic processes. In applied aspects, greatest interest was focused on reports dealing with multivariate statistics.

2. The Division of Earth Sciences of the Academy of Sciences of the USSR conducted a general meeting on December 16, 1965, devoted to mathematical methods. Interest was aroused by the contributions of geophysicists on the use of computers in solving geophysical problems. Problems of actual use of mathematics in geology or of the state of the question as a whole were not considered.

3. During the second half of May 1966, seminars on mathematical methods in geology were planned for the Ukrainian Institute of Geology at Kiev and for the Committee on Sedimentary Rocks at the Division of Earth Sciences of the Academy of Sciences of the USSR at Moscow.

In a number of foreign countries mathematical methods were widely applied in the study of geological problems. At the University of Kansas (Lawrence) work was organized for preparing and publishing computer programs for processing geologic data (under the direction of D. F. Merriam). After a time about 15 programs were published. The programming was set up for FORTRAN, BALGOL, and ALGOL. The same group published a collection of papers on the analysis of cyclicity in sedimentation, using mathematical methods. At the annual fall meeting of the Geologic Society of America in Miami, 1964, a special section was reserved for application of the methods of mathematically modeling geologic processes, and reports by W. C. Krumbein and A. B. Vistelius were scheduled for the program.

In the indicated interval, three symposiums were conducted in the USA on the use of mathematical methods and computational techniques in industry for studying mineral resources. These symposiums were held in Arizona, Nevada, and Pennsylvania. Volumes of papers have appeared in print from the reports at these symposiums. From announcements of these reports it is clear that they touch on a broad spectrum of geological questions.

In March 1966, at the Institute of Mining and Metallurgy of South Africa at Johannesburg, an international symposium was conducted on the use of statistics and electronic computers for evaluation of gold reserves. The papers at the symposium include one by A. B. Vistelius on types of trend surfaces and their use in specific geologic situations.

In the coming years further discussions of mathematical geology are planned, particularly in three large international congresses. There will be first the 7th International Petroleum Congress. A special section will be provided for mathematical methods. The section will include papers on five leading problems. Of these, the one appealing to the most geologists will be "Mathematical methods in geological interpretation," and a report on this theme has been commissioned of A. B. Vistelius by the organizing committee of the Congress.

At the 7th International Congress on Sedimentology (Great Britain, August 1967), one of the leading problems will be dealt with: "The use of models and statistics in sedimentology." Lastly, reports on mathematical methods will be presented at the International Geological Congress in Prague.

The present summary is based on the personal participation of the author at several conferences and on data from the Laboratory of Mathematical Geology of the Mathematical Institute of the Academy of Sciences of the USSR.

In conclusion, the following remarks may be made.

1. The introduction of mathematical methods into geology has been effective whenever the studies have been made by geologists properly oriented in the specific geologic aspects of the problem and well grounded in mathematics. When the problem has not been set up in proper geologic relations, the efforts have accomplished no real purpose.

2. At the present time a tremendous volume of experimental work on mathematical methods in geology has been performed. Without consideration of the opinions of specialists competent in these matters, it is impossible to bring about any development of mathematical methods in geology or to evaluate the status of the problem.

WITHDRAWN